the secrets of codes

weldonowen

the secrets of codes

EDITED BY PAUL LUNDE

weldonowen

weldon**owen**
415 Jackson Street
San Francisco, CA 94111
www.weldonowen.com

President, CEO Terry Newell
VP, Sales Amy Kaneko
VP, Publisher Roger Shaw
Creative Director Gaye Allen
Art Director Tina Vaughan
Assistant Editor Sarah Gurman
Production Director Chris Hemesath
Production Manager Michelle Duggan

Produced for Weldon Owen Inc. by Heritage Editorial
Editorial Direction Andrew Heritage, Ailsa C. Heritage
Senior Designers Philippa Baile, Mark Johnson Davies
Illustrators Andy Crisp, Philippa Baile, David Ashby, Mark Johnson Davies, Peter Bull Art Studio
Picture Research Louise Thomas
DTP Manager Mark Bracey

Consultant editors
Dr. Frank Albo, M.A, M.Phil.
Trevor Bounford
Anne D. Holden, Ph.D.
D.W.M. Kerr, BSc.
Richard Mason
Tim Streater, BSc.
Elizabeth Wyse, BA

Weldon Owen is a division of
BONNIER

Library of Congress information is on file with the publisher.

ISBN 10: 1-61628-462-5
ISBN 13: 978-1-61628-462-6

Printed and bound in China

10 9 8 7 6 5 4 3 2 1

01

THE FIRST CODES

ACPOTNMTHEBNVTJSD

02

SECTS, SYMBOLS, AND SECRET SOCIETIES

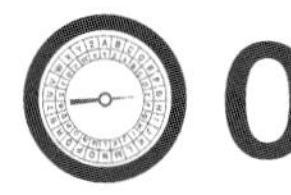

03

CODES FOR SECRECY

04

COMMUNICATING AT DISTANCE

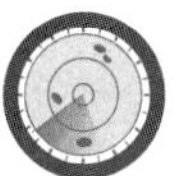

05

CODES OF WAR

ACPOTNMTHEBNVTJSD

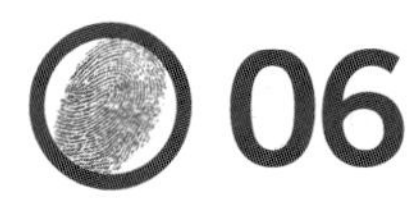

06 CODES OF THE UNDERWORLD

07 ENCODING THE WORLD

08 CODES OF CIVILIZATION

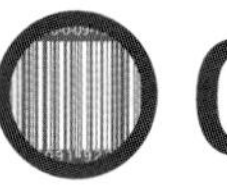

09 CODES OF COMMERCE

CONTENTS

10 CODES OF HUMAN BEHAVIOR

VISUAL CODES

12 IMAGINARY CODES

13 THE DIGITAL AGE

INTRODUCTION

We are all proficient cryptanalysts. We live in a global culture underpinned by a massive number of codes that determine our actions, provide us with information, and provide information about us to others.

Even before learning to speak, children begin to decode their immediate surroundings. They instinctively read expression and gesture and are sensitive to intonation from the beginning. Grasping language is an enormously complex encoded process, not only involving mastery of a set of sounds, but also the rules that govern them, along with all the gestures, intonations, and facial expressions that convey further meaning. We all continue to decode for the rest of our lives, evaluating our environment and assessing those around us, almost always without being fully conscious of what we are doing. We even learn to listen to what is not said, for language is used as much for concealment as it is for communication.

The semantic ambivalence of the term 'code' – a system of rules or laws, as well as a hidden means of communication – is no accident. We speak of a 'dress' code and a 'code of behavior,' meaning a set of rules, but to be operative these must be decoded by the observer. Probably since the beginning of recognizable human behavior, coding has been fundamental to all human groups, just as it is now. The way we dress and act still defines us, and is meant to send specific messages. In traditional societies, these messages can be very complex, indicating age, sexual availability, origin, status, and many other values.

We must also decode our physical environment. Early humans were dependent upon an ability to decode the world, to divide the edible from the inedible, the threatening from the nonthreatening. Survival involved deciphering signs, learning to read the landscape and weather, developing tracking skills, judging the passage of time from the movement of heavenly bodies, and understanding the rhythms of the seasons. Modern urban humans may have lost many of these skills, but our survival still depends on the ability to correctly decode an overwhelming amount of coded information, from advertising billboards to emergency exits and highway signs.

Concealment, too, seems to be endemic in human societies. Secret languages and gestures are characteristic of many human groups, serving as social markers or means of concealing messages from the noninitiated. Children often use 'secret' languages to conceal things from adults, while adults use periphrasis to evade the curiosity of their children. Thieves as well as ruling elites have historically used argots or minority languages to conceal their thoughts from the general public. In some societies, men and women use different forms of the same language.

LIONWTFRNOEDMULCOTWINOENM

The invention of writing systems, the graphic representation of sounds, made it possible to have a durable record of what had formerly been ephemeral. Writing itself is a form of encoding, and the decipherment of some ancient 'lost' writing systems, like Egyptian hieroglyphics or Linear B was only made possible by using cryptanalytical techniques. Codes, in the sense of deliberately concealed messages, are probably almost as old as the writing systems that made them possible. Early states, like their modern counterparts, frequently wished to conceal (while at the same time communicate at a distance) by writing. Military ciphers are known from ancient times, and a bewildering variety of forms of 'secret' writing was used throughout the medieval world.

Cipher systems, the systematic concealment of the meaning of a message by substituting the words or phrases of the original message with another set of characters, began to be widely used in Europe in the 16th century. And each new code spawned new decryption techniques, culminating in the epic story of the breaking of the Enigma system used by the Germans in World War II. The transformation of communication technology has added new opportunities for the cryptographer, and new challenges for the cryptanalyst.

Today, with computerized communication systems, codes have moved from a largely military concern to the province we all inhabit. Every telephone call we make or electronic message we send is automatically encrypted by the systems, machines, and algorithms we have created in order to make these wonders possible. But in turn, every message can be intercepted and read. We live in an age where the most valuable commodity is – as Bill Gates has demonstrated – code, and the second most valuable is the 'inside' information that possession of the appropriate code provides us access to. In order to guard the latter, encryption systems have to be continually updated; just as every code generates its own solution, every encryption system is eventually foiled. The subject is no longer a matter for specialists. How to preserve privacy in an electronic age, and at the same time protect society from assault, is one of the burning questions of our time.

This book surveys the ways in which codes of every sort have been used to convey an enormous range of information. Its thematic structure presents codes of certain types in coherent groups. Although the organization of these thematic sections is broadly chronological, cross-referencing between sections reveals a clear pattern. An awareness of the interconnected mesh of secret languages that surrounds us is now – more than ever – paramount to our survival and success: this book could change your life.

– *Paul Lunde*

01

It is clear that humans developed an ability to understand the meaning of the natural patterns in the world around them from the earliest times; it was a significant factor in their survival and ascent to become the Earth's dominant species.

the first codes

With the growth of the first communities, humans developed their own complex systems – language, counting, writing – that involved processes of abstract thought, organization, and an ability to create symbologies, creating the first ciphers. Working from often fragmentary material evidence, archeologists have sometimes used the techniques of cryptanalysts to unravel how these ancient systems developed and functioned.

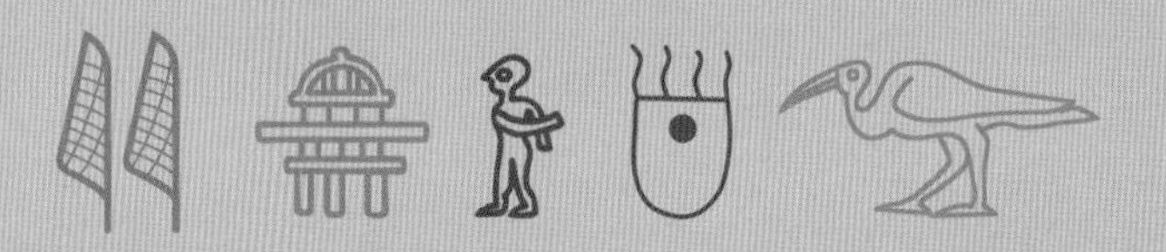

READING THE LANDSCAPE

Caves provided a natural shelter from the weather for early humans, although finding suitable sites and defending them from human and animal rivals could be challenging.

For the first humans, survival was dependent upon their ability to understand the processes that governed the world around them. As primitive social groups developed, their ability to forage and hunt for food, to find shelter, and to avoid calamity became critical. Gathering and interpreting such information was probably the first example of human ingenuity in perceiving hidden patterns in the surrounding world. This became particularly important as they began to adopt a migratory hunting and gathering lifestyle, spreading gradually from the home of *Homo sapiens*, in tropical East Africa, eventually to colonize much of the globe, adapting to contrasting biogeographical regions.

Looking for signs in the landscape

Some of the earliest human remains have been found in the Rift Valley of East Africa. Although the environmental conditions have undoubtedly changed, it is a good starting point in the quest to discover the roots of human ingenuity. How would early humans 'read' such a landscape?

River valleys
Even in an arid environment, dry gulches and wadis indicate the presence of flowing water at certain times of the year, and the likelihood that water might be found below the surface by digging wells. However, these may also indicate hazardous flash floods.

Hills
Provide both a vantage point and a defendable location, and also the possibility of springs providing drinking water.

Cliffs
Indicate the possibility of caves to provide shelter and often seeping groundwater. High ground is also advantageous as a look-out point to detect game or other groups – possibly competitive or hostile – of hunter-gatherers. There is evidence that cliffs were used to inaugurate mass killing of game, when herds would be driven to stampede over a precipice.

Plant life
Shows the presence of groundwater, potentially edible plants, and a gathering point for game animals. Identifying which plants were edible was presumably a matter of trial and error.

Arid conditions
Limited rainfall, unlikely to provide sustenance for an extended period, probably also means confronting daily and nightly extremes in temperature.

The seasons

As humans migrated away from the tropics, they would have had to come to terms with patterns of seasonal change. As migration was inevitably a gradual process, much of this information would have been learned by trial and error over the years. Nevertheless it is clear that, jumping forward several millennia, the first organized agricultural communities developed on major river systems – the Tigris and Euphrates in Mesopotamia, the Nile in Egypt, the Indus in South Asia, and the Huang He or Yellow River in China – where understanding the annual cycle of inundation was critical for planning planting and harvesting, and for the siting of settlements to avoid flooding.

Reading weather signs

Understanding the meaning of cloud patterns and formations would have provided early humans with much useful information about what was to come.

1 Cirrus High, wispy clouds associated with fair weather, although in colder climates if they begin to accumulate with a steady wind they can indicate a coming blizzard.

2 Cumulonimbus Towering, anvil-shaped clouds with increasingly dark gray bases, a definite sign of approaching storms of rain, hail, or snow. Wind from the direction of the clouds will increase and the temperature may drop sharply.

3 Scuds Unstructured vapory clouds running before the wind mean continuing bad weather.

4 A hailstorm In the midst of a deluge may indicate that a storm is going to turn into a tornado.

5 Cirrocumulus High, puffy balls of cloud in groups – fair weather.

6 Cirrostratus (altostratus) A fine unbroken layer of gray clouds. When these follow cirrus, bad weather may be in store.

7 Cumulus Puffy white clouds in a blue sky – stable conditions. Accumulating through the day and becoming taller and fuller they can develop into storm clouds.

8 Nimbus Flat gray clouds covering the sky, usually providing rain showers preceding bad weather.

9 Stratus Low, even, gray clouds obscuring the sky, associated with damp, dank conditions. Extremely low stratus can form enshrouding fogs.

Impending disaster Observing the behavior of animals and birds would have provided many clues. Many species frequently sense an impending disaster. Before earthquakes, earthworms will dig their way to the surface, dogs will stop barking, and often try to hide; larger animals such as horses and antelopes will become restless and may even stampede (as they may also before a storm or wildfire). In the path of a hurricane or typhoon, birds will scatter (or frequently drop dead), and sharks will vacate the reefs they inhabit. The migratory habits of birds and game animals also provide important information concerning changing seasons and weather patterns (*see page 14*).

TRACKING ANIMALS

One of the earliest examples of humans being able to 'decode' information arose from hunting, a talent dating back some 100,000 years or more. The stealth and ingenuity needed to hunt wild animals for food relied in large part on the ability of the hunters to read various signs left by their quarry: simple footprints, tracks, spoor, evidence of eating habits, and other activities. These skills were also vital to avoid predatory animals who may in turn hunt the hunters. These traits are still embedded to some extent in human common memory today.

Baboons are omnivores that live, forage, and hunt in groups, so their distinctive paw prints are rarely found in isolation. Their tracks clearly show the distinction between the hands – with a pronounced opposable thumb – and the feet.

Savannah water holes

Of the many environments in which animal tracks and traces can be found, the water holes of the savannah regions of the Americas, Africa, and Asia are the most fruitful in providing many clues to the animal life in the area. This is due to a number of factors: the watering holes act as magnets for most of the fauna in the area (including humans), not just in search of water, but because of the verdant plant life water holes are likely to support; the damp, muddy shore-line means that most animals will leave tracks of some sort; and in addition, depending on the climatic conditions, it is possible to determine fairly accurately the age of any tracks or spoor.

Rabbit or hare these droppings have been left by herbivores such as rabbits or hares. Too light to leave paw prints here, these telltale signs nevertheless indicate their presence in the area.

Warthog The dung of a warthog is difficult to date, as their slow digestion means that their spoor tends to be compacted and dry upon delivery.

Tracks of hyenas will indicate that game, or freshly killed quarry, is likely to be found in the locality.

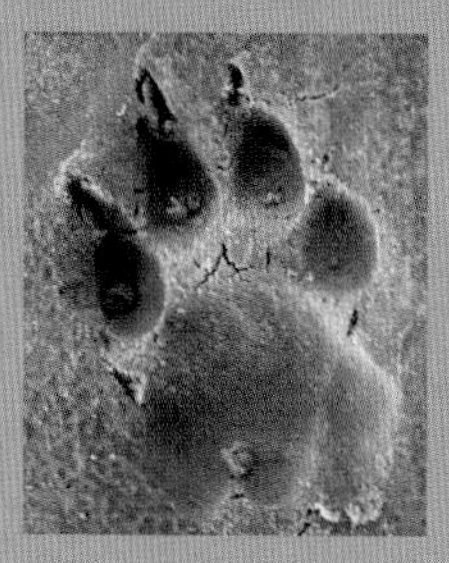

Lion prints are distinguished from most other cat prints by little more than their sheer size. Finding a print of this sort would alert the hunter to the presence of game – but also to the presence of a dangerous predator.

Gnawed grass This is a sign of recent animal activity. The torn stems show that this was grazed by an ungulate, possibly an antelope or zebra.

Contrasting environments

Different biogeographical regions not only support unique animal populations, but also provide a variety of animal tracks in different terrains.

Floodplains and river banks
Frequent inundation by floodwater or tides means that such environments provide a wonderful source of animal tracks when the mud is relatively damp. However, subsequent rises in water or heavy rainfall may swiftly erode these tracks. In the tropics, where inundation might only occur during the rainy season, paw prints and other marks, such as these antelope tracks in Kenya, may be baked dry for months.

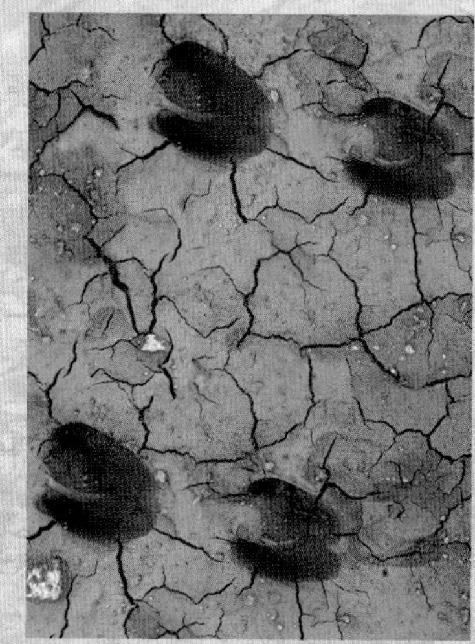

Gopher

Ibis

Zebra

Desert tracks Sandy deserts such as the Mojave or the Great Erg of the Sahara tend, like snow, to show fresh animal tracks extremely well – but only for a limited time, as wind will erase or fill in any tracks quite rapidly. Here sidewinder patterns crisscross with coyote prints in the Mojave Desert.

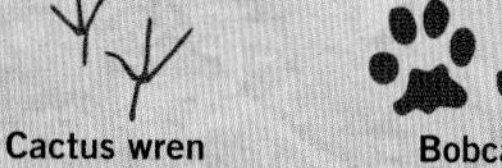

Cactus wren

Bobcat

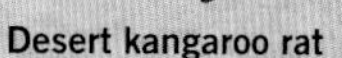

Desert kangaroo rat

Coyote

Snow tracks Fresh snow cover will bear the tracks of even the smallest animal or bird, although the effect of melting and windblown drifting means that these tracks are unlikely to retain their distinctive patterns or even last very long. Nevertheless, unlike in many other environments, snow tracks will show the quarry's route in its entirety. These polar bear tracks were found in Greenland.

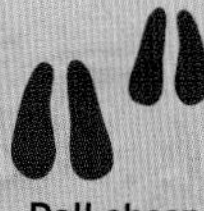

Dall sheep

Arctic fox

Snowy owl

BUSHCRAFT SIGNS

Penan twig codes

The Penan hunter-gatherers of Sarawak, Borneo still use an ancient field message system involving cut twigs.

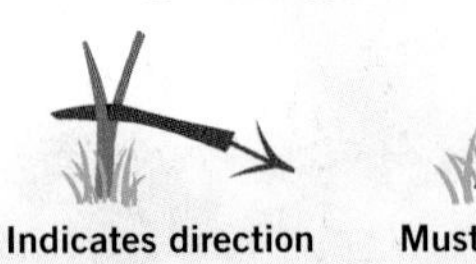

Indicates direction to follow

Must follow

Hurry up

Long way

Be there in three days

No food, but in good mood

Be warned, do not follow us

Although there are various theories about how and when the first spoken languages evolved, little is truly known, although we live today with the 'Tower of Babel' legacy. What is more clear was the necessity among migrant hunting groups for a sophisticated means of silent communication, involving hand signals and body language while stalking, and the ability to provide signals and instructions to others from the same group or tribe concerning their movements. We can find examples of these among many primitive cultures the world over today, and some have been adopted and adapted by modern hunters, armies, and organizations like the Scouts.

Indian signs

One of the most comprehensive systems of signs used whilst stalking game known to us today was developed by the Plains Indians. These involved both complex body language and hand signs, and images that could be drawn (*left*). In addition, the Plains Indians developed a complex signing language which allowed them to overcome the language barrier between tribes (also achieved among the Aboriginal tribes of Australia's Western Desert), and also acted as a primal form of signing for deaf people (*see page 242*).

US military personnel learned Native American hand sign language.

Field information signs

While many hunting and gathering groups such as the San Bushmen of the Kalahari Desert and the Penan of Borneo (*opposite*) developed their own, unique bushcraft signals and ways of leaving messages, it was from encounters with these systems that an internationally recognized vocabulary of bushcraft signs was developed, initially by colonial military troops, and latterly by the Scouts movement. These are designed to provide information for other people or groups in the field, and are closely linked to the vocabulary of modern survival signs (*see page 220*). These signs may be drawn in the sand or earth, or constructed from available materials such as sticks or boulders.

Military signs

In combat or search-and-find situations, silent communication in the field can be a matter of life or death. The US military uses a system of hand and body signaling which closely resembles that used by other armed forces, and is designed to communicate key information to fellow soldiers, and to potential suspects who might not speak English.

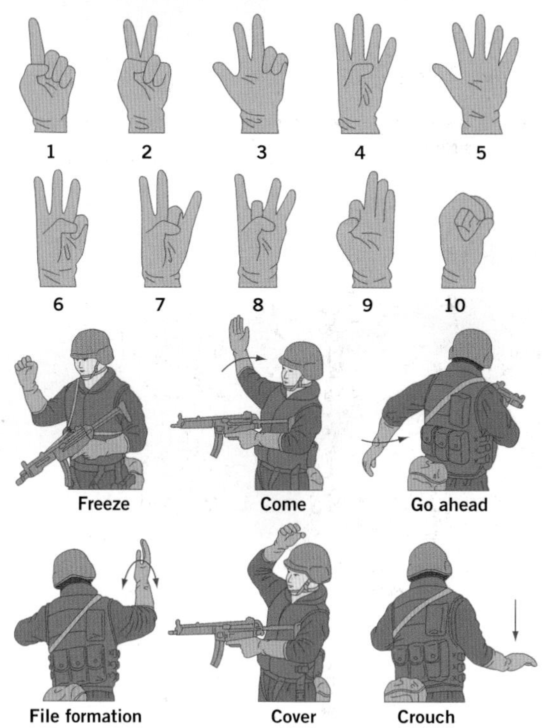

Shadow Wolves

Bushcraft skills are still important today. An elite US police unit, the Shadow Wolves, composed of Native Americans from a number of peoples including the Navajo and Blackfoot, use traditional tracking techniques to hunt down drug traffickers along the US/Mexican border. They have impounded over 45,000 pounds (20,412 kg) of marijuana since 1972, and have traveled to Central Asia and Eastern Europe to teach tracking skills to local police officers.

EARLY PETROGLYPHS

From earliest times, simplified, stylized drawings of the human figure have been used to represent human activity in various forms. Early examples, such as the famous cave paintings at Lascaux in France, dating from about 17,000 years ago, and the carvings shown here from Alta in Norway from about 4500 BC, clearly show human figures – with head, torso, arms, and legs – engaged in hunting varieties of game with spears, and bows and arrows. Although the exact motive for these petroglyphs is not known, this is a clear example of the early development of visual coding to convey specific information.

Pictograms

Early petroglyphs were obviously not intended to accurately depict the human or animal form but were icons (from the Greek *eikon* meaning 'image') combined in specific arrangements to convey a message – forming pictograms. Whether that message was about hunting techniques, the game available in the region, or merely a means of commemorating a particularly successful foray remains unclear. Later examples of hunter-gatherer art from various parts of the world show more sophisticated attempts at embedding meaning in the image.

Cave paintings from El Castillo in Spain include rows of dots, grids, and other repeated geometric forms. These may represent cadastral records, plans, or be clan emblems.

Variety
The differences in scale and illustrated activity could indicate that this petroglyph was completed over a long period, perhaps by successive generations; the colors were probably added much later.

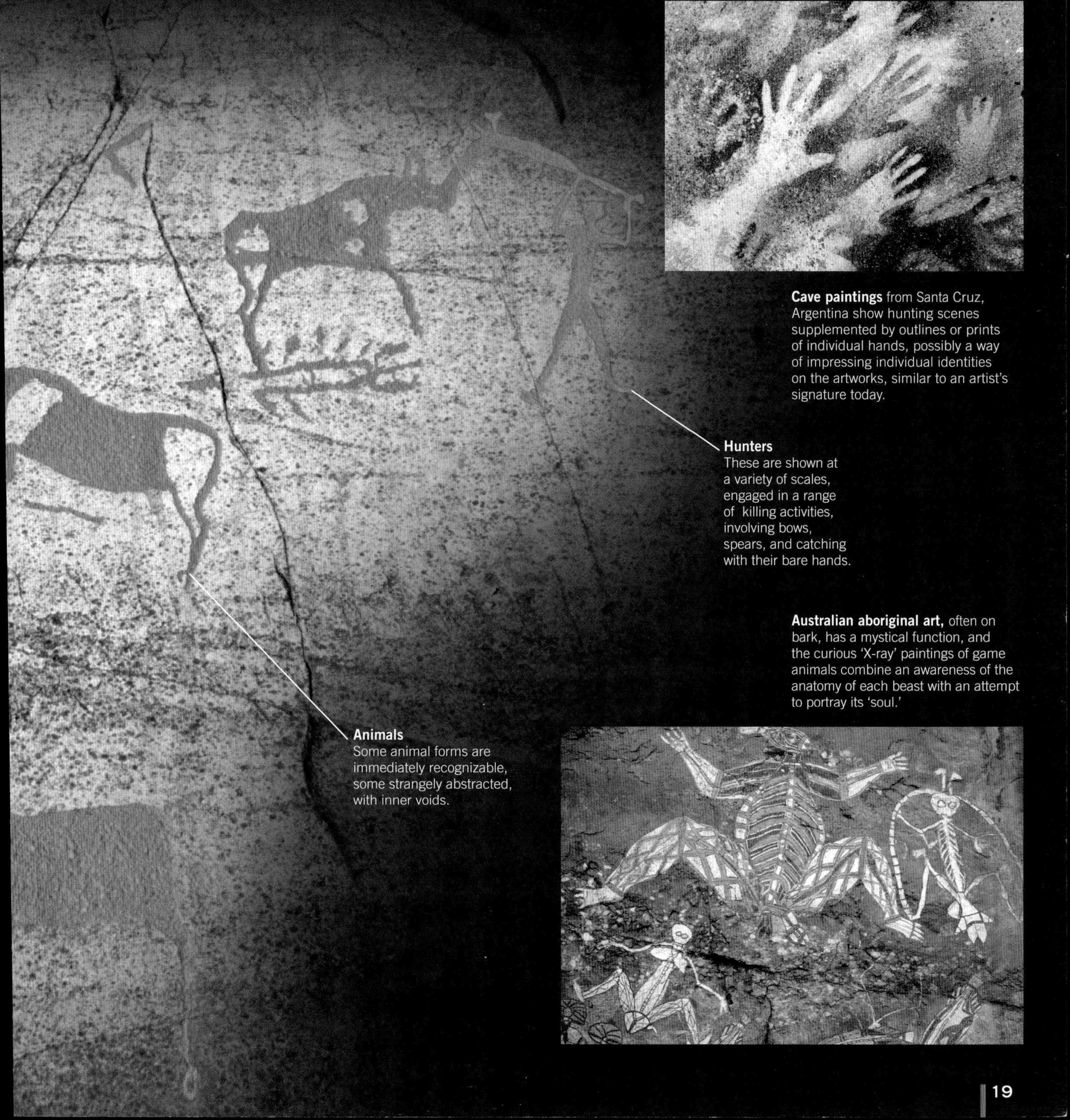

Cave paintings from Santa Cruz, Argentina show hunting scenes supplemented by outlines or prints of individual hands, possibly a way of impressing individual identities on the artworks, similar to an artist's signature today.

Australian aboriginal art, often on bark, has a mystical function, and the curious 'X-ray' paintings of game animals combine an awareness of the anatomy of each beast with an attempt to portray its 'soul.'

FIRST WRITING SYSTEMS

Scribes were so highly prized in ancient Egypt that they were exempt from taxation.

The first systems for encoding messages developed, along with methods for counting (*see page 26*), some 5,500 years ago when early agricultural towns and cities needed to record stored and redistributed goods, livestock, and trade transactions. The earliest examples of records using inscribed symbols were produced in Sumer in Mesopotamia around 3400 BC, in the form of marked clay counting tokens. Slowly, over the next few thousand years similar numerical and writing systems, in various languages, spread throughout West Asia, while separate systems evolved in South Asia, China, and Central America. These achievements represented a huge step in human development: mastering the abstract action of counting and writing led to the development of calendars, weights and measures, coinage, mathematics, geometry, and algebra. Writing enabled governments to communicate edicts and laws, and history and myths to be recorded.

c.3400 BC
Sumerian clay counting tokens. These recorded goods and products, and were sealed in clay envelopes marked to indicate their contents. A pictographic script develops in parallel using icons or pictograms to represent words.

c.3000 BC
Egypt: development of hieroglyphic writing in conjunction with numbering systems.

c.1400 BC
Syria: Ugaritic alphabet using cuneiform for 22 (later 30) consonants; earliest example of alphabetic writing.

c.1400 BC
Syria and Palestine: first Aramaic alphabets.

c.1100 BC
Phoenicia: development and spread of alphabet.

Syria/Palestine

Mesopotamia

3500 BC | 3250 BC | 3000 BC | 2750 BC | 2500 BC | 2250 BC | 2000 BC | 1750 BC | 1500 BC | 1250 BC

Key to writing systems

- pictographic
- hieroglyphic
- cuneiform
- inscriptions
- alphabetic
- probable relationship between scripts

c.3250 BC
Tell Brak, Syria. Earliest finds of writing using cuneiform script on clay tablets, the first known writing system.

c.2400 BC
Mesopotamia: spread of Akkadian, written in cuneiform. First literary texts.

c.1700 BC
Sinai: Proto-Canaanite script.

c.1500 BC
Anatolia and Caucasus: Hittites and Urartians adapt cuneiform.

China

c.1400 BC
Shang China: first inscriptions written on oracle bones. These record predictions made by reading the cracks on heated bones.

c.2600 BC
Indus Valley: possibly influenced by Mesopotamian pictograms, the Harappan civilization produces a unique, and still largely undeciphered, pictographic script. System dies out with decline of the culture c.1800 bc.

India

c.2000-1600 BC
Development of Cretan hieroglyphic writing (Linear A, Linear B), probably influenced by Egyptian hieroglyphs (*see page 28*).

Aegean

An early Sumerian token inscribed with cuneiform.

The evolution of scripts

Writing systems fall into four categories: pictographic, hieroglyphic, cuneiform, and alphabetic, and many developed from one stage to another. Most writing systems were originally pictographic, in which symbols were used for individual words or ideas. However, such systems became cumbersome, requiring a new symbol for each word, and symbols soon came to be used for sounds rather than concepts, the symbols themselves becoming more stylized and less representational. Most scripts used in western Eurasia evolved from Sumerian cuneiform or Egyptian hieroglyphs, although local styles developed relatively rapidly. Chinese writing developed independently, and came to dominate much of East Asia, while Mesoamerican writing too developed entirely independently, but was largely eradicated after European contact in the 16th century.

c.300 BC
Central America: hieroglyphic/syllabic writing emerges with the Maya (*see page 36*).

c.600 BC
Central America: Pictographic writing first associated with the Zapotecs.

Mesoamerica

c.AD 1000
Mesoamerica: Mixtec Nahuatl (Aztec) language written in pictograms.

c.650 BC
Italy: development of Etruscan writing, superseded by Latin. First Latin inscription c.500.

c.AD 250
Northern Europe: Runic script (*above*) appears, with characters in futhark (alphabetic order). Possibly influenced by Etruscan or Roman alphabet (*see page 188*).

c.AD 300
Western Europe: Latin established as main writing system.

Eastern Europe

c.AD 1000
Eastern Europe: Slavonic scripts evolve from Greek.

Western Europe

c.750 BC
Greece: earliest alphabetic inscriptions in Greek.

Europe

c.500 BC
Persia: introduction of Aramaic alphabet.

Middle East

c.300 BC
India: Brahmi alphabetic script develops, possibly based on Aramaic alphabet.

c.AD 450
Arabia: first texts in Arabic alphabet.

India

1000 BC | 750 BC | 500 BC | 250 BC | 0 | AD 250 | AD 500 | AD 750 | AD 1000 | AD 1250 | AD 1500

AD 75
Mesopotamia: last known use of cuneiform.

c.AD 300
Korea: Chinese script spreads north.

c.AD 750
Japan: Chinese-influenced script in use in Japan.

Cuneiform writing

The earliest form of writing developed in Mesopotamia, using a wedge-shaped stylus (*cuneus* in Latin) to impress the characters on clay tablets, and is known as cuneiform. Stylized pictographs were used to make up words, each pictograph representing a syllable, with abstract words being related to nouns ('ear' used for 'hear'). For 3,000 years this form of writing was used on a wide variety of materials to record everything from mundane business transactions to epics like the story of Gilgamesh. It was adapted throughout Southwest Asia to write a number of different, often unrelated, languages, just as our Latin alphabet is today – Sumerian, Akkadian, Elamite, Hurrian, Hittite, Urartian, Ugaritic, and Old Persian. Cuneiform may have influenced Egyptian hieroglyphs. After the conquests of Alexander the Great (334-323 BC), cuneiform gave way to Aramaic alphabetic writing and, like Egyptian hieroglyphs, became a 'lost' language until the 19th century (*see page 22*).

pig | bird | eat | head | walk/stand | ox | pot | hand | day | well | water

Pictographic c.3000 BC

Early cuneiform representation c.2400 BC

Late Assyrian cuneiform c.650 BC

READING CUNEIFORM

From wedge to word
In the 18th century the traveler Carsten Niebuhr, perhaps the first truly scientific traveler to the East who accompanied a Danish expedition, provided scholars with quite accurate transcriptions of the Persepolis inscriptions. He noted that three different forms of cuneiform writing were employed in the inscriptions, confirmed that the script ran left to right, and even succeeded in isolating most of the characters of the simplest of the scripts which, in trilingual inscriptions, always preceded the other two.

The apparently impossible task of unraveling cuneiform is an epic tale. Cuneiform writing was first described by García Silva Figueroa, Spanish ambassador to the Persian court of Shah Abbas. He saw the cuneiform inscriptions at Persepolis in 1618, describing the characters as 'triangular, in the shape of a pyramid or miniature obelisk.' A later 17th-century traveler, Engelbert Kaempfer, was the first to coin the current name for the script, and although it was popularized by Oxford linguist Thomas Hyde, he didn't believe it was writing, but merely architectural decoration.

Finding the key

The key to cuneiform was discovered in 1802 by a German schoolmaster and philologist, Georg Friedrich Grotefend (1775-1853). Concentrating on two short inscriptions transcribed by Niebuhr, he realized that there were too few signs in the inscription for it to be logographic, while at the same time the length of the individual words was too great for the script to be syllabic. He therefore assumed that it was alphabetic, some of the signs indicating short vowels, others long. Grotefend was on the right track.

Knowing that Persepolis (*below*) was the Achaemenid capital, he assumed the inscriptions were probably commemorative, and might contain the names and genealogies of the rulers of the Achaemenid dynasty known from Greek sources. A Middle Persian, or Pahlavi, formula from Palmyra had recently been deciphered with the help of a Greek bilingual, containing the name of the Sasanian ruler Ardashir, followed by the title 'king of kings of Iran… of the race of the gods, grandson of the god Babak, the king.' Using this model Grotefend assumed that the first word in the two short inscriptions was the name of the ruler, and that the next word would correspond to the word 'king,' and 'king of kings' might occur. The repetitions of the word 'king' made it easy to identify the repeated signs for this word, but as the signs that made up the first word in each inscription were different, they must spell out the names of different kings. He then noticed that the first word of the first inscription occurred in line 3 of inscription 2, followed by the word he had identified provisionally as meaning 'king.' It was therefore probable that the first word of inscription 2, assuming that it was the name of a king, was the son of the king mentioned in line 3. The Achaemenid ruler Xerxes was the son of Darius, son of Hystaspes. So the first word in inscription 2 must be the name of the ruler the Greeks called Xerxes, while the name in line 3 of inscription 2, which was the same as the first word in inscription 1, must be the father of Xerxes, Darius.

From script to sound

Grotefend knew of a translation of the Middle Persian *Avesta* which gave several different forms for the name of Darius' father, Hystaspes, the most common being 'Goshtap.' It also gave the Middle Persian word for 'king' – *khšeio*. The first and second words of inscription 2 begin with the same cuneiform sign, which Grotefend could now identify as the sound 'kh.' The second sign in both words was also identical, so this must represent the sound 'sh.' The third letter in the first word of inscription 1, already identified as the name Darius, recurred in the first word of inscription 2, which Grotefend was now sure was the Old Persian form of the Greek name Xerxes. This gave him the value of the sign representing 'r.' Following this procedure, he succeeded in correctly identifying ten of the 22 different signs used in the two inscriptions – the names of Xerxes, Darius, Hystaspes/Goshtap, and two words, 'king' and 'great.' His approach was that of the cryptanalyst rather than the linguist. After this inspired breakthrough, further progress was left to others, as the knowledge of Sanskrit and Avestan languages, closely related to Old Persian, gradually deepened. His achievement was nevertheless remarkable. For the first time in history, an unknown ancient script had been partially deciphered.

Inscription 1
Line 1 Dārayavauš : xšāyaθiya: vazra
Line 2 vazraka : xšyaθiya : xša
Line 3 yaθiyānām : xšāyaθiya:
Line 4 dahyūnām : Vištāapahy
Line 5 ā : puça : Haxāmanišiya : h
Line 6 ya : imam : tacaram : akunauš

Grotefend's translation: **Darius the Great King**, **King of Kings**, **King** of countries, son of **Hystaspes**, an Achaemenian.

Inscription 2
Line 1 Xšayārša : xšāyaθiya :
Line 2 ka : xšāyaθiya : xšāyaθiya
Line 3 nām : Dārayavahauš :xšāyaθ
Line 4: iyahyā : puça : Haxāmanišiya:

Grotefend's translation: **Xerxes**, **the Great King**, **King of Kings**, son of **Darius**, the Achaemenian.

The notes above show the words Grotefend identified in boldface.

Unlocking the past

It took until 1847, and the work of 17 scholars of different nationalities, to fully describe the cuneiform system for writing Old Persian. Grotefend had assumed that the script was alphabetic; in fact the 36 cuneiform characters represent consonants plus a vowel. Apart from three signs for the vowels 'a,' 'i,' and 'u,' they form a syllabary, with 22 signs for consonants plus the inherent vowel 'a,' four signs for consonants followed by 'i,' and seven for consonants followed by 'u.' There are also logographs for the common recurring words king, god, country, and earth.

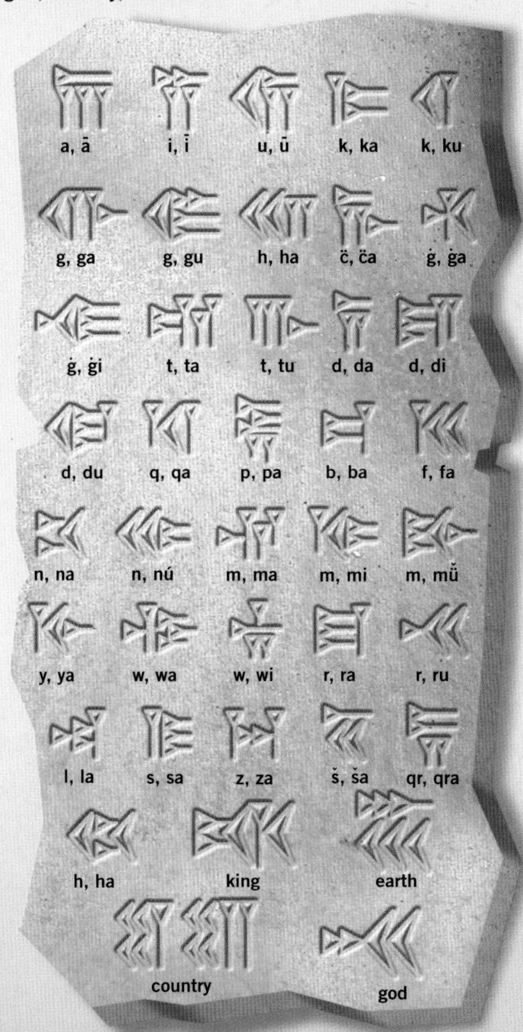

The Old Persian writing system is unique in the ancient world and seems to have been invented at the instigation of Darius himself. It appears to have been used only for the highly formulaic inscriptions of the Achaemenid kings. All the script has in common with the Akkadian and Elamite versions that accompany it is the use of cuneiform. Nevertheless, it was the decipherment of Old Persian that unlocked Akkadian cuneiform and added 3,000 years to the historical record.

ALPHABETS AND SCRIPTS

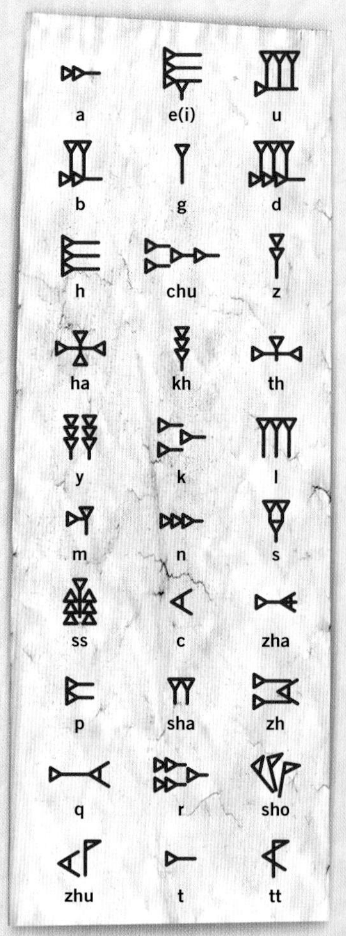

Ugaritic remains the earliest known alphabet. Dating from c.1400 BC, it was written in cuneiform. It originally comprised 22 consonants, but grew to 30.

The earliest consonantal alphabet was written in cuneiform in the city of Ugarit on the Syrian coast c. 1400 BC, but the order of the signs suggests that it was influenced by an alphabet similar to the somewhat later Phoenician, the earliest example of which dates to 1000 BC. The latter was spread throughout the Mediterranean by Phoenician traders. The Greeks perfected the system by adding signs for vowels, while to the east, in India and Southeast Asia, syllabic alphabets, possibly inspired by Aramaic letter forms, were brought to an extraordinary degree of phonetic perfection. Curiously, Akkadian cuneiform and Egyptian hieroglyphics continued to be written in the traditional way for 1,000 years after the invention of this much simpler way of writing.

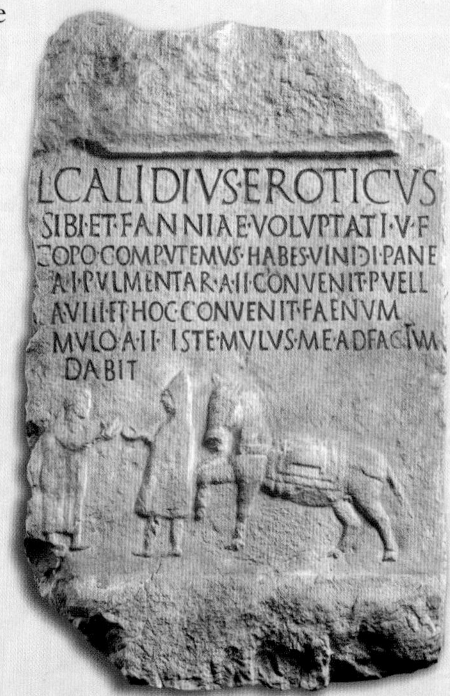

Abjads and abugidas

Ugaritic was closely related to Phoenician, Canaanite, and Aramaic, as well as to Hebrew. Such Semitic alphabets, consisting solely of consonants, are today called 'abjads,' after the first three letters, aleph, beth, and gimel, the sign for aleph representing not the vowel 'a,' but a glottal stop. Almost all scripts used for Semitic languages are abjads. Scripts like Ethiopic, which developed from the South Arabian abjad, but modified the shapes of the letters to indicate following vowels, are known as abugidas (*see Devanagari, opposite*). Most Indian and many Southeast Asian scripts are of this type.

The Greeks adopted the Phoenician alphabet (*below*), but although well adapted to writing Semitic languages, a consonantal script was clearly inadequate for a vowel-rich language like Greek. Signs that represented Semitic sounds not present in Greek were assigned vocalic values, and after much regional experimentation, the first 'true alphabet,' in which every sound of the language could be represented by a single sign, was formed. The Greek versions of the Phoenician names of the first two letters of the Greek alphabet, *alpha* and *beta*, give us our word for 'alphabet.'

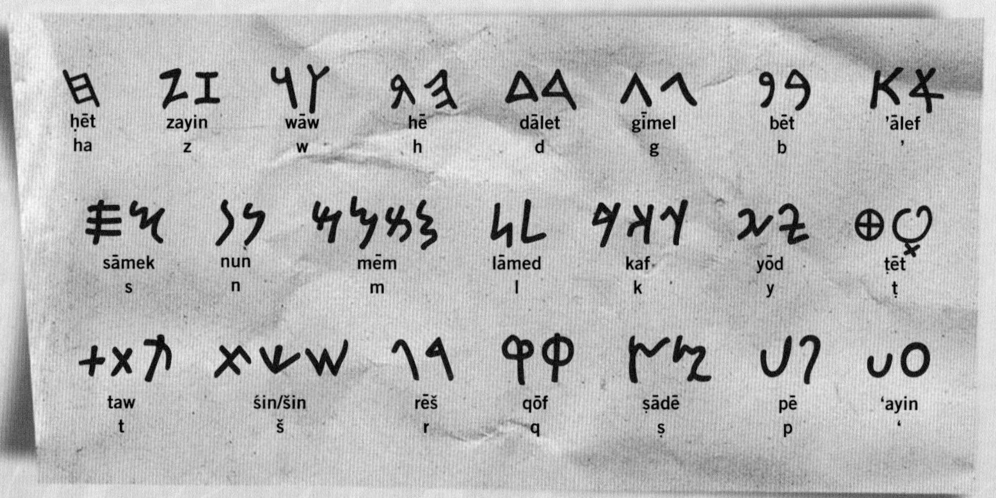

Knowing your Ps and Qs

The Roman alphabet, which forms the basis of the modern Western writing system, first appeared in inscriptions in the 6th century BC, and was probably derived from Etruscan. Originally it comprised only 21 letters, 'V' standing for both the sounds 'V' and 'U' and 'I' standing for both 'I' and 'J.' 'U' was not distinguished from 'V' graphically until the 10th century, and 'W' – originally two V's written side by side – did not appear until substantially later. 'J' was finally assigned its own graphic form in the 15th century. Italian still rejects 'K' for a hard 'C,' preferring the digraph 'CH.' Special letter forms and diacritics have been adopted for certain sounds in Scandinavian and some Central European languages, as well as Turkish.

Syllabic alphabets and syllabaries

The letters of a syllabic alphabet normally indicate consonants plus vowels by modifying the shape of the consonant letter, or by adding diacritics, or both. These alphabets are richly represented in the many complex scripts of the Indian subcontinent. The Bhahmi script is the oldest (c.300 BC), and the Devanagari (*right*) is the most widespread. A true syllabary, with a separate sign for each possible combination of consonant and vowel, would have several hundred characters. Syllabic alphabets instead modify the shapes of the letters depending on which vowel follows or precedes it. Japanese Hiragana and Katakana, and the Korean Han'gul script (*below*), are examples, and such syllabic alphabets are used to write Inuit and other North American Indian languages.

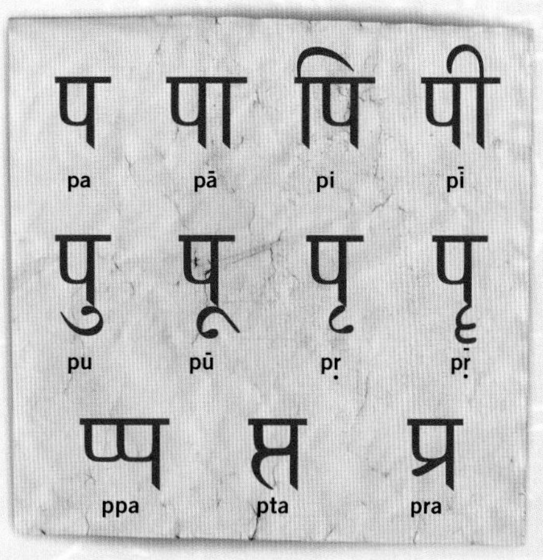

Devanagari script An example of an abugida showing how a single consonant sign is adapted to show its syllabic values.

The Korean Han'gul script is an elegant syllabary, in which the consonants and vowel sounds are treated separately, the vowel sounds acting as modifiers to the consonants.

The Roman alphabet was used not only throughout the western empire, but was carried much further by Christian missionaries in succeeding centuries, which explains its modern predominance. In the Orthodox east, Greek was still used; a new alphabet was developed by Byzantine missionaries in the 9th century which combined features of both Latin and Greek, adapted to translate the scriptures into Old Church Slavonic. Its use was spread by Saints Cyril and Methodius who led missions into eastern Europe and Russia, and where it took root and became known as Cyrillic (*right*). Its 33 letters lend themselves to Slavonic vowel sounds, and it is currently used to write some 50 Central Asian languages across the former Soviet Union.

Chinese script

First appearing as a fully developed script on oracle bones from c.1200 BC, Chinese has developed over the centuries using four fundamental types of character: pictographs, visual representations of objects; differentiated characters, not pictorial, used for various relational and abstract ideas; associative pictographic compounds, made up of two semantic elements – two graphic components whose meanings taken together suggest another word; and phonetic compounds drawn from either semantic or phonetic signs which, when combined, indicate pronunciation and meaning. The latter are used for some 90% of modern Chinese writing. Chinese today comprises some 60,000 characters, although less than 4,000 are normally used.

Writing Chinese requires knowledge of a huge number of characters.

	Pictograph Horse	Differentiated characters Upwards, rising	Pictographic compound Sunset, ending	Phonetic compound Willow
c.1200 BC for divination	馬	上	莫	柳
c.1500 BC for religious purposes	馬	上	莫	柳
221 BC for proclamations or names	馬	上	莫	柳
c.200 BC for official texts or literature	馬	上	莫	柳
c.AD 200 for official texts or literature	馬	上	莫	柳
c.AD 1400 for general use	馬	上	莫	柳
1956 for general use	马	上	莫	柳
c.AD 200, for drafts, notes, and letters	马	上	莫	柳

The Evolution of Numerical Systems

The Ishango tally bone
This four-inch (10-cm) bone was found in the Congo, and has been tentatively dated to at least 6500 BC; several of the groups of incised notches form prime numbers. Its function remains obscure – was it an early form of calculator, or merely a method of recording kills, or days, or other values? Similar tally sticks, some much older, have been found in southern Africa and central Europe, and the San Bushmen of Namibia today count the passing of time on similar counting sticks.

Keeping an account of things necessitated the creation of numbering systems, and these remain probably the first example of codifying an abstract concept, preceding the evolution of writing in most parts of the world. Over 30,000 years ago wooden or bone tally sticks were used by hunting communities, probably to record the number of animals they killed. Clay counting tokens from Sumer in Mesopotamia dating from 3400 BC show the beginnings of a stocktaking and accounting system, used by producers and tradesmen, the basis of the first written numbering system. Humans usually rely on their digits (fingers and toes) for rudimentary counting, hence most systems have a decimal base of ten, although the Maya, Aztecs, and Celts used base 20, while in Mesopotamia base 60 was adopted. Alphabetic counting systems, in which numbers were written using letters, were used by the Greeks, Romans, Hebrews, and later the Arabs – although 'Arabic' numerals are now used throughout most of the world.

Keeping a tally of livestock and crops was essential as sophisticated states such as Egypt developed.

c.3300 BC Sumer
additive, base 60
1 digit 0.75 inch, (1.65 cm)
30 digits 1 cubit

c.3000 BC Egypt
4 digits 1 palm (4 inches, c.7.5 cm)
7 palms 1 cubit
100 cubit 1 khet (rod)

c.1950 BC Crete
additive, base 10, hieroglyphic

c.1800 BC Babylonia
positional, base 60, cuneiform numeric

c.1450 BC China
additive and multiplicative, no base

c.1400 BC Hittites
additive, base 10, cuneiform numeric

4000 — 3000 — 2000

Maths in cuneiform
In Mesopotamia, the Sumerians and Babylonians developed a sophisticated and flexible sexagesimal system, base 60, probably derived from their astronomical observations and calendrical calculations. Sixty is divisible by 2,3,4,5,6, and 10. Despite the prevalence of a decimal base in most aspects of the modern world, the sexagesimal approach remains with us today in our use of 60 seconds in a minute, 60 minutes in an hour, and the division of angles as proportions of a full circle of 360 degrees, each degree in turn comprising 60 minutes.

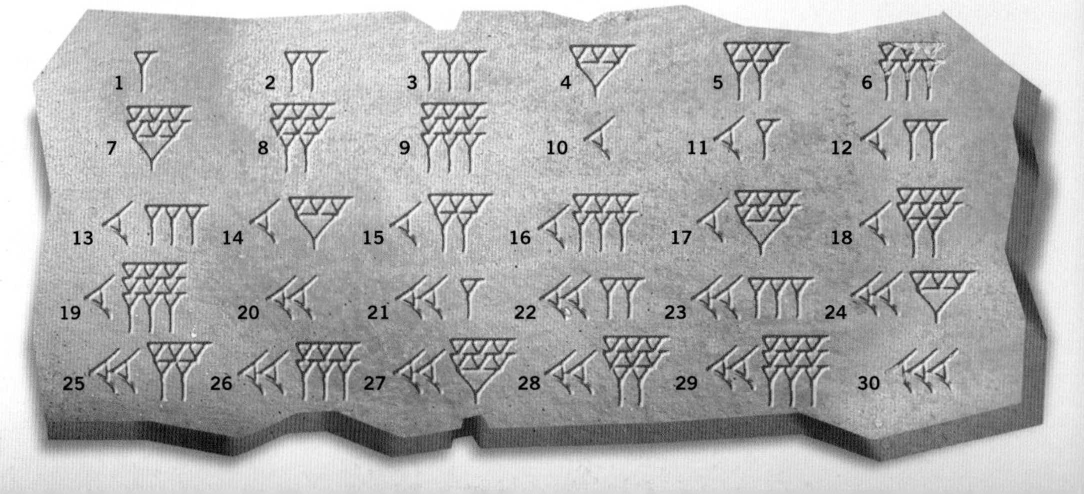

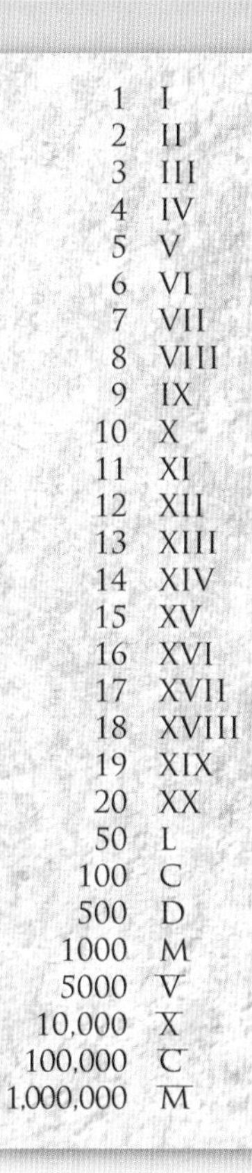

1	I
2	II
3	III
4	IV
5	V
6	VI
7	VII
8	VIII
9	IX
10	X
11	XI
12	XII
13	XIII
14	XIV
15	XV
16	XVI
17	XVII
18	XVIII
19	XIX
20	XX
50	L
100	C
500	D
1000	M
5000	V̄
10,000	X̄
100,000	C̄
1,000,000	M̄

Roman numerals

This is an example of an ancient alphabetic system still in occasional use today, notably in a formal or monumental context, and often to express the date of a copyright. A mixture of positional and additive, it uses a total of seven characters to express any number, the numbers four and nine being shown as five minus one, and ten minus one respectively.

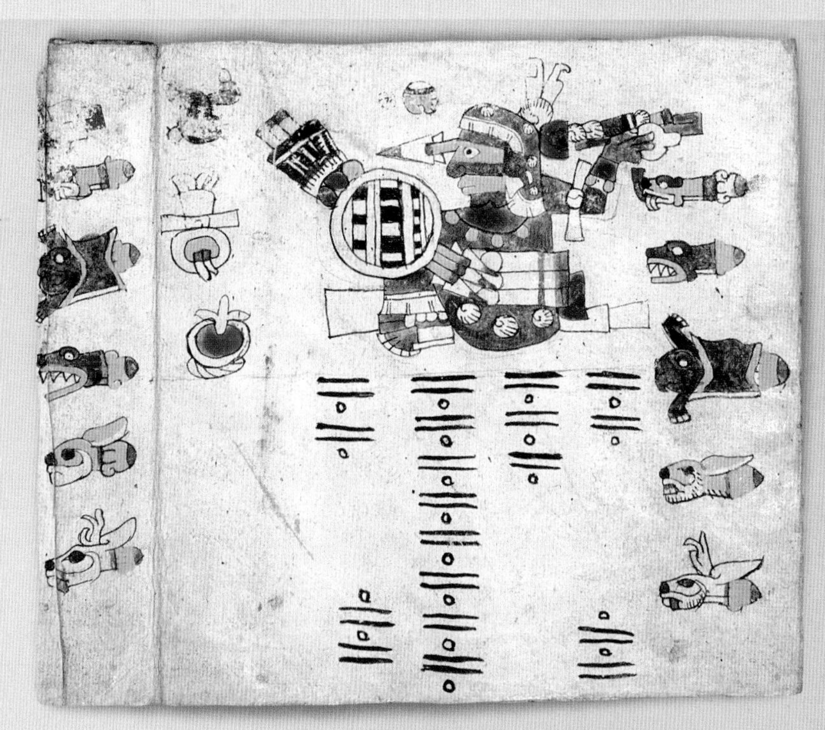

Mesoamerican numerals

The Maya employed an elegant and simple numbering system. The numbers 1-19 were indicated by just two shapes, a dot or circle, and a bar. Twenty was usually indicated by a symbol for 'completion' or 'zero.' This base 20 system was fundamental to the complex and interlocking calendar systems used by the Maya (*see page 152*). The numbers one to four were indicated by the respective number of circles, five by a bar, six by a bar surmounted by a circle, ten by two parallel bars and so on, up to 19, a stack of three bars surmounted by four circles.

c.700 BC Greece
additive, base 10, alphabetic
4 daktyloi (a finger's breadth) 1 palaste (palm of the hand)
4 palastai 1 foot, c.12 inches, (30 cm)
1.5 feet 1 cubit

c. 500 BC Rome
additive, base 10, alphabetic
4 digitii (a finger's breadth) 1 palm
4 palms 1 pes (1 foot, c.12 inches, 30 cm)
5 pedes 1 passus (pace)
1,000 passus 1 mile

c.AD 240 India
positional, base 10, numeric

c.AD 450 Maya
positional, base 20, glyphic

c.AD 1000
Arab world using an adapted form of Indian numeric system

1000 BC 0 AD 1000 2000

c.200 BC Hebrew
additive, base 10, alphabetic

c.AD 1200
Arabic numerals introduced to Europe

Arabic numerals

The numbers recognized and used most widely today are known as 'Arabic' numerals. This is a misnomer, as the system originated among Hindu mathematicians in India, and was the first system to include the idea of zero, and effectively represents a decimal system, as only the digits 0-9 are described, higher numbers being expressed as combinations of these. It is known as a place-value system. It was adopted by the Arabs, who hitherto spelled out mathematical functions, and via them it spread to the West.

Brahmi		—	=	≡	+	𑁖	𑁗	𑁘	𑁙	𑁚
Hindu	०	१	२	३	४	५	६	७	८	९
Arabic	٠	١	٢	٣	٤	٥	٦	٧	٨	٩
Medieval	O	I	2	3	ꝗ	ς	6	ʌ	8	9
Modern	0	1	2	3	4	5	6	7	8	9

Medieval numbers

The Italian mathematician Leonardo Pisano (also known as Fibonacci) encountered the Arabic system whilst in North Africa, and introduced the numerals to Europe in the 13th century. However, over the next few centuries, a range of numeral systems, some of them regarded as esoteric, were in use in Europe, although finger counting predominated (*right*).

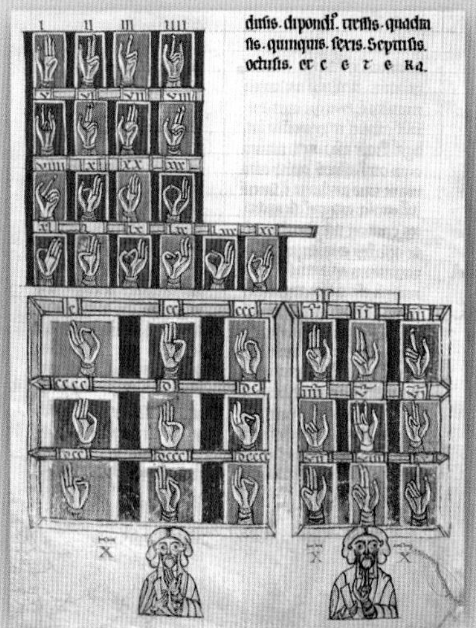

Linear A and Linear B

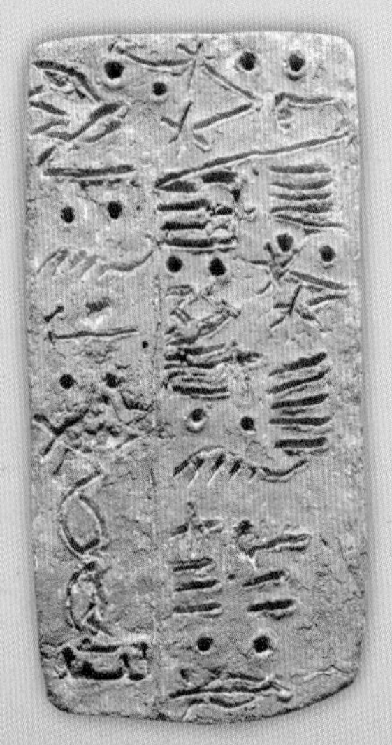

The evidence

The signs used for both Linear A (*above*) and Linear B are quite similar, but there are major differences in rendering fractions, and a sign used in inventories clearly meaning 'total' was different. No sequence of signs was the same in both scripts. It was clear that the tablets were written in different languages.

Evans deduced from the number of signs used in Linear B that the script was syllabic. He determined the direction of writing (left to right), identified a word divider, the use of logograms or pictographs, and a decimal system for counting. The tablets were inventories of some kind, in the main consisting of word(s) plus pictogram plus number. When Evans stopped working at Knossos in 1904 some 3,000 Linear B tablets had been found but few were published until 1953.

In 1939, 600 more tablets in Linear B were unearthed, on the mainland, at Pylos. These were published in 1951, and the editor, E.L. Bennett, established the basic forms of the 87 signs that make up the Linear B writing system.

On April 5, 1900 Sir Arthur Evans discovered a cache of clay tablets in the palace complex of King Minos at Knossos in Crete. They were inscribed with signs drawn with a stylus. Evans had earlier come across seal stones in antique shops in Athens bearing a similar unknown hieroglyphic script. He traced these to the island of Crete, where the local women wore them as charms. In 1902 another deposit of tablets was found at Hagia Triada. These were written in a script similar to the Knossos tablets, but with significant differences. This script Evans called 'Linear A,' and the tablets found at Knossos he called 'Linear B.' He regarded both scripts as 'Minoan,' and assumed they had developed from the hieroglyphic signs he had found on the seals.

Sir Arthur Evans (1851-1941) excavated much of the palace at Knossos (*below*), revealing the rich culture of Minoan Crete.

Before Greek

Seven of the signs of Linear B were similar to those of a syllabic script used on Cyprus. Deciphered in 1871-73 by George Smith and Moritz Schmidt with the help of a Phoenician bilingual, this proved to be an early form of Greek. One of these was the sign for the syllable 'se,' used in the Cypriot writing system both for this syllable and for the final 's' (the most common ending for Greek nouns) in which case the 'e' sound was not pronounced. Yet the sign indicating it, shared by both Linear B and the Cypriot syllabary, almost never occurs as the end of words in Linear B. This convinced scholars that Linear B could not be Greek. There was another, even more compelling reason: the palace of Knossos had been destroyed around 1400 BC, and it seemed inconceivable to Evans that Greek could have been spoken in Crete then, some 800 years before the earliest inscriptions in Greek.

Kober's contribution

The first step in the right direction was taken between 1945-49 by the US scholar Alice Kober who proved, by comparing recurrent sets of signs, the existence of declension. If three five-letter words have the form 'abcde,' 'abcfg,' and 'abchi,' it was logical to suppose that the final groups '-de,' '-fg,' and 'hi' were inflectional endings. She then showed that a sign group occurring before the numerals must mean 'total' and had two forms: one of these occurred on tablets with inventories of women and one class of animals, another on tablets with inventories of men and another class of animals, swords, and tools. She suggested that this indicated that the language represented by the signs possessed gender. These observations were the first positive advance toward decipherment.

Linear B
Many of the tablets are clearly inventories or ledgers recording trade and products. The writing reads left to right, obvious from the alignment of the initial characters.

Lists
The writing in the left-hand column identifies particular items.

Characters
When cataloged, the writing system totaled 87 characters, far too many for an alphabet, meaning that some characters were inflectional, and also probably possessed a gender.

Numbers
The right-hand column includes a decimal counting system, preceded by variations on a symbol for 'total.'

The Ventris insight

A breakthrough occurred in 1952. A young British architect named Michael Ventris (1922-56), interested in Minoan scripts since his boyhood, had published his first paper at age 18 on the Linear B tablets, suggesting that their language might be Etruscan. After serving in the war, he returned to the problem. In 1950 he circulated a questionnaire among scholars working in the field to ascertain current thinking on the script. The general consensus appeared to be that the language was Indo-European, possibly related to Hittite. Ventris still thought it might prove to be Etruscan, a language with no known affinities. In 1951 E.L. Bennett's publication of the Pylos tablets provoked Ventris and others to compile frequency lists to discover which signs most often occurred at the beginning and end of words. This resulted in the tentative identification of the sign for 'a.' Over the next 18 months Ventris circulated 20 'work-notes' among his correspondents. These included four grids, inspired by Kober's methodology. The signs were plotted on the grid, their frequency noted, and tentative guesses as to their accompanying vowel plotted, contributing greatly to the eventual decipherment.

A clue from Homer

But the real breakthrough came through the identification of a place name. Many ancient scripts so far deciphered had given up their secrets through the identification of a proper name (*see pages 22, 34*). In the case of Linear B, no personal names were known. Ventris had noticed a group of signs which occurred on lists of different commodities, and he suggested that they might represent place names. Homer mentions the port of Amnisos, near Knossos. In a syllabic script, this would be written 'a-mi-ni-so;' Ventris had already identified the sign for 'a,' and was sure, from analogy with Cypriot, of the sign for 'ni.' Assuming the values of the second and last syllables to be 'mi' and 'so' gave him the tentative value of two more signs. Ventris was then able to identify the adjectival forms of these two words, 'a-mi-ni-si-ya' and 'a-mi-ni-si-yo.'

Ventris then conjectured that a group of three signs, the third of which could now be read as '-so,' spelled 'ko-no-so' – Knossos. He had learned an important point about Linear B – the final 's' was not written. This explained the 'un-Greek' look of the language. A spelling convention had concealed the language of the tablets.

The tripod clue

Working with the former cryptanalyst John Chadwick, Ventris began to identify other possible Greek words in the script. In May 1953 Ventris (*above*) received a letter from the excavator of Pylos that confirmed his investigations into Linear B in an unexpected way. The letter described a tablet listing different pots, whose shapes were clearly drawn and followed by a number. Applying his values to the signs preceding the drawing of a three-legged pot, Ventris obtained 'ti-ri-po-de;' the four-handled pot gave 'qe-to-ro-we,' the three-handled pot 'ti-ri-o-we,' and the pot with no handles 'a-no-we.' These were all instantly recognizable Greek words – tripod, 'four-handled,' 'three-handled,' and 'no-handled.' The evidence destroyed Ventris' original theory, proving that the language Linear B represented was indeed Greek, possibly imposed on the Minoans by their mainland neighbors. The Minoans had adapted their writing to accommodate the language. The mystery of Linear A, still probably a way of writing native Minoan, remains unsolved.

On June 24, 1953, surely an *annus mirabilis*, Ventris announced the decipherment. It was reported on the same day that Edmund Hillary scaled Everest (*see page 86*).

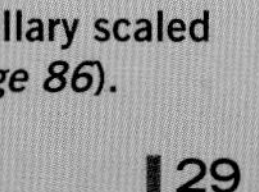

THE PHAISTOS DISC

Discovered in 1908, during an excavation of the basement rooms of a Minoan palace at Phaistos in Crete, the origins and dating of this baked clay disc remain disputed, but it probably dates from the first half of the second millennium BC, the Middle or Late Minoan period. It measures about 6 inches (15 cm) in diameter, and ½ inch (1 cm) thick. The clay disc is decorated on both faces with a total of 241 pictograms (comprising 45 individual tokens or characters) arranged in a continuous clockwise spiral on each face. It remains an example of a script that has proved unreadable.

Spiral
The spiral's lines, which contain the groups of pictograms, were incised by hand, as were the vertical lines that group or divide the characters, presumably into 'words,' or possibly 'sentences.'

Eagle
Appears five times, only on Side A.

Shield
The second most frequent pictogram, occurring 17 times. In 13 instances, it directly follows the plumed head pictogram; in the remaining four it falls at the end of a word.

Carpentry plane
Appears three times, only on Side A.

Side A

It is not clear which is the 'front' or which is the 'rear' face of the disc, or in which order they should be read, so they are simply termed sides A and B. Side A has 31 'words' on it. Each of the pictograms was impressed in the clay using a stamp – an extremely early example of movable type printing.

The Phaistos disc characters

The impressions on the disc have been organized into an analytical table naming the characters, and counting the frequency with which they occur (*right*). Although some similarities have been noted to identifiable Cretan hieroglyphics and related Egyptian hieroglyphs, and also to both Linear A characters and Anatolian hieroglyphs (*see pages 28, 32, 34*), few other parallels have been found. It seems unlikely in the absence of any other archaeological finds using the same alphabet of pictograms that the meaning of the disc will ever be unlocked. The characters have been provisionally assigned Unicode status.

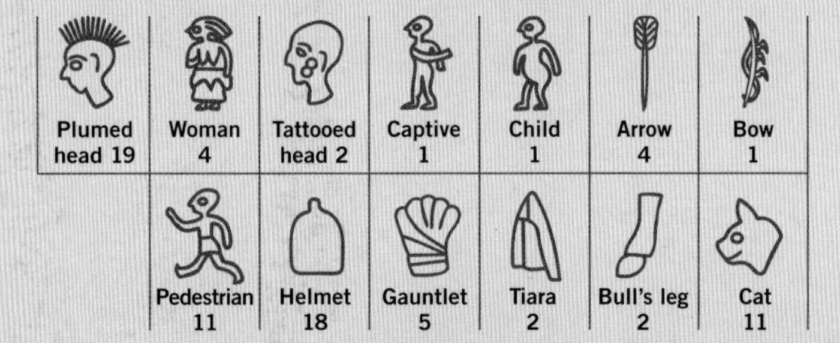

Plumed head 19	Woman 4	Tattooed head 2	Captive 1	Child 1	Arrow 4	Bow 1
	Pedestrian 11	Helmet 18	Gauntlet 5	Tiara 2	Bull's leg 2	Cat 11

Side B

Only 30 'words' appear on this face. It is notable how many of the pictograms used on both sides are immediately recognizable, and function successfully as icons.

The plumed head
This pictogram is the most frequent, appearing 19 times, and only ever occurs at the beginning of a word.

Vine
Appears four times, only on Side B.

Sling
Appears five times, only on Side B.

Indicators
Some letters are accompanied by diagonal strokes, also incised by hand. These may indicate the beginning or ending of a word, depending on the direction of reading, although it is generally accepted that the disc should be read inward, toward the center.

Strainer
One of only nine pictograms which occur only once, known as hapaxes.

Shield 17	Club 6	Manacles 2	Mattock 1	Saw 2	Lid 1	Boomerang 12	Carpentry plane 3	Dolium 2	Comb 2	Sling 5	Column 11	Beehive 6	Ship 7	Horn 6	Hide 15
Ram 1	Eagle 5	Dove 3	Tunny 6	Bee 3	Plane tree 11	Vine 4	Papyrus 4	Rosette 4	Lily 4	Ox back 6	Flute 2	Grater 1	Strainer 1	Small axe 1	Wavy band 6

THE MYSTERY OF HIEROGLYPHS

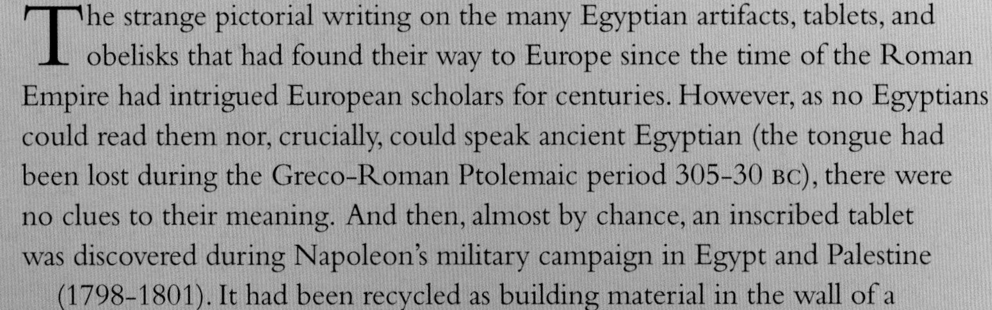

The strange pictorial writing on the many Egyptian artifacts, tablets, and obelisks that had found their way to Europe since the time of the Roman Empire had intrigued European scholars for centuries. However, as no Egyptians could read them nor, crucially, could speak ancient Egyptian (the tongue had been lost during the Greco-Roman Ptolemaic period 305-30 BC), there were no clues to their meaning. And then, almost by chance, an inscribed tablet was discovered during Napoleon's military campaign in Egypt and Palestine (1798-1801). It had been recycled as building material in the wall of a fortress at Rosetta in the Nile Delta. Inscribed in 196 BC, it described a priestly decree in three languages.

Earlier interpretations of hieroglyphs

Ancient Egyptian writing, known as hieroglyphic, had long been assumed to be a pictographic script – that is, one in which the individual pictograms represent an idea, basically a rebus or picture writing. Many imaginative attempts had been made to interpret them, although one problem was that while many of the hieroglyphic characters appeared to represent a recognizable item (a falcon, a plow, and so on) there remained a number of cursive notations which were impenetrable. Were they also highly stylized pictures, or were they merely punctuation marks or linking signs? Also there appeared to be no structural order to the presentation of the script: sometimes characters were presented in rows, sometimes vertically in columns. Nevertheless, at no time did the earlier scholars assume that hieroglyphic was true writing, and that it had, at root, a phonetic function.

Hieroglyphs were first thought to be picture writing.

Napoleon's invasion of Egypt provided the opportunity for French antiquarians to study Egyptian monuments, and to advise Napoleon on what to loot. Among the booty was the Rosetta Stone that, as a result of his defeat by the British, was one of many artifacts which found their way back to England. The arrival of the artifacts caused a wave of interest among antiquarians and scholars.

The Rosetta Stone

This extraordinary find contained a single text in three scripts which, when read against each other, constituted a decoding manual. However, due to its damaged state many parts of the comparable texts are missing. While the Rosetta Stone made it possible to attempt to read the Greek text against fragments of the hieroglyphic text, it was not possible to reconstruct how ancient Egyptian actually worked.

Upper section Hieroglyphic text, but unfortunately the most damaged; only 14 lines remain, which more or less correspond to the last 28 lines of the Greek text, but none is complete.

Middle section Demotic text comprising 32 lines, but damaged top right which, as demotic is written right to left, means that the opening of the first 14 lines is missing.

Lower section Greek text, and the only immediately readable one, comprising 54 lines, of which the last 26 are incomplete. This key text eventually allowed both Egyptian hieroglyphs and demotic to be unraveled.

Unlocking the puzzle

Thomas Young, a gifted linguist and polymath, was fascinated by the Rosetta Stone. Looking for a common link between the parallel texts, he noticed the appearance of several characters within loops or cartouches. Guessing (correctly) that this might be a way of emphasizing something special, he compared these characters to the only pharaoh featured in the Greek text – Ptolemy (or Ptolemaios). Being a proper name, he knew how it was spoken, and so began to build the beginnings of a linguistic alphabet.

Thomas Young (1773-1829) was a linguist and scientist, whose Cambridge sobriquet was 'Phenomenon Young.'

Ptolmis (Ptolemy)

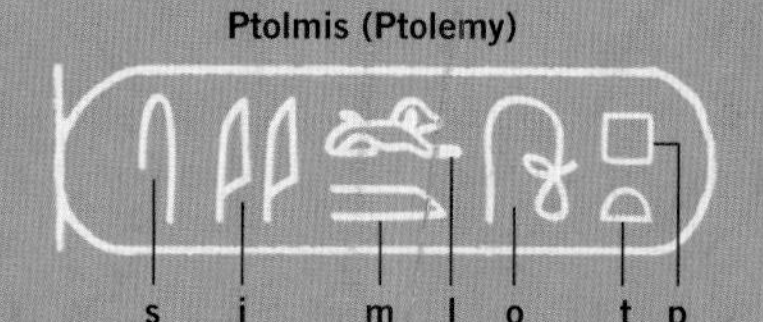

Shown in its original right to left order, the core spelling reveals the name 'Ptolmis,' also spelled Ptolemaios (in Greek) or Ptolemy (in English).

He repeated the exercise with an inscription featuring another Ptolemaic ruler, Queen Berenika. Nevertheless, as the Ptolemaic dynasty was Greek rather than native Egyptian, and characters would have had to have been made up to phonetically spell such a name, he still assumed that in principle hieroglyphics remained picture writing.

Berenika (Berenice)

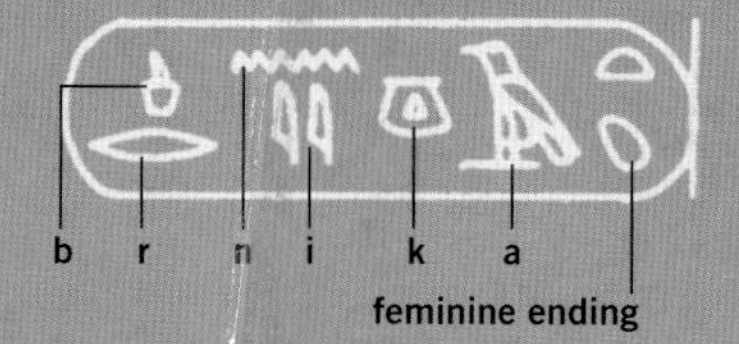

Shown reading left to right, the absence of the 'e' vowel is clear, while the final characters are not to be read, but are a symbol indicating a feminine termination, often included after the names of queens or goddesses.

Clues to the code

After successfully deciphering the cartouche of Ptolemy V (205–180 BC) as displayed on the Rosetta Stone, Young began to build a hieroglyphic dictionary. He also began to recognize certain features of hieroglyphic writing which would help him unravel a further cartouche found at the temple of Karnak at Thebes, that of a later Ptolemaic Greek queen, Berenika, or Berenice, (58–55 BC).

Right to left
One elementary factor was that ancient Egyptian was usually written from right to left, unlike most European scripts, but not uncommon elsewhere.

It looks better like this
In decorative designs involving hieroglyphs, the scribe, painter, or engraver may make arbitrary decisions about the position of certain characters on aesthetic and compositional grounds.

Extra symbols
The cartouche for Ptolemy occurs six times on the Rosetta Stone, repeating the core spelling, but often with additional hieroglyphs which Young assumed were included as describing various titles (such as Ptolemy 'the Great').

Fewer vowels
Egyptian spelling often missed out vowel sounds, possibly reflecting the way in which these names were actually spoken at the time.

HIEROGLYPHS REVEALED

Determination
Champollion (*above*) decided to become the first person to read ancient Egyptian. Trying Young's ideas out on cartouches for Cleopatra, Alexander the Great, and Rameses II, Champollion was able to compare his solutions, allowing him to establish the phonetic value of some signs, and the alphabetic value of others. Using this knowledge, he returned to the Rosetta Stone and concentrated on the text outside the cartouches.

After Thomas Young had made the first inroads to understanding how hieroglyphs worked (*see page 33*), he set his studies aside, distracted by other interests. It was left to an equally talented young Frenchman, Jean-François Champollion (1790-1832), to follow up Young's lead. Working from copies of inscriptions, he established the principles of Egyptian writing and its grammar. Only late in his short life did he travel to Egypt to see the abundance and variety of Egyptian hieroglyphs with his own eyes.

Champollion's success

By applying Young's technique to other inscriptions from the Ptolemaic period, Champollion was able to confirm Young's readings for Greek names enclosed in cartouches. Applying the same method to a non-Ptolemaic inscription from Abu Simbel, known to have been built by Rameses II, he succeeded in identifying and reading the name of this pharaoh, the first purely Egyptian name to be deciphered. Champollion had studied Coptic and realized that this, the liturgical language of the Coptic church, was descended from the language encoded in the hieroglyphic inscriptions. This was a tremendous help in testing new readings. Champollion quickly realized that the hieroglyphic system consisted of logograms – signs standing for a single word or concept – as well as phonograms, signs indicating one to three consonants, and signs which functioned as determinatives, used to distinguish homophones. He finally proved that hieroglyphs were indeed a form of true writing. By September 1822 Champollion was able to write to the Académie des Inscriptions et Belles-Lettres announcing his findings. The news provoked a wave of interest, transforming the study of ancient Egypt.

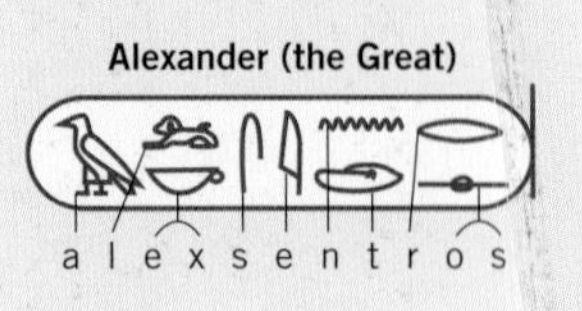

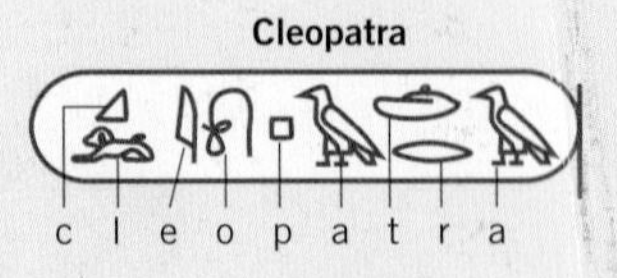

Champollion worked on the cartouches for two Ptolemaic Greek names, Alexander and Cleopatra, before attempting a native Egyptian, Rameses II.

Egyptian figures were often combined with pictograms and hieroglyphs. Here Rameses III is shown with the goddess Isis, with their cartouches identifying them.

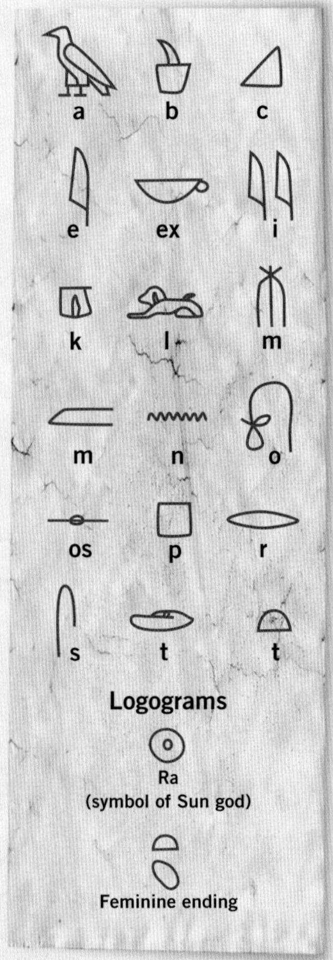

Champollion's first alphabet
From his readings of the royal cartouches, and those of Young, Champollion developed a tentative alphabet. Some assumptions, such as the gender indicator and the logogram for Ra, the Sun god, proved correct; some needed to be adjusted as more inscriptions were unraveled.

How hieroglyphs worked

In total there are over 2,000 hieroglyphic characters, each of which is derived from a pictogram of a common object, animal, or activity in ancient Egyptian life. They could represent letters of the alphabet, a phonetic sound, a gender, numbers, an abstract idea, or a non-phonetic idea linked to the word (a determinative).

The hieroglyphic alphabet The hieroglyphic script had no vowels, so we do not know how the language was pronounced. The signs used for transcribing foreign words, the so-called 'hieroglyphic alphabet,' in fact indicated consonants, but some were arbitrarily used to represent Greek vowel sounds.

Determinatives

Certain characters, largely pictograms, were not designed to be read as part of the alphabetic sentence, but rather to explain the meaning of the alphabetical letters which preceded it. They could have various functions:

Adjectives and adverbs A set of commonly used pictograms were developed to cover a wide range of situations, mostly quite straightforward:

Clarification This would arise when the consonants used to spell a word might produce ambiguity (as would be the case with our words 'duck' and 'deck,' which in Egyptian would both produce merely the hieroglyph for 'D' and the hieroglyph for 'K.' In such an instance a pictogram of a duck might follow the first characters, while a pictogram of a boat might follow 'deck'). In hieroglyphs this was true of the words 'taste' and 'boat': scribes would add an image of a tongue after the former, and a ship after the latter to make the sense of the preceding alphabetic letters clear. The Egyptian spelling of the noun 'fish' was 'rem.' So was the Egyptian spelling of the verb 'to cry.' Both spellings simply involved the Egyptian alphabetic letters for 'r' and 'm.' In order to distinguish which word was meant, a determinative was added.

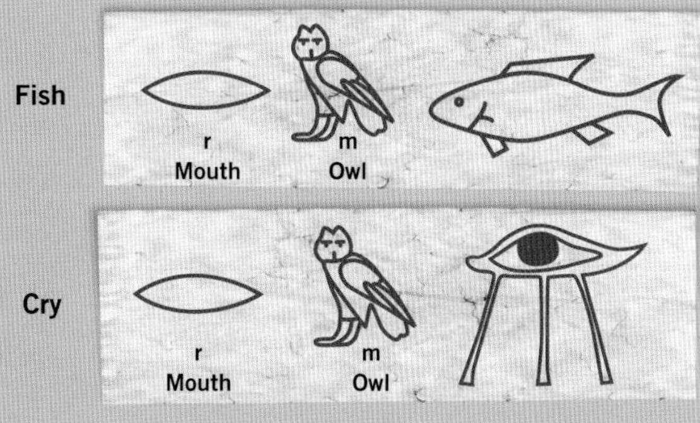

Counting

The Egyptian system of counting was additive and incorporated a variety of signs. Fractions were usually indicated by numbers written beside or beneath the symbol for a mouth, indicating a part.

	1-9	Vertical or horizontal bar
	10-90	Cattle hobble
	100-900	Rope
	1,000-9,000	Lotus
	10,000	Raised finger
	100,000	Tadpole
	1,000,000	God supporting the sky

Abstract ideas

Symbols associated with everyday activities and phenomena could convey otherwise indescribable concepts, as shown below.

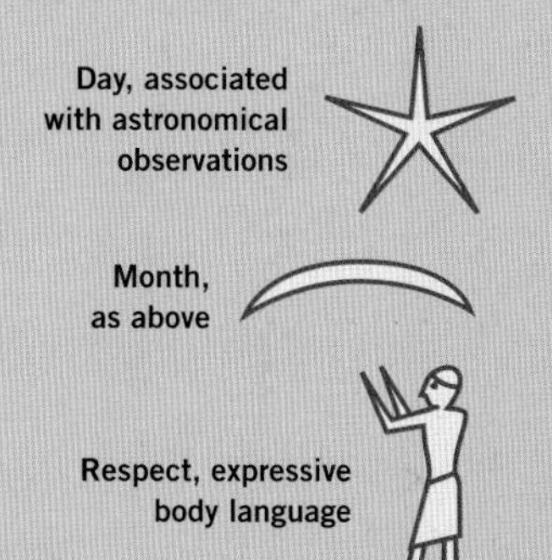

Pictograms often symbolized more than could be communicated in words, and were interpolated among hieroglyphic writing. This relief from the 12th Dynasty mortuary temple of Senwosret I at el-Lisht shows the symbolic union of Upper and Lower Egypt by depicting the gods Seth on the left, and Horus on the right, tied together around the hieroglyphic sign for unity. Both are identified and described in the hieroglyphic writing above them, as is the act of union of the two parts of Egypt.

THE RIDDLE OF THE MAYA

When the Spanish *conquistador* Cortés looted villages on the Gulf coast of Yucatán in 1519, he found books in the houses of the Maya inhabitants. Peter Martyr, the earliest chronicler of the discoveries, described the few copies that had been sent to the Spanish court: "The characters are very different from our own; dice, hooks, loops, strips and other figures, written in a line as we do; they greatly resemble Egyptian forms." Many believed that Maya was just picture-writing, but some scholars persisted, eventually proving otherwise. Maya remains the only Mesoamerican language to have been partly deciphered.

Glyphs in stone

The loss of Mayan books was compensated by the large number of surviving Mayan inscriptions, not only the monumental stone friezes and stelae, but the numerous painted inscriptions on ceramic vessels and even on the walls of caves. These date from c. AD 200 up to the Spanish conquest. Diego de Landa's account of the Mayan calendar meant that their system of time-reckoning was quickly understood, as was the Mayan numerical system (*see pages 26, 153*). Most archaeologists came to believe that the Mayan inscriptions were largely calendrical, and many doubted that the other Mayan glyphs could ever be read, regarding them as a kind of 'picture-writing,' rather than a phonetic script.

Emblem glyphs In 1973 scholars showed how the monumental inscriptions at major sites like Palenque *(below)* could be used to 'read' the architectural monuments in which they were placed and the rituals that were carried out by their rulers.

Lost libraries

Mayan books were written on sheets of whitened bark paper and folded like a Japanese screen, bound between wooden covers – identical in form to some Buddhist manuscripts. Only four survive. In 1562, the Franciscan Provincial of Yucatán, Bishop Diego de Landa, destroyed all the Mayan books he could get his hands on, deeming them unchristian. Ironically, it was de Landa who also provided the essential key to the decipherment of the Mayan writing system, giving an informed account of the Mayan names and symbols for the 20 days of the 260-day 'short calendar,' adding, "These people also used certain characters of letters with which they wrote in their books ancient matters and their sciences, and with these figures and with some signs of them, they understood their matters and made them understood and taught them."

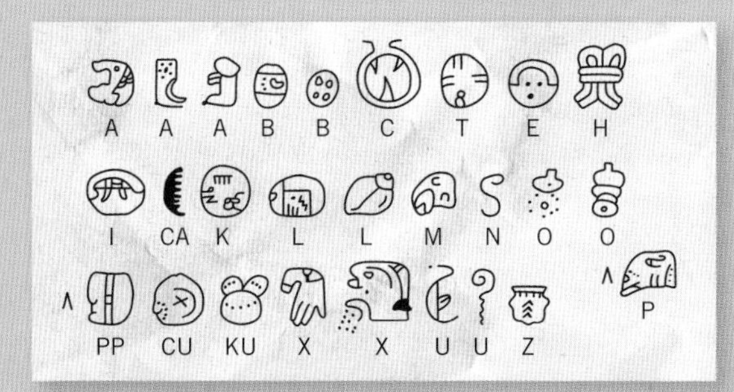

Searching for the alphabet

Theoretically, the decipherment of the Mayan writing system should have been relatively easy. Over 30 closely-related Mayan dialects are still spoken in Mesoamerica, so the type of language that lay behind the glyphs was known. The early Spanish chroniclers gave good descriptions of the physical appearance of Mayan books, establishing the important fact that the script was actually used as a writing system. De Landa actually gave the glyphs for the months and days of the Mayan calendar and explained its workings, as well as supplying the characters of what he called the Maya 'alphabet.' He had asked his Maya informant to write the letters of the Spanish alphabet in the Mayan script. In a classic case of cultural misunderstanding, they transcribed the *sounds* of the Spanish letters of the alphabet according to the principles of their writing system, which was of course not alphabetic. When de Landa pronounced 'a,' they heard the syllable 'ah'; when he pronounced 'b,' they heard 'ba'y.' De Landa asked them for an alphabet; what they gave him was a syllabary. But in this lay the key.

Breakthrough

The de Landa 'alphabet' was central to the work of Yuri Knorosov, a Russian linguist who first cracked the Maya code in 1952. His starting point was simple: writing systems are designed to be read. If we accept that Mayan glyphs are a writing system, then it must contain a phonetic element, or de Landa's account would not make sense. He knew from his study of Egyptian hieroglyphs and Akkadian that both these writing systems made use of signs that were both conceptual (logograms) and phonetic, and that both employed determinative signs for distinguishing homonyms. He postulated that the Mayan system was similar. He proceeded step by step:

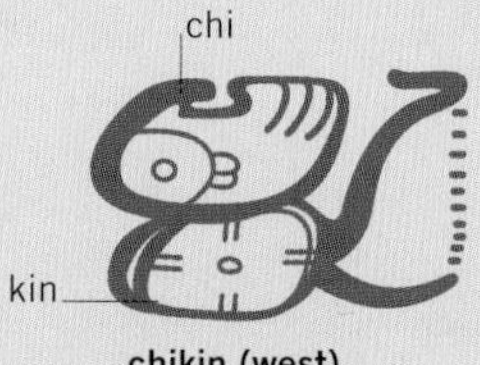

chikin (west)

1 The composite sign for 'west' and its pronunciation, ***chikin***, had been already identified. It consisted of two elements, a 'grasping hand' surmounting the sign for sun, ***kin***. Therefore the 'grasping hand' sign must be the phonetic element ***chi***.

2 The sign for the syllable ***ku*** was known from de Landa's 'alphabet.' Knorosov noticed that the same sign appeared above a picture of the Vulture God in the Madrid Codex, directly above the sign he could now read as ***chi***, spelling the word ***ku-chi***. The word for 'vulture' in Yucatec is ***kuch***. He had incidentally discovered the important fact that the final vowel was silent, common in syllabic scripts.

3 Following the same method, Knorosov found de Landa's syllable ***cu*** together with an unknown sign over a picture of a turkey. The Yucatec word for 'turkey' is ***cutz***, so he had discovered the value of yet another sign, ***tz(u)***. He went on to identify the signs for 'burden,' 'eleven,' 'dog,' 'quetzal,' and 'macaw.'

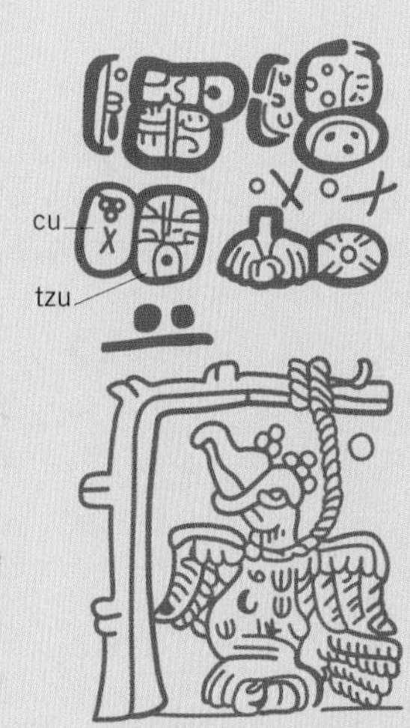

cutz (turkey)

cutch (burden)

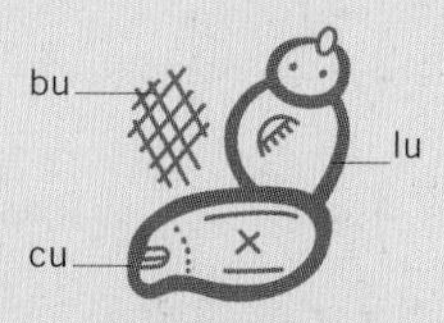

buluc (eleven)

4 However, what became clear was that the glyphs stood for two things: the syllables were derived from logograms that represented words formed from single syllables, such as ***ka*** for 'fish fin.' These in a sense formed the 'alphabet.' Alongside these were a host of genuine logograms representing things or ideas rather than syllables. Thus, the combination of syllables and logograms offered a number of possible ways of expressing a particular word or phrase.

5 It was soon apparent that, like the Egyptians, Mayan scribes and inscribers had a certain amount of design flexibility; while generally following a pattern of writing in columns two glyphs wide to be read left to right, there were no exact grammatical rules about how the individual glyphs might be assembled. This small beginning was the key to the decipherment of Mayan glyphs. More details of the writing system were worked out by many scholars over the next 40 years, and the task is still not complete.

The Maya hieroglyphic word for 'mountain' is ***witz***. It can be written three different ways; as a logogram; as a syllable sign ***wi*** prefixed to the logograph ***witz***, for mountain; or simply spelled syllabically, using two phonetic signs, ***wi*** + ***tz(i)***.

witz

wi-witz

wi-tzi

INDIGENOUS TRADITIONS

There are today thousands of 'lost' cultures, many highly sophisticated, with rich traditions, rituals, and myths, with equally complex means of expressing and commemorating them. Many oral traditions in the Americas, Africa, and Australasia have been eroded by the relentless rise of globalization. However, there remain some enigmatic fragments through which, like Mayan glyphic writing (*see page 36*), a rich but lost past can at least be glimpsed.

A lost heritage

Totem poles are a striking feature of Pacific Northwest indigenous peoples, and are found from southern Alaska to northern Washington State. The word 'totem' is derived from Ojibwa or a related language, and means 'kinship group.' One of the principal functions of the totem pole was to record family and clan legends, lineages, and notable events. When they were created they could be 'read' by the members of the clan or family that erected them, but as they decayed their meaning was usually lost. Their message could simply be to proclaim the successes of a family or individual, commemorate a notable potlatch ceremony, or tell a legendary or historical story. 'Shame' poles were erected as symbolic reminders of unpaid debts, quarrels, murders, and other shameful events that could not be publicly discussed. One such pole was recently erected in Cordova, Alaska depicting the upside-down head of Exxon ex-CEO Lee Raymond.

The carvings would represent the crest of the person or clan concerned, indicating their moiety – Eagle or Raven – and their lineage. The Haida alone, for example, had some 70 crest figures, of which only about 20 were in common use. The following groups of animals are frequently associated with the Eagle or Raven moieties:

Eagle	Raven
Fish	Skate
Amphibians, such as frogs	Sea mammals
Beaver (considered amphibian)	Land mammals (except beaver)

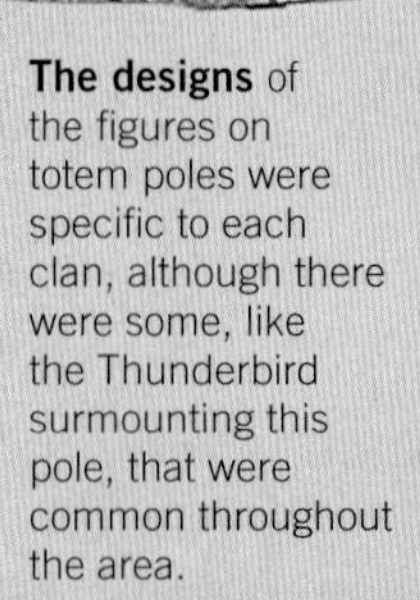

The designs of the figures on totem poles were specific to each clan, although there were some, like the Thunderbird surmounting this pole, that were common throughout the area.

Totemic imagery

The style of totem carving varies among the Haida, Tlingit, Kwakiutl, and other Northwestern and Coastal First Nations, and the style of decoration and iconography appears on all sorts of artifacts: house poles, screens, chests, and canoes, and was also used for identifying tattoos among some groups. The symbolic system was coherent. The universe was perceived as a house and the house itself a reflection of the cosmos. For example, the different parts of the house mirrored the human body:

Front posts	Arm bones
Rear posts	Leg bones
Longitudinal beams	Backbone
Rafters	Ribs
Cladding	Skin
Decoration	Tattoos

The inhabitants represented both the spirit of the house itself and the spirit of their ancestors.

Typically carved of red cedar, totem poles usually did not survive in the rain forest climate for longer than a century, and their original meaning was lost as they decayed.

Adinkra

The Akan of Ghana in Africa have an elaborate traditional system of symbols – *adinkra* – which are not only linked to their proverbs, songs, and stories but also serve to affirm social identity and political views. They are universally recognized by the Akan, and have been for many centuries, but to outsiders they appear simply as decorative motifs. The choice of design is therefore an intensely personal statement available even to those who are illiterate. *Adinkra* appear in wood, paint, and metal, but since the Akan are very much a textile culture, they are most prominent in cloth – for example, the handwoven *kente* or the block-printed *adinkra* or 'proverb' cloths. Over 700 symbols with their associations have now been cataloged. Some *adinkra* are traditional – a wooden comb for beauty and feminine qualities – while others have taken on modern meanings, wealth symbols now standing for a BMW or a television. For example, the symbol of the cocoa tree, introduced in the 19th century, and Ghana's principal cash crop, does not simply refer to the plant or to chocolate, but also to its social effects, bitterly expressed in the proverb: '*kookoo see abusua, paepae mogya mu*' – 'cocoa ruins the family, and divides blood relations.' Again, a pattern which a European might 'read' as a daisy, a generic flower, or the sun is a symbol implying unequal opportunity, linked to the proverb: 'All the peppers on the same tree do not ripen simultaneously.'

Adinkrehene
Chief of *adinkra* symbols:
greatness, leadership.

Denkyem
Crocodile:
adaptability.

Duafe
Wooden comb:
beauty, femininity, hygiene.

Dwennimmen
Ram's horns:
strength, humility.

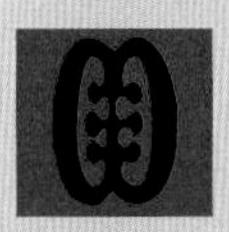

Ese Ne Tekrema
The teeth and the tongue:
friendship.

Funtunfunefu Denkyemfunefu
Crocodiles:
democracy, universality.

Hwemudua
Measuring stick:
inspection, quality control.

Mpatapo
Knot of reconciliation:
peacemaking.

Owo Foro Adobe
Snake climing a raffia tree:
diligence, prudence.

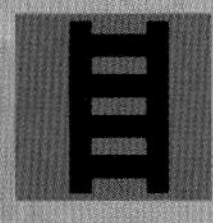

Owuo Atwedee
The ladder of death:
mortality.

Woforo Dua Pa A
When you climb a good tree:
cooperation, support.

Gateways to nowhere

Most monumental architecture – even for religious, ceremonial, or entombment and memorial purposes – has a strong functional element, in addition to its symbolic qualities. A singular exception is the stylized wooden Japanese *O-torii* portal *(below)*, sometimes set as entrances to temples or shrines, which serve to divide the sacred from the profane world. Often freestanding, giving on to nothingness – as is appropriate for Shinto, essentially a nature cult – they are also arranged along paths leading to a shrine.

No one knows the origin of the word – perhaps 'perching place for birds' – but *torii* are traditionally made in three pieces, three being the number sacred to the *kami* or gods. Before passing through the gateway it is traditional to purify oneself by washing at the place provided – *temizu* – and then to bow and clap three times, asking permission to enter the sacred realm. Walking toward the shrine, the center of the path – *seichu* – should be avoided, for that is the walking place of the spirits. These enigmatic gateways are rebuilt on a regular cycle, but of their origin little is known.

02

As human societies developed, often under authoritarian political and religious regimes, so did covert groups who invented disguised ways of communicating a shared belief or activity that would conceal their activities from society at large.

sects, symbols, and secret societies

Unorthodox and persecuted religious groups were outstanding among these, as were those engaged in early scientific research, frequently involving hermetic and arcane practices that continued well into the Enlightenment. Alchemists, necromancers and others drew on a confused abundance of traditions in an attempt to formulate cipher languages that would seemingly unlock the mysteries of creation and existence; their influence persists today among many secret societies.

EARLY CHRISTIANS

The crucifix

The first representation of the crucifixion may be the satirical 'Alexamenos graffito,' found in the remains of a boarding school on the Palatine Hill in Rome. Christ is shown on the cross with an ass's head with a Christian youth praying, and the Greek inscription: "Alexamenos worships (his) God." It has been dated between the 1st and 3rd centuries AD. The cross was thus clearly identified with Christianity at a very early date, but only emerged as the central symbol of the faith in the 5th century.

Christianity in its early years was – literally – an underground sect. Under Rome, it could not declare itself openly, and its adherents adopted secret symbols to express their faith but avoid persecution by the authorities. Many of these coded messages come from funerary remains, especially in catacombs, in Rome and elsewhere, and from Christian secret places of meeting and worship. The faith of the Christian dead was to be declared, but not in such a way that their friends and families would be punished. The cross, now the universally recognized symbol of Christianity, was, however, little used unless disguised. At a time of relentless persecution, it was too dangerous. The first Christians within the Roman empire developed a number of secret signs and symbols, often related to pagan traditions, to identify themselves and each other. These coded messages were fundamental in maintaining the community of belief among members of the early church for several centuries.

The disguised cross

The cross was represented by an anchor, a symbol of safety and coming to rest after the storms of life, or sometimes as a trident; the sword as a symbol of the cross was adopted much later, during the Crusades.

Bread and wine

Grain and grapes were symbols of abundance and joy all across the Roman world, where they were dedicated to Demeter, the goddess of the harvest, and Dionysius, the god of wine. The Christians transmuted them into their central mystery – the Eucharist, the bread symbolizing the body and the wine the blood of Christ Himself.

1ST CENTURY

2ND CENTURY

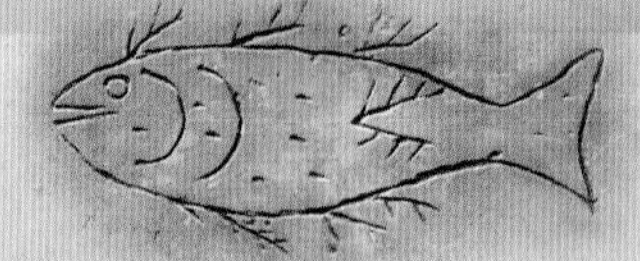

Ichthus

One of the earliest symbols was the fish, an ancient symbol of fertility and of life and continuity, or often two fish flanking a trident. Fish and fishermen are frequently mentioned in the Gospels and were associated with the Eucharist, as a reminder of eternal life. 'Fish' in Greek – *ichthus* – was also used as an acrostic:

Iesous Jesus
CHristos Christ
THeou God's
Uios Son
Soter the Savior

The simple outline of a fish was often drawn in the sand or spilt wine as a means of secretly acknowledging one's faith.

The five-letter Latin words when read in order (either horizontally or vertically) translate as 'he who works the plow sows the seed.'

The Roman square

A symmetrical arrangement of letters found on the walls of certain Roman houses was probably an ingenious early Christian means of identification. This seemingly innocent proverb can be interpreted as an anagrammatic transposition cipher, the letters being rearranged to reveal a hidden message (*right*). The Latin words *pater noster,* meaning 'Our Father,' form a cross, with the spare As and Os representing the Greek words 'alpha' (beginning) and 'omega' (end), which also have a strong Christian significance.

P
A
A T O
E
R
P A T E R N O S T E R
O
S
O T A
E
R

Doves and peacocks

Two further symbols were rooted in the Classical tradition. To the pagan world, the dove was associated with Aphrodite, but for the Christians it represented the Holy Spirit, a pair representing conjugal love, sometimes drinking the water of life from a fountain, while one bearing an olive branch was one of the earliest symbols of reconciliation and peace. Pagans believed peacock's flesh to be incorruptible, and for Christians this was transmuted to represent immortality and the Resurrection.

The chrismon

The cross was often disguised as the chrismon, or Christ's monogram: the two Greek letters *chi rho*. On October 27, 312 these letters changed the Roman world for ever. Two contestants for the Empire, Constantine and Maxentius, were preparing to confront each other at the Milvian Bridge, near Rome. The night before the battle, Constantine had a vision of the *chi rho* blazing against the sky and a voice saying to him "*in hoc signo vinces*" – in this sign thou shalt conquer. Christians in the army told him that it was the emblem of their Redeemer and symbolic of the triumph of life over death. Constantine had the *chi rho* painted on his helmet, his soldiers' shields, and his battle standard. The pagan army had no idea what it meant. Constantine's victory was decisive, and from this date Rome turned towards Christianity. The wreath surrounding the *chi rho* is of palm or bay leaves forming a Roman crown of victory. For Christians, this came to represent the crown of martyrdom.

3RD CENTURY | 4TH CENTURY | 5TH CENTURY

The 'Good Shepherd'

Depictions of a shepherd with a lamb across his shoulders are found from the 3rd century – Christ guarding and protecting His people – but it was also a favorite Classical motif. A lamb by itself, standing for Christ and His sacrifice, would similarly be understood by fellow Christians.

The Orans

The archaic figure of the person praying with lifted hands was a symbol for humans throwing themselves on the mercy of a divinity and not originally uniquely Christian.

The living crucifix

The first true crucifix is from northern Italy, dated to AD 420. The earliest examples show Christ on the cross but living and triumphant, as on the doors of Santa Sabina in Rome, after the Western tradition wearing a loin cloth to proclaim His humanity; in the east Christ wears a tunic, representing His sovereignty.

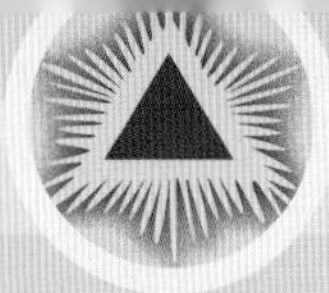

THE PENTANGLE

Among the many arcane symbols regarded as having mystical or magical properties that filtered from various West Asian sources into a number of esoteric disciplines, the pentangle or pentagram probably remains the most significant. From the Greek Pythagoreans onwards it was a design imbued with a plethora of mystical meanings, and remains significant in the modern world for many New Age cults, and as an architectural device for the Church of Jesus Christ of Latter-day Saints, and appears on the flags of Morocco and Ethiopia.

Geometry

The geometric characteristics of the pentagram have attracted many theorists. The ancient Greek Pythagoreans noted many of its most important mathematical principles: it is drawn using only five straight lines of equal length; in doing so it produces a symmetrical five-pointed star, containing eight isosceles triangles, and an inscribed pentagon. The shape can also be inscribed within a circle, but if the points of the pentangle star are joined by straight lines, to form an outer pentagon, a further ten isosceles triangles are produced. Further symbolism can be read into whether the pentangle is used with the single point at the top or pointing downwards.

Babylonian astronomy For the Babylonians, the pentagram was associated with the planets, the points representing Jupiter, Mercury, Mars, Saturn, and most importantly, Venus; ancient astronomers had noted that Venus completed five inferior conjunctions in its orbital cycle around the Sun, its path tracing a perfect pentagram in the sky in the eight years it takes to complete the cycle. Venus was also called the Morning Star, bringer of knowledge, and was called Lucifer by the Romans.

Hebrew lettering Among other mysterious symbols, Lévi included letters resembling Hebrew script.

Echoes of Dr. Dee Lévi incorporated a version of John Dee's ultimate magical symbol (*see page 57*) at the heart of his design.

The golden ratio was first identified by the Pythagoreans (*see page 154*), who recognized that the pentangle was a geometric figure which intrinsically displayed unique properties of proportion, illustrated by the colored lines in this diagram.

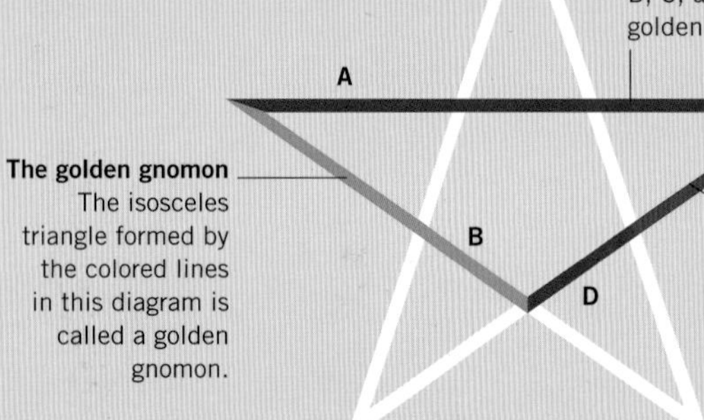

The golden gnomon The isosceles triangle formed by the colored lines in this diagram is called a golden gnomon.

Proportion The colored sections A, B, C, and D are in direct golden ratio to each other.

Equation The properties of the golden gnomon may be expressed to produce phi:

$$\frac{A}{B} = \frac{B}{C} = \frac{C}{D} = \varphi$$

or

$D + C = B$ and $C + B = A$

The 19th-century occultist Eliphas Lévi (1810-75) designed this pentagram as a summary of the 'microcosm,' by which he meant humankind. He incorporated (as with much of his work) an array of symbols and signs drawn from many arcane traditions. Lévi also designed the necromantic Sigil of Baphomet (*right*).

Human form
The upward-pointing pentangle is often associated with the human form with arms and legs outstretched, as in Leonardo da Vinci's Universal (or Vitruvian) Man (*see page 194*).

Pentangles and religion

The properties of five contained within the pentangle came to represent a number of aspects of the Christian faith: the five senses, the five wounds of Christ, the five key stages in Mary's life with Christ – Annunciation, Nativity, Resurrection, Ascension, and Assumption. It was also used as a talisman for health. For some years the pentangle, rather than the Star of David, was the symbol of the city of Jerusalem.

A Greek and Roman tradition of associating the five points of the pentagram with the five classical elements – earth, fire, water, air, and ether (or the idea) – and with the known planets, resonated with medieval alchemists and mystics, who saw the form as having talismanic properties. For them, pointing upwards it was said to represent the spirit dominating the elements; pointing downwards, it was a symbol of evil.

Necromancers, magicians, and satanists frequently use the inverted pentangle within a double circle as a ritual device. In Marlowe's play *Doctor Faustus*, Mephistopheles is temporarily trapped within such a design. With a goat's head design inscribed within the pentangle, accompanied by the five Hebrew letters spelling Leviathan, it is known as the Sigil of Baphomet, representing the pit where the fallen angels are imprisoned. The three downward-pointing arms of the star are said to represent the inversion of the Holy Trinity.

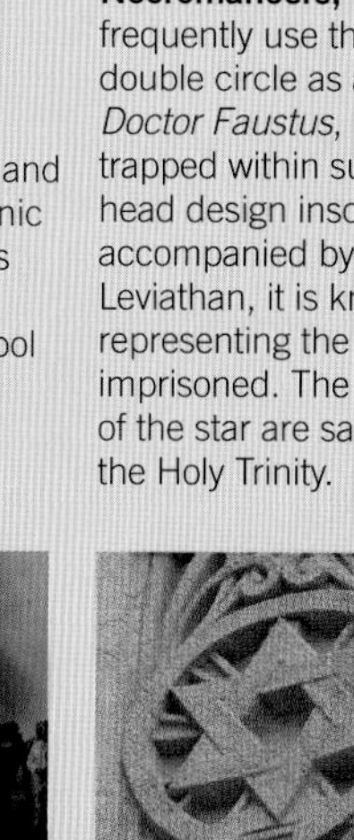

Celtic traditions The pentangle is often confused with a magic pentacle, an amulet which can take many forms, nevertheless the pentangle frequently features in their design. Neo-paganists and practitioners of Wicca use it as a devotional and ritual device, usually with the single point upwards, and inscribed within a circle, denoting completeness, the five points representing the elements, as in the classical tradition.

The founder of the Baha'i faith, Bahá'u'lláh, wrote several texts in the form of a pentagram (or haykal), and the figure is regarded in Baha'i as a manifestation of God. Baha'i also embraces other religious symbols, such as the Star of David and the swastika, and many of these are to be found in combination on Baha'i temples.

Divination

Dissected livers of sacrificial animals were often used for augury in the ancient world, the results preserved in stone.

A variety of mysterious practices developed in the European Middle Ages, often linked to ideas descended from ancient Middle Eastern traditions, all involving esoteric and hermetic laws, rituals, and secret writings, influencing alchemy, necromancy, and Kabbalism, all of which are examined on the following pages. Underlying many of these was the wish to divine the future. From very early times, and in many cultures, those that seemed able to decode the signs and thus forecast events were deemed especially gifted and strove to mask their abilities behind skeins of mystery. Two practices steeped in symbolism, with their roots in an indistinct past and which have persisted to the modern world, are astrology and Tarot.

Varieties of divination

Divination can take many forms, which fall into four main sign-reading categories:

Reading omens The observation, recording, and interpretation of natural events (astronomical, meteorological) on an historical basis. If, for example, a full moon occurred before a significant victory, this might be seen as a precedent for future success. Astrology falls into this category.

Sortilege (or cleromancy) Usually casting of lots (sticks, stones, bones, runes), the pattern of which is interpreted by the 'seer.' In more modern times, popular forms include interpreting the 'spread' of Tarot cards and 'reading' palms or tea leaves. (*See also* I Ching, *page 181*).

Augury The interpretation of patterns in the natural world, such as the flight of birds. 'Reading' the livers and entrails of sacrificial animals was a common form of augury in Greece, Rome, and Mesoamerica.

Spontaneous The ecstatic or inspirational forecasting of events, based on little other than the seer's instinct, 'gifts,' or 'ability.'

Astrology

The observation of the heavens, especially the behavior of the Sun and Moon, the visible planets, and the brighter stars, can be traced to the earliest civilizations. The distinction between astronomy (the scientific observation of planetary and astral cycles) and astrology (divination derived from this) emerged first in ancient Babylonia and Egypt, but the disciplines overlapped until relatively recently.

The ancient Greeks learned astrology (Chaldean wisdom) after Alexander the Great's conquest of Alexandria in Egypt, and it is from them that the Western tradition was transmitted via Rome. The Hindu astrological cycle is notably similar to the Babylonian/ Western tradition; separate astrological cycles were developed in the Chinese oecumene. In the West, most people today are familiar with their 'birth sign.'

For Western astrologers, each of the 'Houses' of the zodiac (*left*) is seen to represent aspects of humans' character and condition. The art of divination is based on the influence of one zodiac sign upon another, depending upon related significant calendrical dates in the solar and lunar cycles. Knowledge of astrology was regarded as an essential attribute for doctors in medieval Western Christendom.

The Devil

Decks of cards began to be mass-manufactured soon after the advent of printing in Europe. The most popular was the Marseilles deck. In this variation on the Marseilles design Le Mat (The Fool) is numbered 21 rather than his usual value of zero, making Le Monde (The World) number 22.

The Sun

The Magician

Tarot

Tarot cards have an obscure origin. They are known to have been used in Mamluk Egypt in the 11th century, and seem to have spread to Europe by the 15th century, the earliest European example being the Visconti-Sforza deck. Essentially a pack of playing cards comprised four suits of numbered cards 1-10, and the four court cards (the 'minor arcana'), plus 22 trump cards (the 'major arcana'). It is the latter, often linked to the Hebrew alphabet, that have attracted a sense of mystery. Tarot cards in various forms and designs are still used widely in Mediterranean countries for regular card games, and it was largely among the north Europeans that a divinatory property was attributed to them. Among many variations, there are three principle designs: the Marseilles (dating from the 15th century); the 19th-century Rider-Waite-Smith deck – Waite was a member of the Hermetic Order of the Golden Dawn (*see page 254*) – used widely in North America; and Aleister Crowley's Thoth Tarot deck, which drew on his studies of arcane principles.

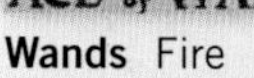

Wands Fire

Swords Air

Cups Water

Pentacles (or coins) Earth

The four suits are frequently linked to the elements, as here in the Rider-Waite-Smith deck.

The Tarot trumps

0 Le Mat (The Fool) Value zero.
1 Le Bateleur (The Magician, Magi, Juggler) Intelligence, quick wit, ability; lying, confusion, inconsistency.
2 La Papesse (The Papess, Priestess) Patience, intuition, knowledge; seclusion, inflexibility, inaction.
3 L'Impératrice (The Empress) Fertility, emotion, reward; dependency, sterility, self-devotion.
4 L'Empereur (The Emperor) Stability, firmness, power; tyranny, self-pride.
5 Le Pape (The Pope, Hierophant) Righteousness, faith, fusion; narrow-mindedness, remoteness, vanity.
6 L'Amoureux (The Lovers) Desire, union, choice; pathos, conflict, temptation.
7 Le Chariot (The Chariot) Sacrifice, determination, conquest; relentlessness, ardor.
8 La Justice (Justice, Adjustment, sometimes No.11) Impartiality, integrity, judgment; intransigence, mercilessness.
9 L'Hermite (The Hermit) Wisdom, abstinence, self-sacrifice; alienation, mysticism.
10 La Roue de Fortune (Wheel of Fortune) Change, destiny, evolution; instability, lack of control.
11 La Force (Strength, Lust, sometimes No.8) Willpower, domination; inhibition, constraint, domination.
12 Le Pendu (The Hanged Man) Reckless confidence; fate, helplessness.
13 Treize (Thirteen, Death) Sometimes called the 'Death' card, it represents a state that goes beyond normal concepts. Metamorphosis, purification; the inevitability of loss, disillusionment.
14 Tempérance (Temperance) Balance, coordination; imbalance, volatility.
15 Le Diable (The Devil) Desire for money, possessions, and physical attainment; greed, extreme ambition.
16 La Maison Dieu (Tower, Temple) Disruption, change, freedom; imprisonment, negativity.
17 L'Etoile (The Star) Hope, renewal, spiritual love; self-doubt, stubbornness.
18 La Lune (The Moon) Imagination, psychic abilities; secrecy, self-delusion.
19 Le Soleil (The Sun) Contentment, health, happiness; failure, arrogance, broken agreements.
20 Le Jugement (Aeon) Decision-making, changes, improvement; stagnation, delay, fear of death.
21 Le Monde (The World, Universe) Fulfillment, accomplishment, achievement, completion; frustration, inability to resolve or complete.

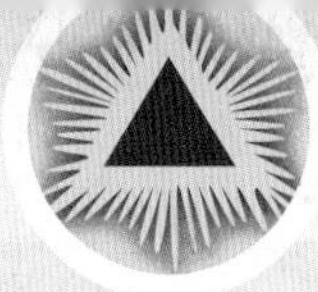

HERESIES, SECTS, AND CULTS

As any faith or system of belief begins to spread across a wide geographical area, encountering other traditions and beliefs, so local adaptations and modifications will occur, often disapproved of by the orthodox. While in Hinduism and Buddhism this process rarely caused problems, and indeed enriched and expanded the nature of those faiths, for Christianity, and latterly Islam, the label of mortal heresy for those at the fringe was all too often applied.

Christian symbols

Folk symbols

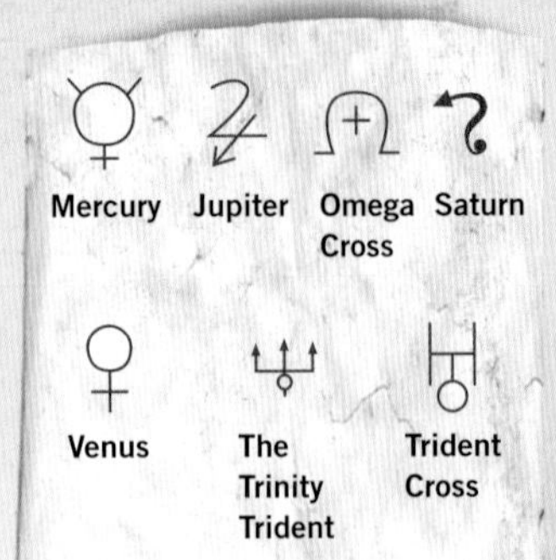

Magic/classical symbols

The interaction between Christianity and other, often older, belief systems is apparent in the mysterious decoration of traditional houses in Apulia in southern Italy (*right*), where pagan or classical symbolism jostles alongside Christian imagery (*above*).

The Atbash cipher

One of the earliest ciphers concerned with religious secrecy and mysticism was the so-called Atbash cipher used by certain Hebrew scholars as a means of secret writing and as a formula to apply to the text of the Torah (the first five books of the Old Testament, or Pentateuch) to reveal hidden meanings. It is a simple monoalphabetic transposition cipher, similar to the Rail Fence cipher, and like that, has only a single key.

The cipher has been used by many secretive groups, but is not very secure, having a single key, with none of the flexibility of the Kama Sutra or the Caesar Shift monoalphabetic transposition ciphers (*see pages 67, 103*). The use of Atbash on Hebrew texts prefigures activities of the Kabbalists and those interested in the Bible code (*see pages 60, 256*).

The principle of the Atbash cipher is to reverse the positions of the first and last letters of the alphabet in successive order until you reach the middle of the alphabet. In Hebrew this swaps the first letter, 'aleph,' with the last letter, 'tav,' the second letter, 'beth,' with the penultimate letter, 'shin,' and so on.

Using the Roman alphabet this produces:

plaintext
a b c d e f g h i j k l m n o p q r s t u v w x y z
ciphertext
z y x w v u t s r q p o n m l k j i h g f e d c b a

This may be recast as a two-way table, each letter in the plaintext to be read against its counterpart:

a	b	c	d	e	f	g	h	i	j	k	l	m
z	y	x	w	v	u	t	s	r	q	p	o	n

An example:

plaintext	the enemy at the gates
ciphertext	gsv vmvnb zg gsv tzgvh

Gnosticism

As Christianity spread, under the Roman and Byzantine empires, towards Asia, it came into contact with the principal faith of Persia, Zoroastrianism. This creed saw Creation as a constant struggle between the equal forces of light and darkness, and good and evil. Closely resembling the Christian belief in the eternal combat between God and the fallen angel, Satan or Lucifer, nevertheless the difference lay in the notion of an equality of power between the two forces. And therein lay heresy. For Christians influenced by these ideas, the Supreme Being was counterbalanced by the Demiurge, who was seen to have created the Earth, and his various emissaries or 'archons.'

A further heresy resided in the notion of 'knowledge,' with which the serpent had corrupted Eve in the Garden of Eden. This produced a number of Christian sects under the general name 'Gnostics,' from the Greek *gnosis* or 'knowledge,' with links also to pre-Christian mystical concepts such as Pythagoreanism. Various forms of Gnosticism spread across Europe at the time of the Crusades, notably the Bogomils in Bulgaria and the Balkans, and the Cathars or Albigensians in southern France. Humble, peace-loving, and spiritual, their beliefs nevertheless were seen as a direct threat to the authority of the Papacy, and they were ruthlessly eradicated.

On Bogomil tombstones a recurrent symbol is the Gnostic cross (*top*). Moons, stars, and crescents also occur, which led to an assumed association with Islam.

> "Kill them all,
> the Lord will recognize His own."
>
> ORDER GIVEN BEFORE RANSACKING A TOWN POPULATED BY BOTH CATHARS AND CHRISTIANS.

Gnostic symbols

Gnosticism adopted a variety of forms, such as Manicheanism and Paulicianism, all methodically condemned by the Catholic church. Associated with these ideas were a variety of coded symbols which remain today uncomfortably ambivalent in meaning.

The Gnostic Cross Derived from the ancient Egyptian symbol for the god Ogdoad, this was adopted by Gnostics to represent the eight Aeons, the eighth being the Messiah who will return. It also appears in Catholic symbolism as the Baptismal cross, the spokes representing the eight days between Christ's arrival at Jerusalem and the Resurrection.

The Messianic Seal Supposedly worn by the followers of Jesus in the 1st century, a menorah surmounts the Star of David, from which depends the Christian fish. It reflects the early Messianic practice of certain Jews who converted to Chritianity, and attempted to convert others.

Iao Sabaoth Derived from the Greek for 'Lord of Hosts,' an Old Testament name for God; for Gnostics it also represents Abraxas, a Sun deity, and the seven 'archons' or spirit agents of the Earth's creator, the Demiurge.

The Ouroboros From the Greek for 'tail-swallower,' the ouroboros was a symbol for the Sun in ancient Egypt, and in Gnosticism represented eternity and the Sun deity Abraxas. The snake had further symbolic value as a symbol of self-birth (because of its skin-shedding), self-birth being associated with God. The snake also refers to the giver of knowledge, the serpent in the Garden of Eden.

The Serpent Wheel This combines the Gnostic Cross with a simplified serpent, uniting the eight Aeons with self-birth, meaning that it is the symbol of the Gnostic Messiah.

The Crucified Serpent Stemming from Moses' use of brazen serpents as magical charms, for Gnostics this brings together the central Christian image of the Cross and the provider of knowledge in the Garden of Eden. This symbol was also used by later alchemists to represent the elixir of mercury, and is related to the modern symbol of the medical profession.

Abraxas The Gnostic Sun deity is usually presented as a warrior with serpents for his lower body. He is often shown driving a four-horse chariot, representing the seasons.

Knightly orders of the Crusades

There has been much speculation concerning the knightly orders – especially the Knights Templar (*below*) – established during the various Crusades in the first half of the second millennium AD. Closed communities, especially those straddling the sacred and secular worlds, attracted suspicion, and by the 14th century the political, moral, and economic vulnerability of the Papacy made the Church of Rome their prime antagonist. The accumulation of vast wealth by these orders, exemplified in the ornate Templar church at Tomar in Portugal (*above*), also attracted attention and envy, as did their unique independent legal status. Although there seems to have been some contact with Gnostic ideas (the Sun deity Abraxas appears in some Templar imagery), there is little evidence to support the accusations of heresy, hermeticism, and Freemasonry that have been leveled against the Templars. On Friday October 13, 1307, the Templars in France were rounded up, their assets seized, confessions of heresy extracted under torture, and many condemned to the stake. Persecution spread. The sudden demise of the order and rumors of hidden treasure have provided rich ground for conspiracy theorists.

Rosslyn Chapel

Brought to global attention by *The Da Vinci Code*, this extraordinary church, encrusted with a superabundance of detailed carvings, weaves together a wealth of eccentric symbolism around its Christian core, ranging from Norse and Celtic references and vernacular myths, tales, and proverbs, to allegedly Masonic iconography. It was founded in 1446 as the Collegiate Chapel of St. Matthew by William St. Clair, Earl of Orkney, in Roslin, a few miles south of Edinburgh, Scotland. It was one of 37 collegiate churches built in Scotland at the time, often extravagantly decorated, as secular foundations intended to spread spiritual and intellectual knowledge within a Christian ethos.

Carved in solid stone, the spectacular arched roof is decorated with many symbols: squares, five-pointed stars, ball flowers, roses, tablet flowers, and a dove bearing an olive branch. The barrel-vaulted roof of the main aisle carries a number of carved decorative motifs, arranged in series. Some, like the five-pointed star, are ambiguous, partly because of their position: seen upwards it represents aspiration, knowledge, enlightenment; presented downwards, it is evil and is associated with witchcraft.

Masonic connections

Although undoubtedly many masons were involved in the creation of the chapel (and it bears a large number of masons' marks) there is scant evidence of a direct link with either the Knights Templar or the Freemasons. The St. Clair family testified against the Templars in 1309, although interestingly descendants of the founder, the Sinclairs of Roslin, held the position of Masters of the Grand Lodge of Scotland.

Rosslyn contains many proverbial allusions. Here the tumbler represents chance or fate, acting as an intermediary between a crowned Death and his inevitable earthly victim.

The Apprentice Pillar Local legend tells that the master mason who began this carving left for Rome seeking guidance in its completion. He returned to find his apprentice had finished the work and, in a rage, struck him dead. The base of the column is formed by eight dragons from whose mouths grow the vines which wrap the pillar. Both the monsters and the vine have strong roots in Norse mythology, in which the Tree of Knowledge, Yggdrasil, held up the heavens, while dragons gnawed the roots. Images of the Celtic Green Man also recur throughout the carvings.

The 'musical' cubes protrude from arches springing from the Apprentice Pillar, and also feature on the ribs of the barrel vault (*above*).

A sequence of 213 square blocks protrude from the pillars and arches, with a variety of patterns on them. One interpretation of their meaning is that they form a musical score, as the motifs resemble cymatic or Chladni patterns, the physical manifestation of certain sound waves.

ALCHEMY

Alchemy is an ancient quasi-science of obscure origin concerned with the transmutation of base metals into gold and the mystical art for human spiritual transformation. By compounding abstruse chemical formulas, the alchemists' aim was to create a magical substance called 'The Philosopher's Stone.' Alchemy was not merely the capricious precursor to chemistry, however, but an important catalyst in the development of modern ideas. Although often portrayed as the epitome of irrationality, alchemy attracted some of the greatest natural philosophers of Western thought, including Robert Boyle, Gottfried Leibniz, Isaac Newton, and the Swiss psychologist C.G. Jung, who interpreted alchemical symbolism in psychological terms. Ranging over two millennia, alchemy aroused the cupidity of kings, the blind fear of mobs, and the occult aspirations of artists, scientists, philosophers, and countless secret societies.

Origins

Alchemy emerged from the cultural melting pot of Hellenistic Egypt in the third century BC, blending an eclectic mix of Aristotelian theories of matter, Gnosticism, and ancient metallurgical (and magical) techniques. The legendary semi-divine sage Hermes Trismegistus is credited with writing the first library of books on alchemy, magic, astrology, and philosophy, and gave his name to 'Hermeticism.' The earliest known alchemical writer, Zosimus of Panopolis, lived in Alexandria about AD 300, and wrote a cryptic theological and practical handbook of the alchemists' craft. Western alchemists turned their attention to gold-making, the discovery of panaceas, elixirs of longevity, and the attainment of a spiritual 'gnosis' or knowledge. They also observed an ultimate aim, discussed in Plato's *Symposium*, of reunifying the two sexes into a prelapsarian, ideal being.

The *Tabula Smaragdina*

The *Emerald Tablet* (1614) is a short but canonical text; it was purported to be a copy of an ancient inscription. It contained the celebrated alchemical axiom: 'That which is below is like that which is above, and that which is above is like that which is below.'

Development

Alchemy flourished throughout the Arabic world in the early Middle Ages. The most famous name of the Arabic alchemical writers was the learned Muslim polymath Jabir ibn Hayyan (c.AD 721-815) who wrote treatises on numerology, astrology, talismans, and the invocation of spirits (and so inscrutable were his writings that it is thought the word 'gibberish' derived from his name). By the 12th century, translations of Arabic alchemical texts were flooding Europe, attracting medieval theologians Roger Bacon (c.1220-92) and St. Albertus Magnus (c.1200-80), and the art began to be enthusiastically patronized by princes, nobles, and emperors. In the 16th and 17th centuries, alchemy was an intrinsic part of the pre-scientific order, and proliferated as a legitimate means of investigating the manifest world.

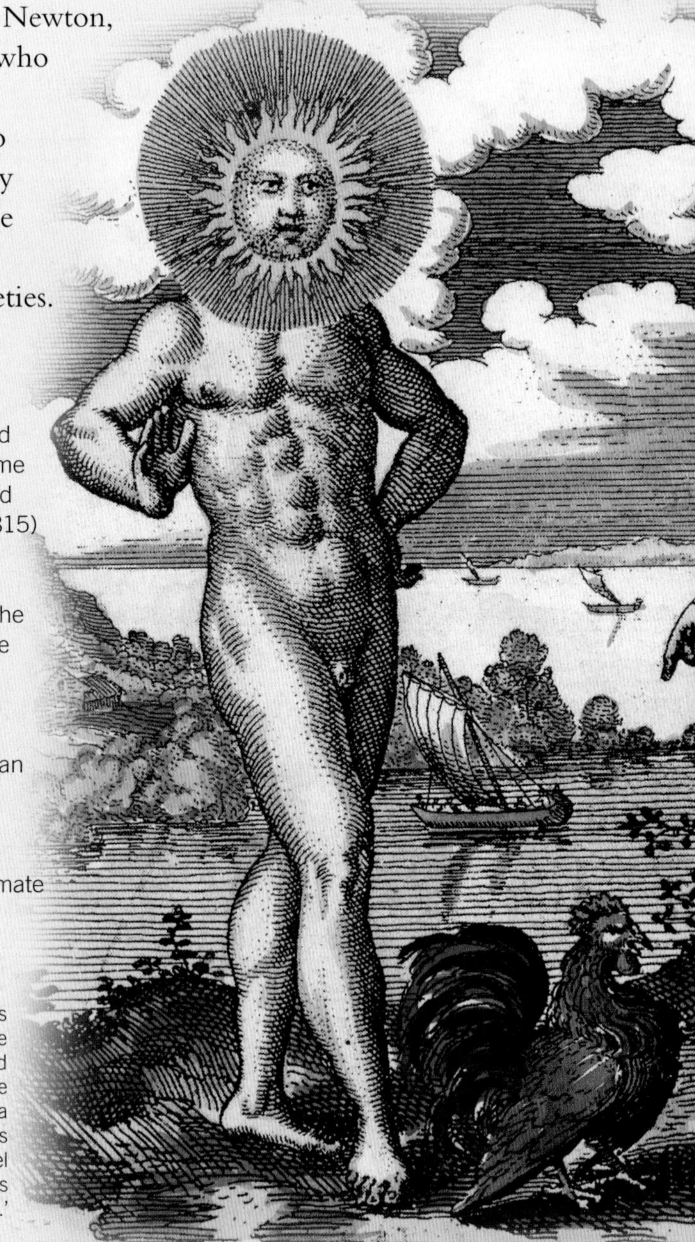

Alchemical emblems often picture this conjunction of opposites, culminating in the 'chymical wedding' of an allegorical king and queen. The Marriage of the Sun and Moon is the 30th 'emblem' from *Atalanta Fugiens* (1617), a comprehensive survey of alchemical techniques and imagery by the Lutheran physician Michael Maier (1568-1622). He noted 'Luna is as requisite to Sol as a Hen is to a Cock.'

AVTRE ALPHABET, PA'R lequel certains Alchimistes ont voulu secrettement couurir & descrire les reigles, & secrets de leur science, faisants d'icelle plus grande estime, qu'elle n'est digne ny merite.

Heinrich Cornelius Agrippa published an alchemical alphabet in his *Three Books About Occult Philosophy* (1531-33, *see also page 57*).

Symbolic codes

In his laboratory, the alchemist endeavored to re-create the act of creation, and God was regarded as an archetypal alchemist. For the alchemist, the analogies or correspondences connecting the diverse skeins of reality were best expressed by cryptic symbols. The alchemist envisaged each stage of the alchemical process being heralded by a color change and a 'contest' with certain animals. For instance, when the lion was depicted clashing violently with another animal it signified the manufacture of sulphuric acid, vitriol, created by distilling green crystals of iron sulphate in a flask. The mysterious language used among alchemists was designed to disguise their activities, and protect their knowledge; one aspect of this language however was the attribution of parallel chemical and characteristic values to substances and other aspects of the natural world. The planets especially attracted their attention.

Planet	Substance
Mars	Iron
Mercury	Mercury (quicksilver)
Jupiter	Tin
Saturn	Lead
Sun	Gold
Moon	Silver
Venus	Copper

Symbols	Meanings
Yellow lion	Yellow sulphides
Red lion	Cinnabar
Crow	Black sulphides
Salamander	King of animals as it was believed to live in fire, and represents in alchemical texts the purification of gold.

Paracelsus

The original Baron Frankenstein, Paracelsus' foray into the occult allowed him to make major advancements in microchemistry, antisepsis, homeopathy, and surgery. A notorious peripatetic iconoclast, Paracelsus (1493-1541) conducted scores of controversial experiments, including an attempt to create an artificial human being through a strange combination of alchemy, earth, blood, and sperm. Paracelsus developed a mysterious 'alphabet of the Magi,' similar to Hebrew script, with which he engraved the names of angels on talismans to imbue them with magical and healing properties.

The alphabet of the Magi

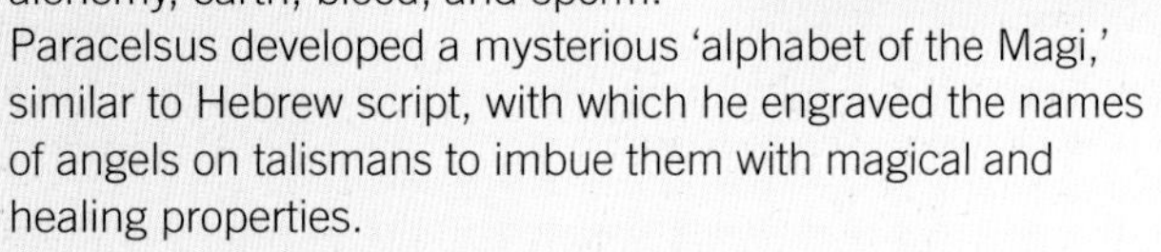

Scientist or magus?

In 1678, an English physicist, astronomer, and natural philosopher surreptitiously jotted down notes about 'hermaphrodite,' a mysterious chemical compound associated with alchemy, and other cryptic references like the 'green lion,' and the 'blood of the sordid whore.' By day, Isaac Newton (1643-1727) was a distinguished Member of Parliament and later the President of the Royal Society; by night, he became a magus of occult knowledge and Hermetic formulae. Newton spent countless hours scouring ancient Greek myths and biblical references for hidden truths about nature and the universe in search of encoded alchemical recipes.

Ironically, Newton's knighthood in 1705 was not granted for his groundbreaking scientific and mathematical treatises, but for his role as Master of the Royal Mint from 1699, supervising the transformation of bullion into coinage.

His obsession with alchemy yielded nearly a million words contained in his private papers which were later deemed of no scientific value. After his death, his body revealed exaggerated quantities of mercury – quicksilver – a key alchemical element.

Kabbalism

Origins

The word 'Kabbalah' means 'to receive,' and developed in 12th-century Provence, culminating in the classical work *Sefer ha-Zohar*, or *Book of Splendor* (c.1300), which records revelations regarding the divine mysteries. Kabbalah is not monolithic, but rather constitutes a complex and highly systematic collage of disparate doctrines loosely grouped into two categories: theosophical Kabbalah, which deals with visual contemplation of the Tree of Life, and ecstatic Kabbalah, which is based on the practice of the recitation of divine names hidden in the Torah, the first five books of the Old Testament, leading to ecstatic states of consciousness and mystical union.

This bookplate of *Portae Lucis*, or *Portal of Light* (1516), included the first depiction of the Sefirotic Tree of Life, the Kabbalistic representation of the divine world with its ten abstract qualities.

Kabbalah is a Jewish system of theosophical speculation concerning God and the creation of the universe. It is conceived as a divine science and seeks to understand the rules by which God administers the existence of the cosmos through the Tree of Life – ten divine emanations that trace God's descent into the material realm and the channels through which our souls must travel on our return journey back to God. In Kabbalah, the Bible is a cosmic recipe book in which each letter represents a primordial essence of reality, a cosmic periodic table of elements, which was orally transmitted in secret enclaves since the time of Moses. Kabbalah is fundamentally a quasi-science in which the Bible is treated as a dense, and infinite, network of potential coded messages; it is in effect a decoding system seeking a code, but with its amulets, grimoires, and number mysticism, Kabbalah exerted a strong influence on the history of biblical speculation.

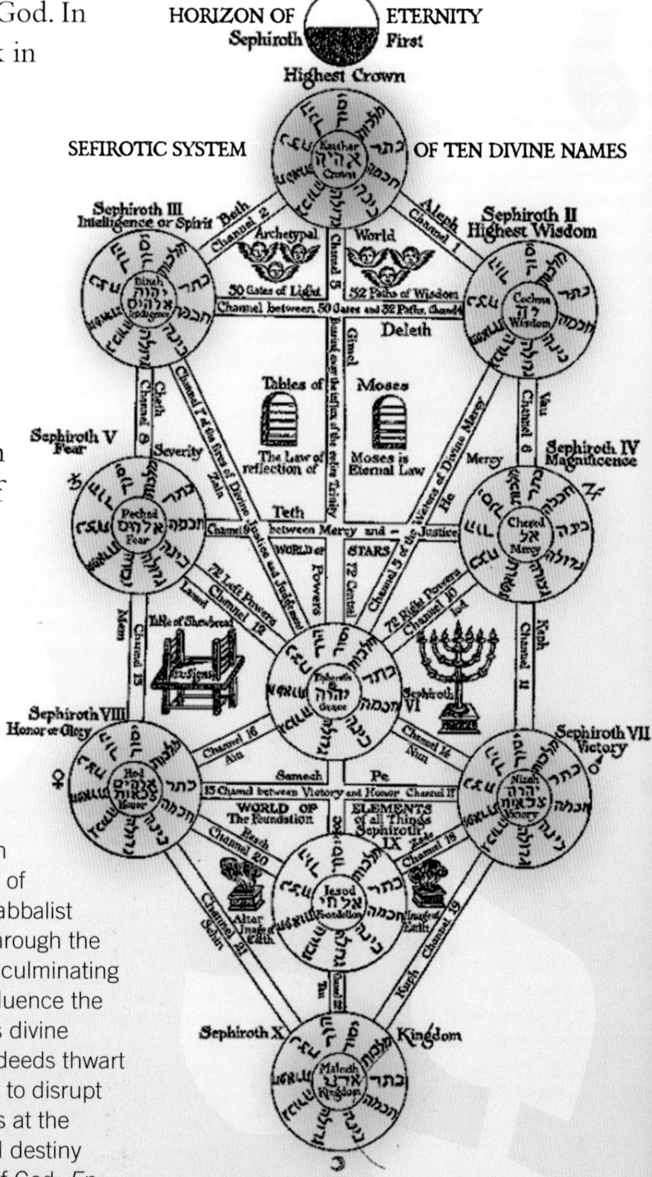

The Sefirotic Tree of Life *Sefirot* are not static transcendental powers, but rather divine potencies which are in a constant state of flux and are capable of being influenced by human actions and deeds.

The art of Kabbalah

In Kabbalah there are two facets of the Godhead – God as manifested in creation, and God (Hebrew *En Sof*), the ineffable and immeasurable, who is beyond all human comprehension or understanding. *En Sof* takes part in the manifest world through a series of emanations in the Tree of Life (*right*) through the ten *Sefirot*, ten divine hypostases of the Supreme Being of God from which all creation flows. The role of the Kabbalist is to tap into the unlimited aspect of the Godhead through the ten *Sefirot*, which descend throughout the universe culminating in the human soul. Benevolent actions positively influence the harmonious matrix of the *Sefirot* thus allowing God's divine grace to flow freely in creation. However, malicious deeds thwart God's divine grace by transmitting baneful impulses to disrupt the harmony of the *Sefirot*. For the Kabbalist, man is at the center of creation with dominion over the future and destiny of the world. The link between the limitless aspect of God, *En Sof*, and the ten *Sefirot* is Adam Kadmon, Primordial Man. Adam Kadmon is the most perfect manifestation of God that the human mind can comprehend and is identifiable with the Messiah, the incarnation of God.

Kabbalah and Christianity

Kabbalah's popularity among the Diasporic Jewish communities in Europe in the 15th century coincided with the period's enthusiastic study of Hermetic philosophy. Notably, the alchemist and necromancer Heinrich Cornelius Agrippa included the gematria table in his *Three Books About Occult Philosophy* (1531, *see page 57*). It also led to a syncretic fusion of Christianity, called Christian Kabbalah. The two chief exponents of this school of thought were the Italian polymath, Giovanni Pico della Mirandola (1463-94), and the German humanist Johannes Reuchlin (1455-1522). Their studies deeply inspired the Kabbalistic friar Francesco Giorgio (1466-1540) whose theories of harmony, numerology, and sacred geometry had a profound impact on the architecture of Freemasonry.

The German polymath Agrippa included Kabbalistic ideas, notably the gematria, in his survey of Hermetic practices.

A modern Messiah

Shabbetai Zevi (1626-76) was a Jewish leader whose followers thought was the incarnation of *En Sof*. Shabbetai rode into Jerusalem on horseback declaring himself the Messiah.

Before being excommunicated, imprisoned, and eventually converting to Islam (*right*), Shabbetai produced such apocalyptic fervor throughout the Jewish Diaspora that broadsheets and pamphlets circulated widely in Europe declaring the advent of the Messiah.

The American singer, actress, and pop superstar, Madonna is one of the many modern glamor icons embracing the mystical tenets of Kabbalah. Some controversy occurred when it was revealed that Kabbalah was being taught in junior high schools in the US.

In the beginning was the Word...

The theosophical aspect of Kabbalah deals with the combination of the letters of the Hebrew alphabet and their values. A cosmic and sacred language, Hebrew is considered to be the source from which all other languages emerged. In this system of thought each of the 22 Hebrew alphabetical letters is seen as a fundamental component of reality, a kind of irreducible atomic element alive with memories and hidden meanings. This allowed for free linguistic creativity: it altered, decomposed, and recomposed the literal stories of the Bible to unveil inner hidden meanings.

Gematria
One key to unlocking these meanings is the gematria, by which a word (or group of words) is assigned a number according to the value ascribed to its constituent letters. Selected letters, words, and sentences can thus be converted into numbers or geometric forms, which in turn have mystic properties. For example: the Hebrew for 'love' is Ahebah אהבה, (aleph-he-beth-he) which adds up to 13. The word for 'unity' is Achad אחד, (aleph-cheth-daleth), which also adds up to 13. Thus, users of gematria see a direct correspondence between love and unity.

Mystic numerals
By adding the five 'sofit' or word-final alternate forms to the 22 letters of the Hebrew alphabet one reaches 27, which is the number of numerals required to express every number from 1-999.

Mystic forms
One version of these properties is the link to the 22 solid geometric figures composed of regular polygons; these comprise the five Platonic solids, the four Kepler-Poinsot solids, and the 13 Archimedean solids.

Letter	Value	Final	Value	Name
א	1			aleph
ב	2			beth
ג	3			gimel
ד	4			daleth
ה	5			he
ו	6			vau
ז	7			zayin
ח	8			cheth
ט	9			teth
י	10			yod
כ	20	ך	500	kaph
ל	30			lamed
מ	40	ם	600	mem
נ	50	ן	700	nun
ס	60			samekh
ע	70			ayin
פ	80	ף	800	pe
צ	90	ץ	900	tzaddi
ק	100			qoph
ר	200			resh
ש	300			shin
ת	400			tau

NECROMANCY

Strictly, necromancy is the purported practice of communicating with or conjuring the dead for the purpose of extracting information or to predict the future. Beginning in ancient Egypt and Babylonia, necromancy was also widely practiced in Israel, China, and throughout the Greco-Roman world. According to the Christian Church, necromancy amounted to diabolical commerce with unclean spirits, rites of criminal curiosity, and the forbidden invocations of departed souls. But, persisting throughout the Middle Ages, necromancy achieved unprecedented popularity among a clerical underworld who enthusiastically consulted magical grimoires to communicate with subterranean spirits and even angels. Other ancient 'sciences' such as geomancy and theurgy were also avidly studied. Among alchemists and the clergy alike, the notion that specially-coded alphabets and other symbolic devices, derived from their studies of arcane sources, could place humans in contact with the 'other world' became increasingly popular.

Magic in the Bible

Despite the Bible's repeated condemnation of magic, in one peculiar episode King Saul visits a witch from the Canaanite city of Endor and cajoles her into summoning the recently deceased prophet Samuel in order to consult him about the threat of the Philistines. While from the 5th century the orthodoxy of Christian dogma became increasingly set in stone, the possibility of alternative ritual and magical paths to communion with the spirit world held its appeal.

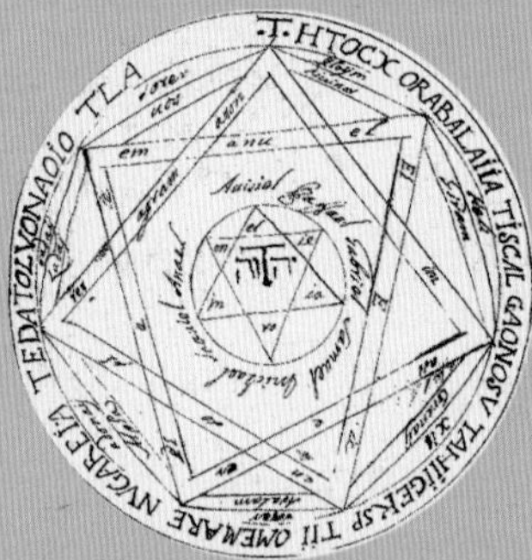

The *Key of Solomon* is one of the most notorious medieval handbooks of magic (attributed to King Solomon himself). The book contains magic circles (*above*), instructions for necromancy, invocations, and binding spells to conjure and constrain infernal spirits.

Ancient origins

Necromancy rituals and ghost expulsion/exorcism texts were commonplace throughout the ancient Near East. In Egypt, as early as the second millennium BC, consultation with deceased royalty was sponsored by the state for public benefit. The central rite usually consisted of the rubbing of magical salves onto the necromancer's face or onto the figurine of the spirit to be consulted. In ancient Turkey, ritual communication with infernal spirits and deities was achieved by means of pits dug into the ground which served as a portal for the chthonic deities to ease their passage between worlds. From this rich bed of mysterious rites, and from the great wave of translations of Arabic magical texts during the 12th and 13th centuries, literate, well-educated members of the European clergy mined necromantic texts which contained a synthesis of astral magic and exorcism techniques combined with Christian and Jewish teachings. The systems of magic depicted in these writings differed radically from the petty sorcery characteristic of the earlier centuries, becoming in effect an erudite method of conjuration and invocation.

Tricks of the trade

The esoteric implements of the medieval necromancer included magic circles, conjurations, sacrifices, swords, and prayer; magical alphabets were also highly significant. Circles were traced on the ground, often accompanied by various mystical symbols drawn from a mixture of Christian and occult ideas. At the opportune time and location, sacrifices and animal offerings were often provided to propitiate ethereal beings. The most important medieval writings on the occult were by the Benedictine abbot Johannes Trithemius (1462-1516, *see page 73*), and his pupil, Heinrich Cornelius Agrippa (1486-1535). The latter's *Three Books About Occult Philosophy* (1531-33) included coded formulae for alchemy, Kabbalism, and the Theban alphabet (*right*), a table for communicating with the spirit world.

Agrippa has often been identified as the model for the priest/sorcerer in Christopher Marlowe's play *Doctor Faustus* (c.1589) who sells his soul to an emissary of the Devil.

The Theban alphabet

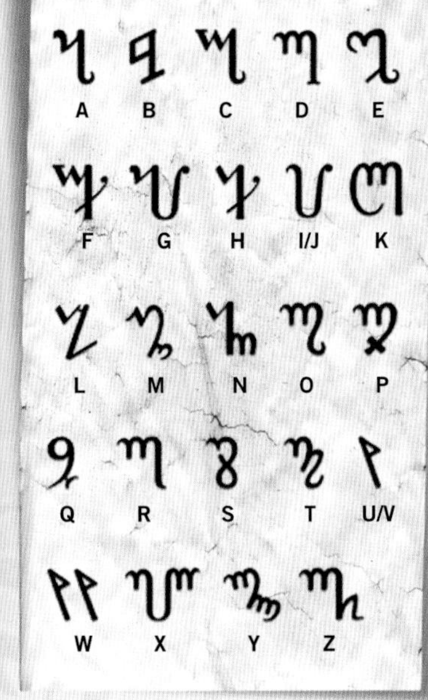

Sword
The origin of the magician's wand.

Magic circles
Marked with mystic alphabetic symbols, these provided a sacred space protecting the necromancer.

An imaginative representation of a ceremony to ward off entities from the spirit world. Protected within a magic circle, the necromancers here are protecting themselves from a demon released by their excavations by citing arcane texts, while the central figure transmits the message using a sword.

John Dee

The most famous magician of his age, an astrologer/alchemist/necromancer, and cryptographer, Dr. John Dee (1527-1608) was revered in his time as the most learned man in all Europe. He was the personal astrologer to Queen Elizabeth I of England, and the model for Shakespeare's Prospero in *The Tempest*. He visited many courts in Europe, along with the mountebank Edward Kelley (1555-97), to find funding for his exercises in divination and the occult. It is unsurprising that Dee made many enemies during his time in the court, several of whom continually brought charges of witchcraft against him. Ultimately, Dee would bring about his own downfall: despite his powerful position and prodigious intellect, his occult preoccupations overwhelmed him and he died in extreme poverty in 1608, reviled and pitied as a madman.

John Dee condensed all magic into a single symbolic equation, the occult equivalent of Einstein's $E = mc^2$. In his *Monas Hieroglyphica* he and Edward Kelley produced the Enochian alphabet, a purported means of communicating with the spirit (or angelic) world.

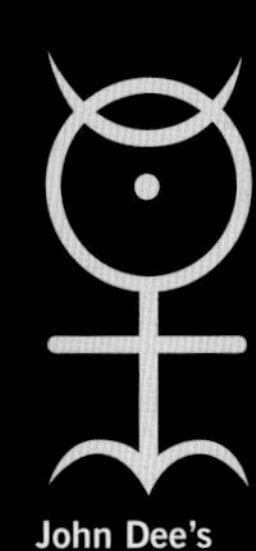

John Dee's single equation

The Enochian alphabet

Pa b	Veh c, k	Ged g, j	Gal d	Or f
Un a	Graph e	Tal m	Gon i	Gon with point w/y
Na h	Ur l	Mals p	Ger q	Drux n
Pal x	Med o	Don r	Ceph z	Van u/v
Fam s	Gisg t			

Rosicrucians

The various threads of medieval Hermeticism – alchemy, Kabbalism, and necromancy – plaited together in the 16th century with Reformational impulses and the resurgence of Catholicism during the Counter-Reformation, in the esoteric cult of the Rose Cross. It was a cult that crossed the boundaries of religious allegiance, and permeated Hermetic thought for the next four centuries, becoming an iconic source of ideas and imagery for mystics and magicians until the present day. Essentially, Rosicrucianism drew together a number of ideas, symbols, and images from a wide range of sources, and attempted to make a coherent magical/spiritual system of them, in the face of the empirical skepticism of the Age of Reason. In doing so, the Rosicrucians have provided conspiracy theorists with a rich fund of themes and ideas.

The Rose and the Cross

When Martin Luther nailed his 95 Theses to the door of Wittenberg's Castle Church, sparking off the bloody division of the Christian Church into Roman Catholics and Protestants, he was voicing only the most potent of many calls for a different interpretation of the religion outside the strict rules of Vatican dogma. Luther's own seal or coat of arms combined the Rose and the Cross (*above*). The Cross had, of course, long been a Christian symbol (*see page 42*), while the Rose had been associated with both Mary the Virgin and Mary Magdalene.

Gold und Rosenkreuzer

Founded in Prague in the early 18th century by Samuel Richter, this hierarchically structured group provided a channel whereby many of the Hermetic interests of medieval and early Renaissance times were transmitted to the 19th century, during which there was a revival of (often non-Christian) spiritualism (*see page 254*). Rosicrucianism has flourished as a mystical cult to the present day, especially in North America.

Origins

Three documents were published anonymously in the early 17th century: *Fama Fraternitatis* (1615), *Confessio Fraternitatis* (1615), and the *Chymical Wedding of Christian Rosenkreuz* (1616), the latter describing the experiences of a 14th-century German pilgrim in the Middle East – a Christian exposed to the influences of Eastern occultism who as a result developed a new ecumenical spiritual world view. These formed the basis of the Rosicrucian cult. Together, they constituted a fable into which many influences could be woven, and relied on a Pythagorean understanding of numbers. The symbolism of the Rose Cross comprised a mixture of symbolic imagery, including references as diverse as the Tarot and divination.

Metaphorical pilgrims seeking 'insight into the physical universe and the spiritual realm' at the mount of knowledge, ideas at the heart of Rosicrucian thought.

A geometric rose-pentangle in a cross with mystic letters and symbols of the Evangelists.

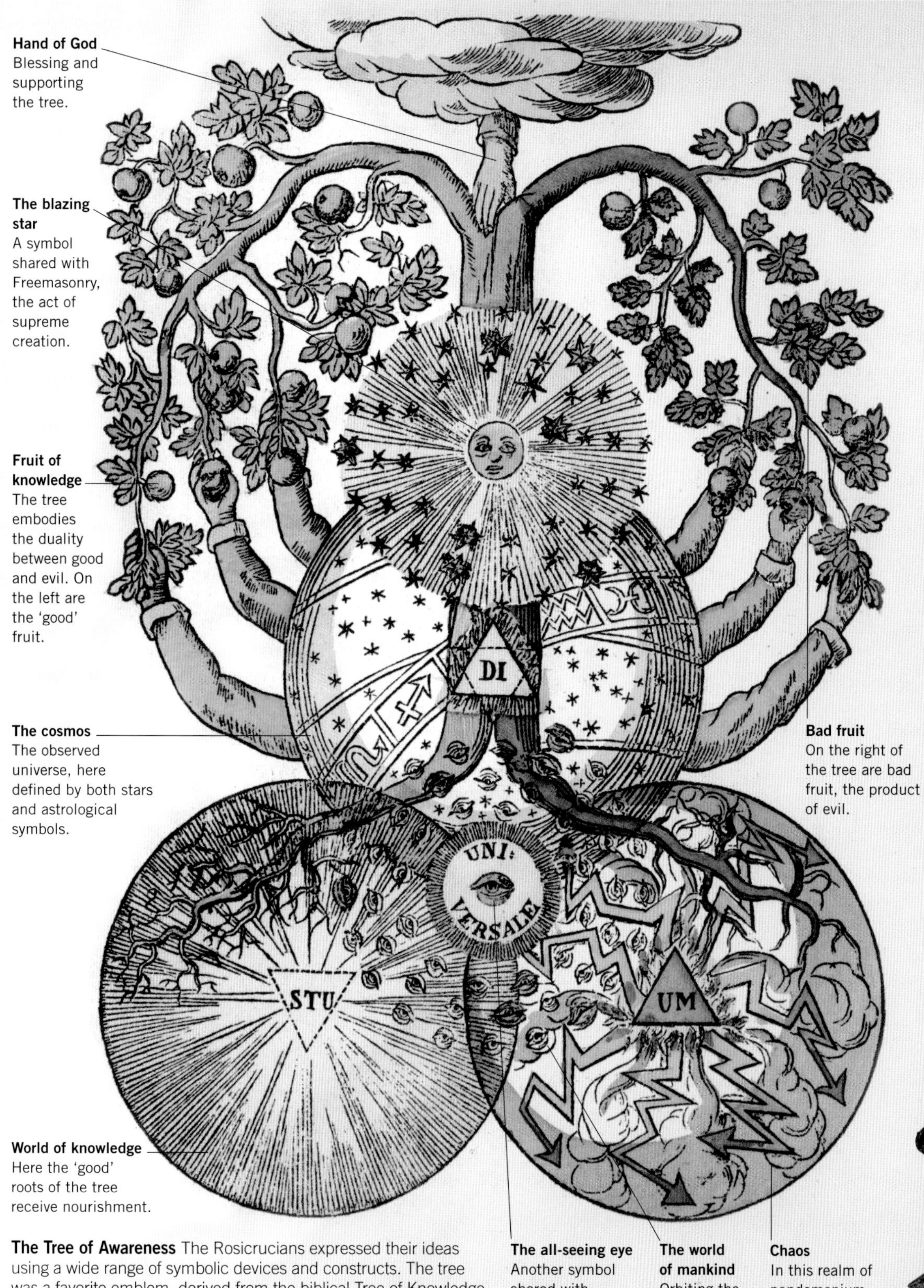

Hand of God Blessing and supporting the tree.

The blazing star A symbol shared with Freemasonry, the act of supreme creation.

Fruit of knowledge The tree embodies the duality between good and evil. On the left are the 'good' fruit.

The cosmos The observed universe, here defined by both stars and astrological symbols.

Bad fruit On the right of the tree are bad fruit, the product of evil.

World of knowledge Here the 'good' roots of the tree receive nourishment.

The Tree of Awareness The Rosicrucians expressed their ideas using a wide range of symbolic devices and constructs. The tree was a favorite emblem, derived from the biblical Tree of Knowledge in the Garden of Eden. One key Rosicrucian tenet was that Eve, in tasting its fruit (which led to the loss of innocence and expulsion from Eden), had condemned mankind to an eternal quest to acquire the appropriate knowledge in order to rebuild mankind's covenant with God. This image, from 1785, attempts to define the cosmographic nature of the spiritual realm.

The all-seeing eye Another symbol shared with Freemasonry, this rests at the intersection of the cosmos, the world of knowledge, and the world of chaos.

The world of mankind Orbiting the all-seeing eye are symbols representing humans with access to 'knowledge.'

Chaos In this realm of pandemonium, the tree's roots wither.

The Illuminati

Rumors have long circulated about a secretive, select band of brilliant, gifted savants, scientists, and artists who mysteriously control the world around us, dedicated to the creation of a New World Order (NWO). While the Masons would seem to adequately fulfill part of this role, various Rosicrucian organizations seem to have embraced and promoted the idea for their own ends. Such organizations are known to have existed, although quite how powerful they were in influencing the affairs of the world is a matter for debate. The Bavarian Illuminati was founded on May 1, 1776 by Jesuit-trained lawyer Adam Weishaupt (d. 1830), and was very much a product of the Enlightenment; it was fundamentally a loose association of intellectual freethinkers, rather than a conspiracy. Numbering some 2,000 members across Europe, it disintegrated in a dispute about succession. On the other hand, organizations such as the Rosicrucians, the Martinists, and the Masons all embraced the notion of illumination and secret brotherhoods, and there was a considerable crossover of ideas, symbolism (*below*), and doctrine.

freemasons

Freemasonry is one of the world's oldest and largest fraternal societies, with over five million members. Its origins are shrouded in mystery. Some claim it descended from the Knights Templar, or the builders of King Solomon's Temple, or the mystery religions of ancient Egypt. Others claim it developed from medieval stonemasons guilds with their arcane rituals culled from Hermetic philosophy and Renaissance mysticism. To the layman, Freemasonry evokes fears of world domination by an elite coterie who enjoy privileged access to wealth and power. Its rigid secrecy has spawned countless conspiracy theories and denunciations, including no fewer than 16 papal pronouncements condemning Freemasonry as being depraved and perverted. Despite being outlawed in various countries, Freemasonry spread rapidly throughout the world, admitting such luminaries as Mozart, Voltaire, Frederick the Great, Benjamin Franklin, George Washington, and Winston Churchill.

Origins and rituals

The principles of Freemasonry are more than 800 years old. Freemasons are taught its precepts through a series of ritual dramas, which use customs and tools of the builder's craft as allegorical guides for moral betterment. Secret signs and passwords have their origins in the practices of the medieval stonemasons responsible for the great castles and cathedrals of Europe. Like other craftsmen, they belonged to a trade union or guild in which craft secrets were cautiously guarded. The term 'Freemason' first emerged in England in the late 14th century, and is an abbreviation of the term 'freestone mason,' a highly skilled and learned artisan who could 'freely' sculpt stone in any direction. The modern fraternity of Freemasonry began in 1717 when four London lodges coalesced to form a central governing body called the 'Grand Lodge.' As early as the mid-17th century, Freemasonic lodges had begun to admit non-operative (or speculative) Masons who were attracted to the legends and mysteries about the divine origins of the Craft. For generations, Freemasons taught that the Order had been established among the workmen who built Solomon's Temple. Destroyed over 2,500 years ago by the Babylonians, Solomon's Temple is the central leitmotif of Freemasonry.

Apprentice rite The candidate enters the lodge with his chest exposed, blindfolded, shoelaces untied, and right knee exposed. The Grand Master is seated and surrounded by fellow brethren.

Floor cloth Masonic ritual is illustrated by the floor cloth, an emblematic representation of the Temple of Solomon differentiated according to the three degrees of initiation: Fellowcraft, Apprentice, and Master Mason.

Mozart in the Viennese Masonic Lodge This painting illustrates several Masonic ceremonies and features Mozart seated on the extreme right.

The Pigpen cipher

This monoalphabetic substitution cipher has been used in various forms for many centuries, and is still used by children today; during the 18th century a version of it was used by the Freemasons for their archives. The key to the encipherment is simple to remember: a symbol is substituted for each letter of the alphabet, the symbols being derived from the position of each letter on the grid below:

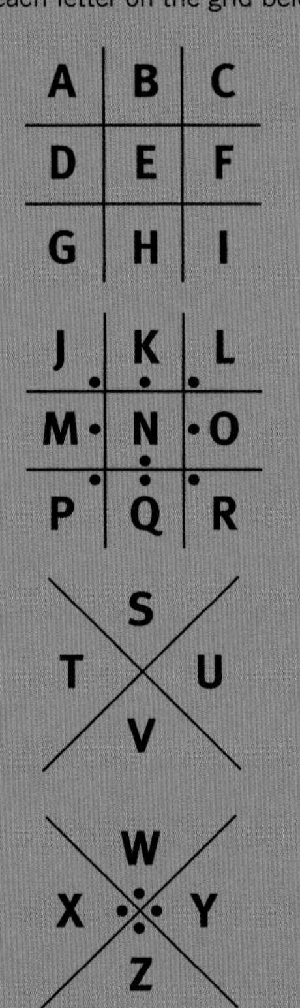

Using this system, 'The Temple of Solomon' would be enciphered as:

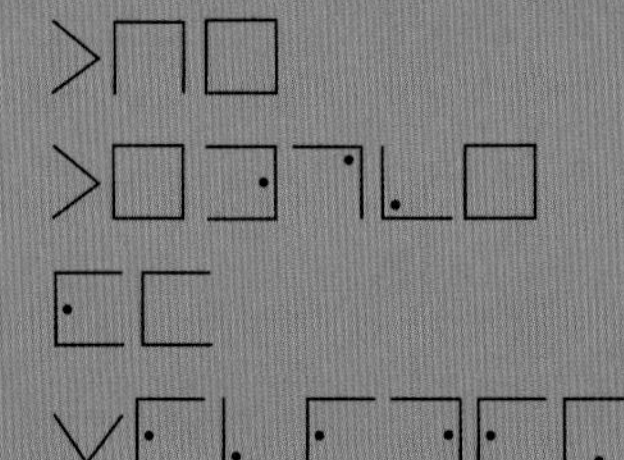

Masonic symbolism and tenets

Dedicated to self-improvement and charity, Freemasonry is often described as a peculiar system of morality, veiled in allegory and illustrated by symbols. Freemasonry is open to all men, exclusive of race and religion, and is guided by the three principles: Brotherly Love, Relief, and Truth. The myth and lore of Freemasonry incorporated the geometric knowledge of the medieval masters with theosophical knowledge inherited from the mystagogue, Hermes Trismegistus. In early Masonic lore Hermes is revered for the invention of everything known to the human intellect and for preserving the knowledge of the Mason's craft and transmitting it to mankind.

Blazing Star
An essential emblem of a Masonic Lodge, the Blazing Star represents either the Star of Bethlehem, Venus, or the Sun.

G – for geometry
For Freemasons, geometry is reputed to be an exclusive and secret science handed down by God to Hiram Abiff, the builder of Solomon's Temple.

Masonic apron The often highly decorated apron is a central Masonic ritual vestment.

Jachin and Boaz
The Two Pillars, Jachin and Boaz, are among the most pre-eminent symbols of Freemasonry and are identified with the twin columns erected at the entrance to Solomon's Temple.

Solomon's Temple
The primary Masonic symbol, representing the origins of the movement, what has been lost, and an inspiration for what may be recovered in a Mason's personal development.

Compasses
As the kernel of all advancement, truth, and mystery, geometry is considered to have the sacred power to re-create the Divine in form – a veritable blueprint for the mind of God.

George Washington lays the cornerstone of the US Capitol. Dozens of American temples and public buildings were consecrated by Masons, including the most famous ceremony ever: wearing a Masonic apron and wielding a silver trowel, George Washington dedicated the US Capitol Building on September 18, 1793.

The American Empire

Freemasonry is closely allied with the foundation of America. Nearly a third of all US presidents were members of the Brotherhood, and Freemasons played pivotal roles in the seminal events of the nation's history, such as the Boston Tea Party, and the Declaration of Independence. Masonic tropes and themes are also observable in the scriptures and rituals of one of America's fastest growing religions, The Church of Jesus Christ of Latter-day Saints. Today, Masonic symbols have become entrenched in the iconic images of American empire, from the Great Seal on the one dollar bill to the Statue of Liberty.

Great Seal of the United States The all-seeing eye features above a thirteen-stepped pyramid, beneath which a scroll proclaims the advent of a 'new secular order.'

Few of the earliest secret messaging and cipher systems seem that secure today, but the many variations on the substitution cipher proved reasonably reliable for over a millennium.

codes for secrecy

The advent of mathematically and linguistically based decrypting techniques, such as frequency analysis, rendered most substitution ciphers vulnerable, while in turn inaugurating a new wave of ingenious methods of encipherment, many of which provided the principles for modern, computer-based cryptographic systems.

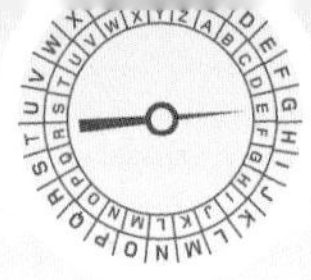

THE ART OF CONCEALMENT

The need to disguise a message, especially in times of war or state security, dates back many centuries. Before alphabetic or numeric codes were devised, various ingenious techniques were developed – some still in use today – for concealing a secret, its location, or the way to find it, or for conveying a piece of information across hostile or enemy lines. Although not strictly coded, the principal of hiding a message using 'secret writing' is similar in intent to cryptography, and is today known as steganography.

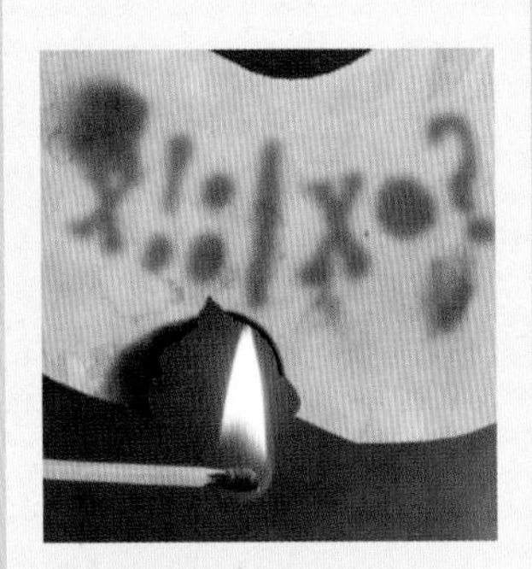

Invisible ink

In the 1st century AD the Romans were aware of invisible inks such as the juice of the thithymalus plant, which is transparent when dry, but when gentle heat is applied turns brown. Other substances have been used, including lemon juice and urine. Giovanni Battista della Porta, the 16th-century scientist, described how a solution of alum in vinegar could be used to write a message on a hard-boiled egg, which appeared to be invisible, but which penetrated the shell leaving its message on the white albumen inside.

The wax technique

In around 480 BC, Demaratus, an exiled Greek living in Persia, observed the buildup of Persian military power and felt he should warn the Greeks. He acquired a pair of wooden writing tablets, typically covered with wax so they could be reused. Having scraped the wax off, he wrote a message about the Persian plans and then added a new layer of wax. The Greeks, receiving two blank wax writing tablets, guessed that a message lay under the wax. It enabled the Greeks to prepare for the Persian onslaught.

The shave technique

Sending a messenger through enemy lines is tricky. But if time was not too pressing, then Histaiaeus's idea might work: he shaved the head of his messenger, tattooed his secret message on the scalp, waited for the hair to grow, then dispatched his messenger, who was aware he was carrying a message, but not what it contained. Arriving at his destination, he simply shaved his head.

'Blood chits' were supplied to airmen in wartime, requesting shelter, food, and medical aid in the local language should they be shot down.

Silk durability

The durability and compactability of silk has allowed it to be used for smuggling information in various ways. The Chinese would write a secret message on the material, roll it into a tiny ball, coat it in wax, then the messenger would swallow the wax tablet, nature being allowed to take its course upon arrival at the destination. During World War II, silk squares had maps printed on them; these were compressed to fit into the heel of a pilot's flying boot so that, if shot down, he could try to reach safety using the map.

Theseus and the Minotaur

One of the earliest examples of elaborately hiding a secret lies in Greek mythology with the tale of Theseus and the Minotaur. The legendary monstrous half-man, half-bull, offspring of King Minos of Knossos in Crete, who needed to be regularly served sacrificial virgins from the Peloponnesian mainland, was confined and concealed in a labyrinth beneath the palace. Excavations at Knossos in the late 19th century revealed a complex of more than 1,000 small, interlocking rooms, which initially suggested the existence of a maze-like lair. Theseus, in tracking down and slaying the beast, ensures he can then escape from the labyrinth by unwinding a cord or 'clew' as he finds his way through the dark tunnels, which he can then follow back to daylight when the deed is done; this is the origin of our word 'clue,' an essential aid in detection.

Theseus and the Minotaur, set within a labyrinth, was a popular motif in classical mosaics and wall paintings.

An elaborate labyrinth on the floor of Amiens cathedral, France.

In the *Hypnerotomachia*, the hero interprets this 'Egyptian' inscription as "Military prudence or discipline is the strongest bond of the empire."

Rebuses

A more playful variant of encipherment, and one that is still popular in children's literature, is the technique of replacing words with pictures or objects, a form of substitution cipher. Early Christians were adept at using such ideas to disguise but communicate their shared faith (*see page 42*). Early Egyptologists assumed incorrectly that hieroglyphs were rebuses, and the rebus idea was often used to create punning images in heraldry (*see page 228*). The fantastic dream-text *Hypnerotomachia Poliphili* (1499) includes several imagined inscriptions 'in hieroglyphs' which the hero manages to somehow interpret. Picture letters became popular among correspondents in the 18th and 19th centuries.

A typical 19th-century children's picture story, an amusing way of developing literacy skills.

This microdot camera is shown here slightly larger than life-size.

The microdot

The technology for photographically compressing an image, usually text, to a tiny slab of film (a microdot) which could then be glued to a document, was developed in the early years of the 20th century. Usually punctuation marks or advertisement artworks were used to camouflage the microdot Advantages: it might not be noticed. Disadvantages: held under light, the microdot tended to show up its reflective surface.

Lost and found

Both Greek and Roman writers describe labyrinths designed as garden ornaments, known as mazes, and such designs are found too in other areas, such as China and South Asia. They demonstrate a fascination with geometric artifice. Labyrinthine carvings, mosaics, or tile patterns are also found in the paving of many medieval edifices across both the Christian and Muslim worlds. For the most part, these are single-track geometric conceits, albeit ones that are designed to demonstrate the complexities of life's road. Mazes became a very popular feature of formal garden design from the Renaissance onwards, where the idea of dead ends and feints became increasingly important to delight and confuse the visitor.

For Your Eyes Only

The first known codes that involved encryption in order to conceal the content of a message or document from prying eyes date back over 2,000 years. These can be divided into two styles: transposition ciphers in which the letters of a message are merely rearranged, and substitution ciphers. The Spartan scytale (*see page 102*) is an early example of a transposition cipher, achieved mechanically. One of the most famous and earliest examples of a substitution cipher is the Caesar Shift (*see page 103*). Although apparently confusing – unless the recipient is aware of the method used for encryption (known as the algorithm) – these codes proved vulnerable to frequency analysis (*see page 68*). Nevertheless, variations of such methods of encryption proved successful for almost 750 years, and have continued to be used, displaying a remarkable breadth of ingenuity.

Anagrams

The simplest way of disguising a word or message by transposition is by merely replacing the order of the letters; this known as an anagram, much loved by cryptic crossword compilers today, but a word game and secrecy system dating back over two millennia. Anagrams themselves are limited by the number of letters in the plaintext: a three-letter word like 'dog' can be rearranged only five times:

ogd / odg / gdo / god / dgo

However, a more complicated message could involve hundreds, indeed millions, of variations – and, possibly, many different solutions that do not match the one intended. Within a simple sentence of 30 letters, the number of possible ways of rearranging the letters exceeds 50 billion, billion, billion. Even the simple phrase 'mind what you say' produces over 350 different words, which could be assembled in a variety of combinations, proving how inefficient anagrams are.

Often anagrams have an inbuilt clue to their structure and intended meaning, indicating how to unlock them, such as the Roman square (*see page 42*).

Another example of this type, often used in cryptic crosswords might be:

'marred dour film'

'marred' indicates that the letters should be rearranged;

'film' indicates what the answer might be;

together these could lead you to the answer:

'dial m for murder.'

Transposition ciphers

One of the earliest forms of transposition cipher is based on a relatively simple algorithm. It is effectively a way of creating an anagram using simple mathematical principles. The Rail Fence system, still widely used today among schoolchildren, provides a puzzling ciphertext, although understanding the algorithm means that it is easy to decrypt and not particularly secure.

Take a plaintext message like:
CAREFUL YOU ARE BEING FOLLOWED
Taking out the word spaces, and then rearranging the alternate letters of the plaintext on two separate lines will produce the railfence pattern:

The letters should then be reassembled as a single ciphertext, strung together line by line. The recipient, aware of the algorithm (the key), can easily rearrange the letters to read the plaintext message.

The Rail Fence system can be made more complicated: for example, by arranging the plaintext on three lines rather than two, then stringing the three lines of letters together, or by switching the order of each pair of letters throughout the ciphertext, so that the first and second letters swap places, then the third and fourth, and so on.

Monoalphabetic substitution ciphers

The next step beyond transposition ciphers are substitution ciphers. The simplest of these rely on the substitution of each letter of the alphabet by either another letter of the alphabet, or by numbers or symbols (and sometimes a mixture of all three). These are known as 'monoalphabetic' substitution ciphers. The best known examples of this method of encryption are the Caesar Shift cipher, which merely involves offsetting the normal alphabet against an alphabet of the normal order, but with a different starting point (*see page 102*), the Kama Sutra cipher (*opposite*), and the Babington Code (*see page 74*).

Algorithms

The component parts of a monoalphabetic substitution cipher are the plaintext (the message to be sent), the key (specifying the details of the particular encryption) and the general encrypting system, known as the algorithm.

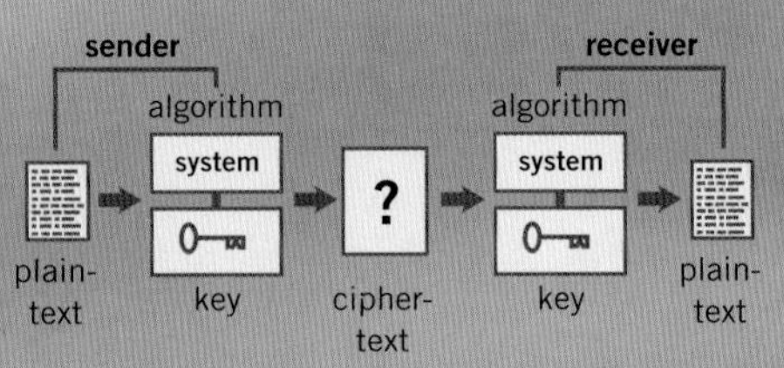

The success of the system relies on the key remaining a secret known only to the sender and the receiver.

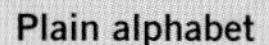

The Kama Sutra cipher

The Kama Sutra was a manual, written by the Brahmin scholar Vatsyayana in South Asia in the 4th century AD, which drew together a wide range of advice for eligible young ladies covering 64 subjects as diverse as dress, cooking techniques, versatility at games and crafts and, notoriously, erotic skills. One of the subjects included was secret writing to be used in order to disguise details of their lovers' trysts. The formula proposed was to encrypt a secret message by creating a random substitution alphabet, thereby avoiding the logical shift of the Caesar cipher.

Plain alphabet

A B C D E F G H I J K L M N O P Q R S T U V W X Y Z
R M E S Z W N A L Y B T F I Q X J U D V K H G O P C

Cipher alphabet

UNDER THE BANYAN TREE TONIGHT

KISZU VAZ MRIPRI VUZZ VQILNAV

Using this particular substitution system, a plaintext message such as the scroll on the left, would appear as the scroll on the right. The fact that such a system can rely on any rearrangement of the alphabet to produce the cipher alphabet means that there are 400,000,000,000,000, 000,000,000,000 potential keys – an impossible task for a cryptanalyst attempting to decode such a message.

A shorthand solution

Memorizing the key to a Caesar Shift cipher is quite simple, while memorizing the exact form of a Kama Sutra cipher could present more problems. A different sort of key can make both systems both simpler to use and simpler to remember. Choose an easily remembered key phrase, such as HELLO MY FRIEND. Remove any repetitions and spaces, and place the phrase at the beginning of the ciphertext, then complete the alphabet in normal order, avoiding any further repetitions.

Plain alphabet

A B C D E F G H I J K L M N O P Q R S T U V W X Y Z
H E L O M Y F R I N D A B C G J K P Q S T U V W X Z

Cipher alphabet

ARE YOU DOING WELL?
HPM XGT OGICF VMAA?

Using this method, the plaintext (*top*) would result in the ciphertext (*above*). The longer the key phrase, the more effective the algorithm for encryption.

Using symbols

Many monoalphabetic substitution keys used symbols to replace certain letters of the alphabet, which had the effect of making the encrypted message seem even more confusing. This meant that memorizing which letters had been substituted by which symbols added a further layer of complexity for both the sender and the receiver, although the underlying principle remained the same.

A simple example might be to use a six-letter Caesar Shift, combined with replacing every vowel within the plain alphabet with a symbol:

Plain alphabet

A B C D E F G H I J K L M
N O P Q R S T U V W X Y Z

Cipher alphabet

✪ G H I ✿ K L M ➾ O P Q R
S ♣ U V W X Y ❤ A B C D E

Encrypting a simple phrase using this ciphertext will begin to disguise the message.

HOW WAS IT AT THE ZOO?
M♣B B✪X ➾Y ✪Y YM✿ E♣♣

Most ciphers will avoid any punctuation marks, especially ?, as it will provide the cryptanalyst with a valuable clue.

M♣BB✪X➾Y✪YYM✿E♣♣

Further elaborations could include eliminating any word breaks in the ciphertext.

M♣B+B✪X+➾Y+✪Y+YM✿+E♣♣

Adding a separate symbol to indicate word breaks, such as +, will create a more complex ciphertext. But once again, the inclusion of word breaks is usually avoided, as such a regularly recurring symbol might be guessed to be a word break.

M♣B3✪X ➾Y ✪Y3M✿ E♣3

Using another unique symbol, for example 3, to indicate a repeated letter can add a further layer of confusion.

Frequency Analysis

Al-Kindi

Al-Kindi (AD 801-73) was an Arab mathematician, scientist, philosopher, astronomer, psychologist, and meteorologist. He was noted for his attempts to introduce Greek philosophy into the Arab world, and translated many classical texts while working in Baghdad. Alongside these and many other advances in the fields of medicine, physics, and musical theory, he is known today for the first known explanation of cryptanalysis based on frequency analysis, which can be found in his *Manuscript on Deciphering Cryptographic Messages*. This text was only rediscovered in Ottoman archives in Istanbul in 1987, but it seems likely that the concept that al-Kindi describes (for use obviously with the Arabic alphabet) was relatively widely known in the Arab world by the 9th century, whether al-Kindi himself invented it, or was merely among the first to describe the technique.

Frequency analysis is a method of decrypting substitution ciphertexts if the original language of the message is known. It relies upon each language's use of its alphabet, and the frequency of certain letters and combinations of letters commonly found within that language. Such frequencies vary from language to language (in German for example, 'e' has a frequency of around 20%, while Italian has three letters with a frequency over 10% and nine letters of less than 1%). Frequency analysis can only be used effectively for decoding substitution ciphers, and is not foolproof; but, by its very nature, the larger the ciphertext, the more effective frequency analysis will be, as many letters have a similar and mathematically measurable frequency.

Frequency analysis in English

Although different sample texts will produce slight variations, the five most frequent letters in English are normally regarded to be 'e,' 't,' 'a,' 'o,' and 'i.' These vary a little, but 'e' is always the most frequent, while 'n,' 's,' 'h,' 'r,' and 'd' generally make up the top ten most frequently used letters in English. This table shows one such order of frequency, in average percentage of use.

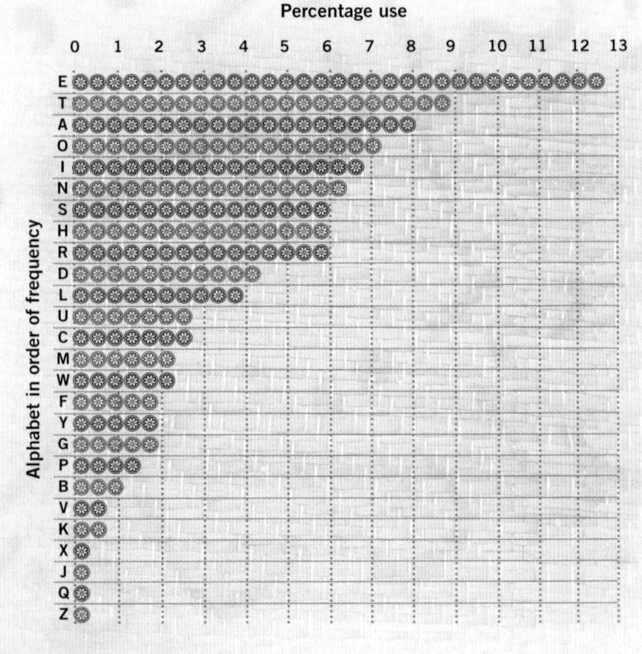

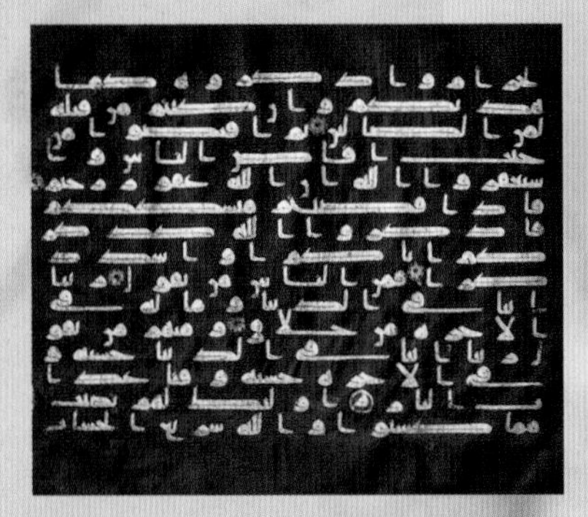

The concept of frequency analysis was developed in the Arab world. Arabic has 28 basic letters, but each letter can have four distinctive forms (initial, medial, final, and isolated), plus letter combinations, making the cryptanalyst's task much more difficult.

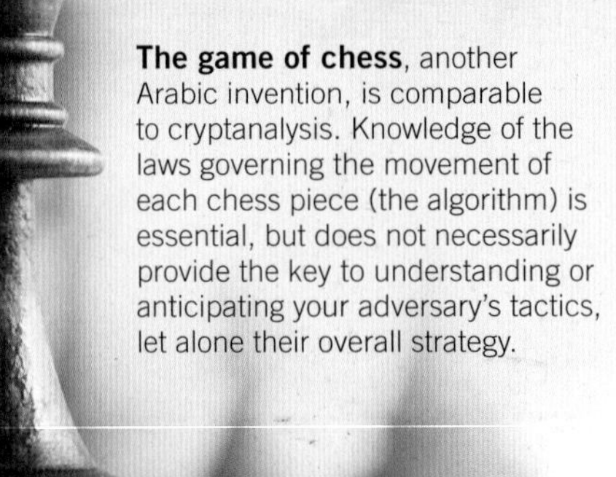

The game of chess, another Arabic invention, is comparable to cryptanalysis. Knowledge of the laws governing the movement of each chess piece (the algorithm) is essential, but does not necessarily provide the key to understanding or anticipating your adversary's tactics, let alone their overall strategy.

Application

Taking a ciphertext such as the one on the right, the cryptanalyst would draw up a table to analyze the frequency of each letter or character, and would then compare the findings with a standard table of letter frequency (*opposite*), and try to identify correspondences. An analysis of the frequency of each letter in this ciphertext is shown on the table (*right*).

DIWHU OXQFK, WKHLU ZDON WRRN WKHP GRZQ IURP WKH
LQQ WR WKH ORFDO PDUNHWV. WKHB PDUYHOHG DW WKH
VHOHFWLRQ RI JRRGV RQ RIIHU, HLJKW RU QLQH VWDOOV MXVW
VHOOLQJ IUXLW, WKH VDPH IRU YHJHWDEOHV, ILVK, DQG PHDW.

Percentage use
0 1 2 3 4 5 6 7 8 9
Alphabet in order of frequency
H W R D O K Q V U I L P G J F N X Y Z E M B S T A C

The cryptanalyst now has a basic mechanism to work with. Using this table they would identify the most common letter, in this case 'H,' and apply the likely decrypt 'e.' Then they would match the next most frequent letter 'W' with 't,' and so on. In this case a Caesar Shift of three has been used, so the cryptanalyst would soon recognize a pattern, and solve the puzzle quite rapidly. However, when a more complex cipher algorithm has been used, they would have to keep going until it became clear that some letters could not possibly occur next to each other. From here, the cryptanalyst would have to retrace their steps, changing a letter here and there, before continuing.

Alongside working through the plaintext letter by letter, the cryptanalyst might search for the most frequent combinations of two letters (digrams) or of three letters (trigrams, *see right*). In this ciphertext, the most common letter 'H' is assumed by the cryptanalyst to represent 'e.' The most common trigram in the ciphertext is 'WKH,' so it might be guessed that 'W' is 't' and 'K' is 'h,' forming the most frequently used trigram in English – 'the.'

'After lunch, their walk took them down from the inn to the local markets. They marveled at the selection of goods on offer, eight or nine stalls just selling fruit, the same for vegetables, fish, and meat.'

No guarantee

The plaintext below shows that no single-letter frequency order can always work for every message. The longer the text, the better frequency analysis works, but it is entirely dependent on the plaintext in question. A plaintext like this will produce a distorted analysis (*right*):

'Sixty-six ex-zookeepers from Zimbabwe and Zambia met in Zanzibar, Tanzania, to discuss the Zulus' attitude to zebras.'

Even in this rather extreme plaintext example (*above*), 'e,' 't,' 'a,' and 's' remain relatively frequent, although the uncommon frequency of 'z' and 'i' might throw the cryptanalyst off the track.

Percentage use
0 1 2 3 4 5 6 7 8 9
Alphabet in order of frequency
I A E Z S T O B M N R U X V D F H K L P W Y C G J Q

Looking for letter combinations

In addition to single-letter frequency analysis, it can help enormously to determine the frequency of digrams and trigrams (commonly occurring two- and three-letter combinations). These might form words or just parts of words.

In detecting words, the most common trigram word in English is 'the,' so once 'e' has been identified, if it comes at the end of a frequent trigram in the ciphertext, then a preceding 't' and 'h' might reasonably be guessed. If 'a' comes at the beginning of the second most common trigram ('and'), then 'n' and 'd' might also be assumed. The most frequently used three-letter sequences in English are 'est,' 'for,' 'his,' 'ent,' and 'tha.'

Taken further, a search might extend to the most common pairings of letters in English – 'th,' 'ea,' 'of,' 'to,' 'in,' 'it,' 'is' – some of which are also words in their own right. Another tactic might be to look for the most commonly repeated letters such as 'ss,' 'ee,' 'tt,' 'ff,' 'll,' 'mm,' and 'oo.'

Although it can be a time-consuming job, frequency analysis, along with perseverance and a good knowledge of the language in question, will almost invariably decrypt most substitution ciphers, even those substituting symbols or numbers for letters or phrases.

Disguising Ciphers

Many attempts were made to create ciphers that would defy, or at least frustrate, attacks using frequency analysis. Before the development of polyalphabetic ciphers (*see pages 72, 104*) the most common were a group of ciphers that used homophones, in which plaintext letters are represented by more than one ciphertext character, number, or symbol. Such cipher systems, in various forms but often of considerable complexity, were widely used for sending and protecting secrets until the 19th century. There were other methods that added further layers of transposition or substitution, progressively transforming the original message in cloaks of disguise.

Nomenclators

Popular by the late 14th century, this style of cipher required a relatively simple basic codebook but a large number of homophones, often using pre-agreed symbols as homophones, with a single symbol being used to represent a word or phrase. The name of the system derives from the courtly official who read out the full titles of visiting dignitaries, their elaborate formal nomenclators encoded in a handbook of easily remembered characters and symbols. Due to its flexibility, and the possibility of adding innumerable homophones, this system remained the stuff of most daily diplomatic code signaling for 400 years. However, Mary Queen of Scots discovered to her cost that such ciphers were hardly impregnable (*see page 74*).

Eliminating the obvious

Deleting gaps between words, and any punctuation marks, will produce a continuous string of ciphertext, which is in itself daunting for the cryptanalyst. Using a transposition table of numbers rather than alphabetical letters also helps to confuse. One technique also used from medieval times to throw the cryptanalyst off the trail was to break the ciphertext up into regular packages of five or six characters (or 'bits,' a technique that became more common after the invention of the electric telegraph in the 19th century, when it became widely used in order to ensure accurate transmission of an encrypted message).

For example, a simple Caesar Shift of 12, substituted in turn by numbers, might look like this:

plaintext	A	B	C	D	E	F	G	H	I	J	K	L	M	N	O	P	Q	R	S	T	U	V	W	X	Y	Z
ciphertext	L	M	N	O	P	Q	R	S	T	U	V	W	X	Y	Z	A	B	C	D	E	F	G	H	I	J	K
numeric ciphertext	12	13	14	15	16	17	18	19	20	21	22	23	24	25	26	1	2	3	4	5	6	7	8	9	10	11

Using this algorithm, the following message and encryption would occur:

plaintext	t h e l a n d o f d r a g o n s
encryption	5 19 16 23 12 25 15 26 17 15 3 12 18 26 25 4
5-bit packages	51916 23122 51526 17153 12182 6254x

Nulls
The 'X' at the end of the cipher is known as a 'null,' included to make up the requisite number of ciphertext characters to complete the 5-bit packages.

Court officials developed a shorthand system of 'nomenclators' to record the elaborate official titles of visiting dignitaries.

Homophones: flattening frequency

The homophonic cipher can be used in a variety of ways to disguise substitution ciphers. Typically, the characters with the highest frequency are given the greatest number of homophones, thereby 'flattening' their apparent frequency. The homophones are used in a regular cycle. As more characters will be needed than are available in the 26-letter alphabet, other characters need to be introduced. Colorful and fanciful symbols might be added; existing alphabetic letters might be modified by using upper or lower case options; and finally, a more straightforward solution is to use almost endless numeric substitution.

This proved an effective disguise in principle and, taken to an extreme, as in the Great Cipher of Louis XIV (*see page 106*), where numerous homophones were used for every letter of the plaintext alphabet, it could prove almost impregnable.

plaintext	A	B	C	D	E	F	G	H	I	J	K	L	M	N	O	P	Q	R	S	T	U	V	W	X	Y	Z
ciphertext of 12-place Caesar shift	L	M	N	O	P	Q	R	S	T	U	V	W	X	Y	Z	A	B	C	D	E	F	G	H	I	J	K
Numerical homophones for common letters	3				1				5					6	4					2						
	9				7				11					12	10					8						
	15				13				17					17	16					14						

A table might be constructed along these lines, combining a Caesar Shift algorithm of 12, with three homophones for the six commonest letters in English (e, t, a, o, i, n), here using numbers.

plaintext	every	man	must	be	seen	to	have	done	his	duty
encryption	PG1CJ	XLY	XFDE	M7	D13P6	2Z	S3G1	O4127	STD	OF8J

Although in this plaintext message the letter 'i' was only used once, and so did not call upon the use of a homophone, the commonest letter 'e' occurred seven times, and ran through the homophone cycle in the ciphertext as 'P,' '1,' '7,' '13,' 'P,' '1,' and '7.'

Straddling checkerboard cipher

This ingenious technique provides a way of converting plaintext into digits and flattening frequency using homophones.

1 Draw up a table of 11 columns and four rows, then number the first row 0-9, leaving the first column blank.

2 In the next row, again leaving the first column blank, place the eight most frequent letters (in English, e, t, a, o, i, n, s, r) in no particular order leaving two blank spaces.

3 In the third and fourth rows add the rest of the alphabet, again in no particular order. As there are 44 spaces on the grid, there will be two further blanks, which can go anywhere in the third or fourth row.

4 In the left-hand column add the numbers corresponding to the two blank spaces on the second row. The sender and receiver must work from the same grid.

5 Compressed into a single stream of numbers the ciphertext 400431371913968 may be sent as it is, or put through a second stage of encryption by using a key such as 3455, which may be repeated, adding the numbers (with no carrying) of the ciphertext to the key to produce a new ciphertext.

6 The new ciphertext can then be converted back into letters using the straddling checkerboard table, remembering that any instance of numbers 3 or 6 should be paired with its following number.

It is clear from this that the most frequent letters have been disguised (e, a, and so on), as have repeated letters like t.

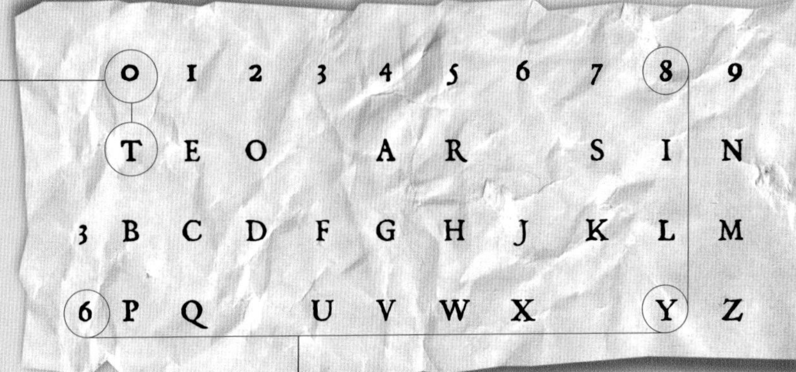

	0	1	2	3	4	5	6	7	8	9
	T	E	O		A	R		S	I	N
3	B	C	D	F	G	H	J	K	L	M
6	P	Q		U	V	W	X		Y	Z

Encryption
The letters on the second row are simply encrypted using the single digit numbers above them.

Double digit numbers
The letters on the third and fourth rows are given double digit numbers, derived from their row number then their column number.

plaintext	A	T	T	A	C	K	E	N	E	M	Y
ciphertext	4	0	0	4	31	37	1	9	1	39	68

ciphertext	4	0	0	4	3	1	3	7	1	9	1	3	9	6	8
+ key	3	4	5	5	3	4	5	5	3	4	5	5	3	4	5
new ciphertext	7	4	5	9	6	5	8	2	4	3	6	8	2	0	3

new ciphertext	7	4	5	9	65	8	2	4	36	8	2	0	3(0)
letter ciphertext	S	A	R	N	W	I	O	A	J	I	O	T	B

Decryption
As there is no ambiguity in the basic grid, this is achieved by simply reversing the process, step by step.

MEDIEVAL CIPHER SYSTEMS

Roger Bacon

The English Franciscan friar and natural philosopher Roger Bacon (1214-94), in addition to advising Pope Clement IV on how to combine empirical science with the largely theological educational syllabus at the time (which later attracted accusations of heresy), was one of the first Western scholars to investigate the mathematical potential of secret writing. In his *Epistle on the Secret Works of Art and the Nullity of Magic* he expressed astonishment that cipher systems were not used more widely for private correspondence.

His two great works (*Opus Majus, Opus Minus*) display an enormous breadth of enquiry and this, combined with his interest in cryptography, has led him to be attributed as the author of the Voynich Manuscript (*see page 168*), a remarkable catalog of illustrated observations accompanied by a coded text using a secret alphabet which has resisted all attempts at cryptanalysis.

The turmoil of the European Middle Ages, with constant internecine warfare among the emerging nation states, and the various Crusades to defend, or extend, Christendom, was played out against a background of emerging humanism and learning. Contact with the Arab realms was not exclusively violent, and many texts from classical times, preserved in Arab libraries, as well as original Arabic works, filtered into Europe, stoking a desire for enlightenment. Many of the concepts developed in the medieval monasteries and courts of Europe appear to have been anticipated by Arab scholars.

> "Mathematics is the gate and key of the sciences."
>
> **ROGER BACON, *OPUS MAJUS*, 1266.**

Alberti and polyalphabetic substitution

The concept of a series of mathematical shifts which would allow the sender of a coded message to constantly alter the encryption algorithm by using different alphabets was noted by Italian Renaissance humanist, philosopher, and architect Leon Battista Alberti (1404-72) in *De Cifris* (1466). He described a device with an outer disk (*stabilis*) with the letters of the Italian alphabet plus the numbers one to four arranged around the rim, and an inner movable disk (*mobilis*) with the letters of the alphabet in a random order. Alberti also devised a codebook of over 300 phrases with numerical values, although the examples below merely require pre-agreed index letters, and letters that trigger an index shift.

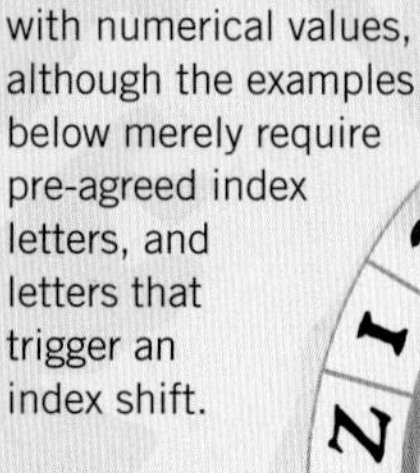

Encryption

Either disk can act as an index, and two methods can be used:

1 Method 1

An initial index letter is agreed, as are two or more further letters that trigger a change in the index. If the initial index letter was 'g,' then 'g' on the inner disk would be aligned with 'A' on the outer disk. The trigger letters may be embedded in either text. It is the trigger letters which make this a polyalphabetic cipher system.

2 Encipherment would then start, reading from outer to inner, until one of the trigger letters in the plaintext is reached (in this example 'T'), at which point the 'g' on the inner disk is realigned with the 'T' on the outer disk, and encipherment then proceeds using the new cipher alphabet, until the next trigger letter is reached, and the process repeated.

Trigger letter

When 'T' appears in the plaintext, the inner disk is rotated.

Realignment

Index letter 'g' rotated to align with 'T.'

Method 2

A simpler method uses the outer disk as the index. Again an initial index letter is agreed, here once again 'g,' which is aligned with the 'A' on the outer disk. No pre-agreed trigger letters are needed. Encipherment proceeds until one of the plaintext letters is matched with one of the four digits on the outer disk. At this point that letter, in this instance 'b,' coinciding with '1,' is shifted to align with the 'A' on the outer disk, and enciphering continues using a new cipher alphabet, until another plaintext letter coincides with a digit, and the process is repeated.

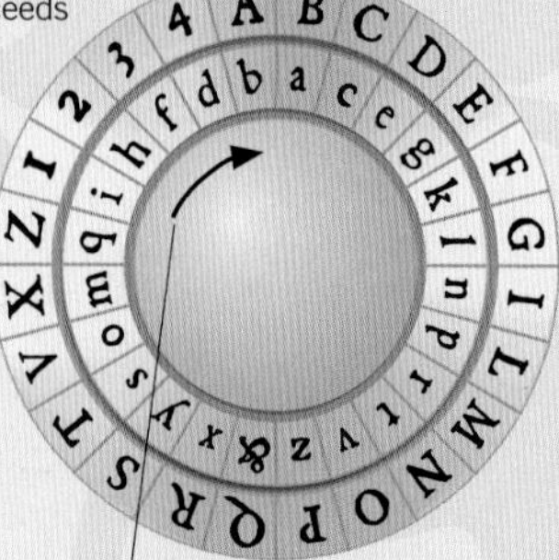

Trigger letters

When 'b,' 'a,' 'c,' or 'e' appear in plaintext, inner disk is rotated to align the letter with 'A.'

Decryption

The coded message for either method can be untangled by simply reversing the steps taken in encryption.

AB	a	b	c	d	e	f	g	h	i	l	m
	n	o	p	q	r	ſ	t	u	x	y	z
CD	a	b	c	d	e	f	g	h	i	l	m
	t	u	x	y	z	n	o	p	q	r	ſ
EF	a	b	c	d	e	f	g	h	i	l	m
	z	n	o	p	q	r	ſ	t	u	x	y
GH	a	b	c	d	e	f	g	h	i	l	m
	ſ	t	u	x	y	z	n	o	p	q	r
IL	a	b	c	d	e	f	g	h	i	l	m
	y	z	n	o	p	q	r	ſ	t	u	x
MN	a	b	c	d	e	f	g	h	i	l	m
	r	ſ	t	u	x	y	z	n	o	p	q
OP	a	b	c	d	e	f	g	h	i	l	m
	x	y	z	n	o	p	q	r	ſ	t	u
QR	a	b	c	d	e	f	g	h	i	l	m
	q	r	ſ	t	u	x	y	z	n	o	p
ST	a	b	c	d	e	f	g	h	i	l	m
	p	q	r	ſ	t	u	x	y	z	n	o
VX	a	b	c	d	e	f	g	h	i	l	m
	u	x	y	z	n	o	p	q	r	ſ	t
YZ	a	b	c	d	e	f	g	h	i	l	m
	o	p	q	r	ſ	t	u	x	y	z	n

The Italian connection

Partly due to intense rivalry between its jigsaw of states and dukedoms, and partly because of the rise of cities such as Florence, Venice, and Genoa as international trading and banking centers, cryptography was in high demand by the 15th century in Italy. Following Alberti (*opposite*) Giovan Battista Bellaso (1505-c.1580) published three tracts on cryptography which examined various polyalphabetic cipher systems. In the first of these he created a form of *tabula recta* (or reciprocal table) and introduced the notion of a key, or 'countersign.' He split the alphabet (then in Italian only 22 letters) in two, spelling out a-m in the usual fashion, and arranging a series of random counterparts below this using the remaining letters. Each arrangement has an index, formed of two adjacent letters of the alphabet. Using this tableau, a plaintext is enciphered following the key or countersign to indicate which row in the index should be used for enciphering that particular letter of the plaintext, the rows being read either way depending on the location of the relevant letter to be enciphered.

Using the tableau (*left*) a plaintext such as 'engage enemy at first light' with the counterpart '*et in arcadia ego*' would encipher like this:

countersign:	e	t	i	n	a	r	c	a	d	i	a	e	g	o	e	t	i	n	a	r	c	a	d
plaintext:	e	n	g	a	g	e	e	n	e	m	y	a	t	f	i	r	s	t	l	i	g	h	t
ciphertext:	q	l	r	r	t	u	z	a	z	x	l	z	b	p	u	c	h	c	y	n	o	u	a

Giambattista della Porta (c.1535-1615), an aristocratic dilettante, reproduced a version of Bellaso's ideas in his *De Furtivis Literarum Notis* (1563) effectively scooping his predecessor, and is now frequently credited with the invention of the polyalphabetic tableau system. Many of the separate threads which led to the development of the polyalphabetic cipher were drawn together and published by the French diplomat Blaise de Vigenère in 1586 (*see page 104*) who in turn now has his name indelibly associated with the idea.

Steganographia

The Benedictine abbot of Sponheim, Johannes Trithemius (1462-1516) wrote a three-volume work *Steganographia* in around 1499. It was not published until 1606 (when it was placed on the Vatican's *Index Librorum Prohibitorum*). It appeared to be about necromancy, purporting to use angels and spirits to convey messages over great distances. In fact, it is a complex work of cryptography, describing a variety of systems for secret writing. Like many of his contemporary clergy, Trithemius was interested in the occult, although it has become clear that his references to angels and spirits in fact refer to different enciphering algorithms. Although many were simply formulae for substitution or transposition ciphers, Trithemius seems to have independently developed the *tabula recta*, a later technique proposed by Bellaso (*see above*), and popularized by de Vigenère (*see page 104*). His system was simple, involving a single progressive Caesar Shift through the tableau: the first letter of the plaintext would be enciphered using the first row of shift ('a' becomes 'B'), the second letter using the second row of shift ('a' becomes 'C') and so on. A slightly later book elaborating Trithemius' ideas on cryptography, *Polygraphiae*, became the first European printed book on the subject (1518).

Johannes Trithemius appears to have developed the idea of the polyalphabetic cipher system independently of Alberti (*opposite*). The style of language Trithemius used places him in the tradition of the clerical mystics of the period, familiar with elements of alchemy and necromancy, which put him in danger of accusations of heresy by the church authorities.

Math and mysticism

Many texts, such as Oswald Kroll's *Basilica Chymica* (*above*), blurred the considerable overlap between the purely mathematical and pragmatic concerns of cryptography and more mystical practices. Some 50 years before de Vigenère published his version of the polyalphabetic cipher tableau, Cornelius Agrippa (a pupil of Trithemius) produced a similar Caesar Shift tableau for his 'tables of commutation,' intended to provide ways of enciphering magical alphabets to produce formulae for summoning and communicating with angels or demons. Alchemists, necromancers, and Kabbalists alike were fascinated by the magic potential of numbers (*see pages 52-56*) as had been the Pythagoreans before them (*see page 154*).

Polygraphiae The first European printed book on cryptography.

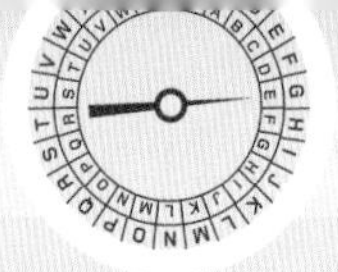

The Babington Plot

England, February 1587: Queen Elizabeth I, head of the Protestant Church of England, has been on the throne for 29 years, a throne threatened by the Catholic powers of Rome and its powerful allies, Spain and France. Elizabeth has just reluctantly approved the death warrant of her cousin, the Catholic monarch-in-waiting Mary, Queen of Scots, to be executed for treason. Aside from natural family ties, Elizabeth is hesitant to aggravate Catholic ill-feeling within her realm by approving her cousin's execution, but the evidence of Mary's guilt, as presented by Elizabeth's spymaster, Francis Walsingham, is overwhelming. He has apparently decoded a series of letters that prove that Mary, long but comfortably imprisoned in England as a security measure, is plotting a coup to overthrow Elizabeth.

Mary, Queen of Scots (1542-87) was a devout Catholic and thus a threat to Elizabeth.

Elizabeth I of England (1533-1603), whose staunch defense of the Anglican faith after her accession in 1558 led her to confirm Mary's execution for treason.

Religious persecution

After the death of Henry VIII in 1547, and his son Edward VI, the reign of his daughter the Catholic zealot Mary (1553-58) saw the mass victimization and execution of Anglican and other Protestants throughout England, including the burning of some 300 'dissenters.' When Elizabeth ascended to the throne upon Mary's death, the tide turned. To be openly Catholic was dangerous, but many chose to demonstrate their faith covertly. Some Catholic families built hidden messages into their coats of arms, and some even into the design of their houses, such as the extraordinary triangular estate house at Rushton, Northamptonshire, UK. Built by the devout Catholic Sir Thomas Tresham in 1593, its triangular form represents the Holy Trinity, and the triform rebus recurs everywhere: there are three floors, and three windows in each wall, and three gables on each facade, surmounted by the Tresham family trefoils. The exterior is further decorated with emblems, biblical passages, and other symbols linked to his avowed faith. If challenged, however, Tresham could always argue it was simply an architectural folly.

Mary's courier

Mary's letters had been intercepted by her jailers for some time. Catholic supporters, among them the dashing Catholic Anthony Babington and Thomas Gifford, managed to smuggle letters to her (in the bung of a beer barrel) from her supporters on the Continent. A trained Catholic priest, Gifford became Mary's regular secret courier, and had the confidence of the French embassy in London and other Catholic cells in England. Gifford, however, was a double agent.

Enter Walsingham

Francis Walsingham was Elizabeth's Principal Secretary and head of security, a ruthless and able man with extensive spy networks throughout Europe. He had been approached by Gifford, and had recruited him in 1585 upon Gifford's return from Rome. When Babington sent Mary details of a planned coup, involving the assassination of Elizabeth, a military invasion of England, and Mary's installation on the throne, Gifford showed it to Walsingham. The message was encoded, not as a straightforward substitution cipher, but as a combination of a cipher alphabet with code words and signifiers (*opposite*). Walsingham had long been aware of the importance of ciphers and decryption, and had employed one of Europe's finest cryptanalysts, Thomas Phelippes, as his cipher secretary.

Elizabeth's devious spymaster Francis Walsingham (c.1532-90) intercepted the so-called 'Casket Letters' and commissioned forgeries to trap Mary.

Mary approached her execution with dignity and resignation (*right*).

Phelippes' solution

Phelippes (1556-1625) was a past master of frequency analysis – his first line of attack. Using this framework, he tried successive guesses at recurring symbols to build a cipher alphabet, tentatively identified the nulls, and boiled the problem down to interpreting the code words, often understood by their context.

An extract from the reply to Mary, forged by Phelippes, which led to her downfall.

Having disentangled Babington's original description of the plot, Walsingham bided his time. To fully expose the plotters and Mary's involvement, he needed replies from her to Babington in order to demonstrate her complicity, and more details of those involved. Gifford continued his work as a courier, while Phelippes (also an outstanding forger) was asked to add a postscript to one of Mary's replies, asking for the names of the "six gentlemen which are to accomplish the design." The ruse worked, although the core cause of the plotters' downfall was their assumption that their cipher was secure, encouraging them to be less circumspect in what they actually said.

Close of play

Babington was intercepted whilst leaving England to finalize details of the invasion, captured a few days later, and he and his confederates were half-hanged, drawn, and quartered for treason. Elizabeth approved Mary's trial on the same charges in October 1586. Despite her outright denials (which have led some to believe that the entire scenario was fabricated by Walsingham) she was found guilty and beheaded at Fotheringay Castle, Northamptonshire, in February 1587.

The Babington code

A nomenclator was created, using symbols in place of 23 alphabet letters (j, v, and w were not commonly included then); a further 35 symbols stood for words or short phrases; in addition, there were four nulls (red herrings, meaning nothing, but placed to confuse the cryptanalyst) and a signifier which doubled ('dowbleth') the following letter.

Nomenclators stood in for alphabetic letters and a number of recurring phrases.

a	b	c	d	e	f	g
h	i	k	l	m	n	o
p	q	r	s	t	u	x
y	z	nulls	nulls	nulls	nulls	dow-bleth
and	for	with	that	if	but	where
as	of	the	from	by	so	not
when	there	this	in	which	is	what
say	me	my	with	send	ire	receive
bearer	I	pray	you	meet	your name	mine

THE DA VINCI CODE?

The notebooks of the Italian High Renaissance artist and engineer Leonardo da Vinci (1452-1519), which are now divided among several major collections throughout the world, have attracted considerable attention, not least because of their subject matter and his use of apparently coded notes and annotations. The content of the notebooks ranges from sketches from everyday life to anatomical drawings and fantastic weapons of war, and encompasses detailed sketches for artistic commissions as well as mere doodles.

Secret writing

The notes da Vinci wrote on nearly every page of his notebooks appear inscrutable, but are in fact simply in 'mirror writing'; whether this was because he felt the need to disguise his notes from unfriendly eyes, or because, being left-handed, he found it easier to write in this manner, remains a mystery. There is little doubt, however, that da Vinci was concerned that his notes – often for good reason – remained private, or at least obscure to the casual viewer.

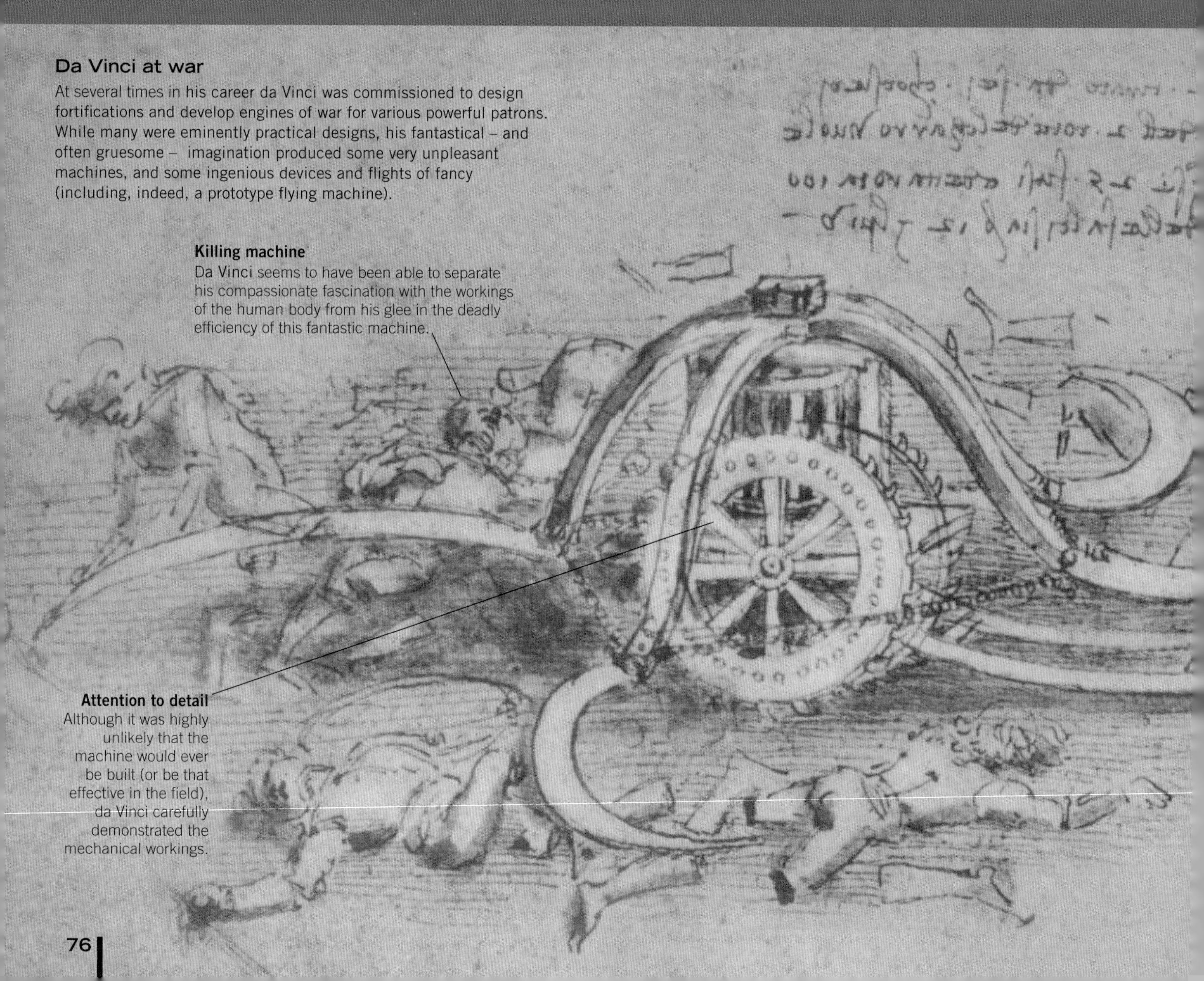

Da Vinci at war

At several times in his career da Vinci was commissioned to design fortifications and develop engines of war for various powerful patrons. While many were eminently practical designs, his fantastical – and often gruesome – imagination produced some very unpleasant machines, and some ingenious devices and flights of fancy (including, indeed, a prototype flying machine).

Killing machine
Da Vinci seems to have been able to separate his compassionate fascination with the workings of the human body from his glee in the deadly efficiency of this fantastic machine.

Attention to detail
Although it was highly unlikely that the machine would ever be built (or be that effective in the field), da Vinci carefully demonstrated the mechanical workings.

Inside the human body

Da Vinci's anatomical investigations undoubtedly involved the flaying and dissection of cadavers, a practice which could have attracted the unwelcome attention of the Church authorities. To modern eyes, his work is informed and authoritative, and certainly the product of practical scientific inquiry.

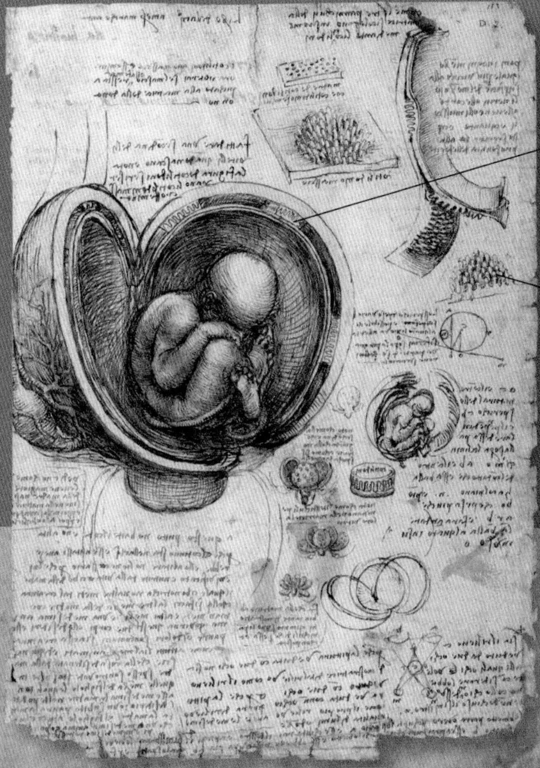

Life after death
This beautiful rendering of a child in the womb belies the fact that it was only possible as a result of dissection. It is surrounded by da Vinci's observations.

Origins
Da Vinci has added various sketches with commentaries explaining his ideas about the progress of the reproductive cycle from fertilized egg to fetus.

How does it work?
Da Vinci's text here provides detailed measurements and descriptions of the mechanism which activates the scythes.

Horsepower
Da Vinci was interested in the science of mechanics and power, and saw the horse as the motor-force which would propel this particular machine into action.

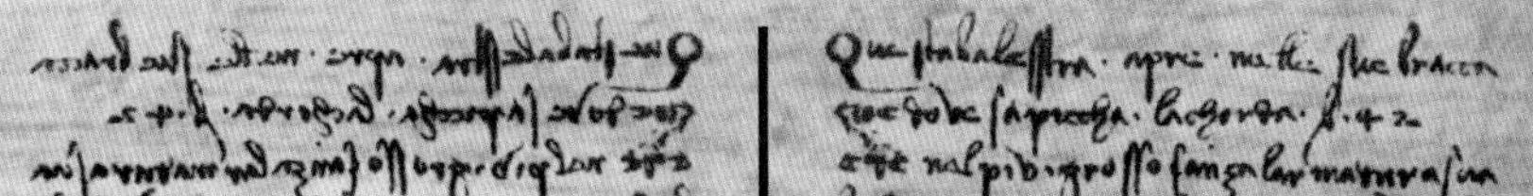

Mirror writing

Although naturally written in classical Italian, da Vinci's elegant hand remains clear, even when written, as it originally was, in reverse (*left*). When photographically transposed (*right*) the accuracy of his writing becomes clear. His secrecy has given rise to numerous far-fetched theories concerning his membership of arcane secret societies – unlikely in a man so interested in investigating the practical properties of the world around him, unlike his contemporary alchemists (*see page 52*).

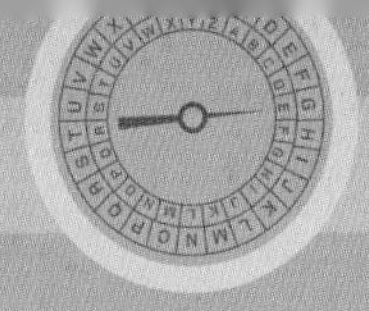

CIPHERTEXTS AND KEYS

A number of cipher systems were developed from the 16th century – including one with its roots in classical Greece – that although originally fairly easy to use were made successively more sophisticated over the next 400 years; some were still being developed for use by the early 20th century, during World War I. Despite the escalating ingenuity of these cipher systems, they were all dependent upon a shared key, and the problem was keeping the key concealed.

The Polybius square

Described by the Greek writer Polybius in the 2nd century BC, this tableau provides a way of encrypting the alphabet by placing it in a 5x5 grid, and converting the letters into two-number coordinates, read first by row then by column. Polybius saw this method as a pioneering way of sending messages from lighthouses using flashing lights, preceding semaphore and Morse code by 2,000 years.

	1	2	3	4	5
1	A	B	C	D	E
2	F	G	H	I/J	K
3	L	M	N	O	P
4	Q	R	S	T	U
5	V	W	X	Y	Z

Using the Polybius square, the word 'code' becomes 13 34 14 15.

The system has often been used by prisoners, notably anarchists incarcerated by czarist Russia, and by US POWs during the Vietnam War, where a number of taps represents each letter, earning it the name Tap (or Knock) code. It was not intended to disguise a message, but merely as a means of communicating from one cell to another.

Developing the *tabula recta*

Many cryptographers sought a less complicated and time-consuming system than that of the de Vigenère tableau (*see page 105*). Looking to the Polybius square, and some of the permutations produced by the more sophisticated grilles (*see page 80*), not to mention the Playfair cipher (*see page 109*), produced some ingenious systems of encipherment.

As the German army prepared for its last major push in World War I, the Ludendorff Offensive, much attention was paid to secure communications. Experiments with grilles (*see page 80*) were abandoned, and a more sophisticated version of their ADFGX cipher was developed.

The Bifid cipher

Invented by Félix Delastelle in 1901, this system involves three stages and combines the basic concept of a Polybius square with transposition and fractionation – the conversion of the ciphertext to numbers.

1 A random mixed alphabet Polybius square combining I and J is established.

2 The plaintext is converted to coordinates, with the numbers stacked in two rows.

3 The two rows of numbers are then strung together.

4 The numbers are then divided into pairs (fractionated).

5 The fractionated numbers are then converted into ciphertext via the Polybius square.

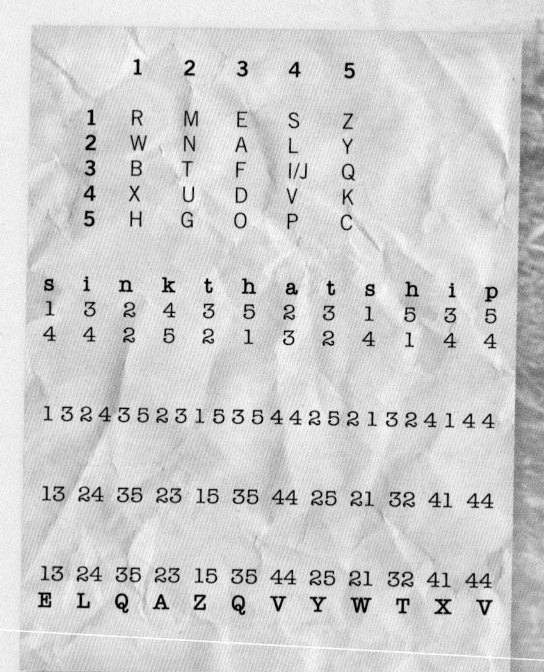

	1	2	3	4	5
1	R	M	E	S	Z
2	W	N	A	L	Y
3	B	T	F	I/J	Q
4	X	U	D	V	K
5	H	G	O	P	C

s	i	n	k	t	h	a	t	s	h	i	p
1	3	2	4	3	5	2	3	1	5	3	5
4	4	2	5	2	1	3	2	4	1	4	4

132435231535442521324144

13 24 35 23 15 35 44 25 21 32 41 44

13	24	35	23	15	35	44	25	21	32	41	44
E	L	Q	A	Z	Q	V	Y	W	T	X	V

6 The receiver merely has to reverse the process step-by-step, although they have to be aware of the key, which in this case is how the random alphabet has been organized.

The ADFGX cipher

Developed by the German military, this system relied on a Polybius square with a random alphabet, transposition, and a key word.

1 Using the same random alphabet square as at left, the rows and columns are denoted by letters. These were chosen as being the most distinct when transmitted in Morse code, and give the name to the cipher.

	a	d	f	g	x
a	R	M	E	S	Z
d	W	N	A	L	Y
f	B	T	F	I/J	Q
g	X	U	D	V	K
x	H	G	O	P	C

s	i	n	k	t	h	a	t	s	h	i	p
AG	FG	DD	GX	FD	XA	DF	FD	AG	XA	FG	XG

2 From this square the message is converted to fractionated pairs of letters, as above.

3 These letters are then placed in another grid, reading left to right, accompanied by a key word or key phrase, which determined the length of the rows.

4 The columns of letters are then transposed following the alphabetical order of the key word.

H	E	A	D
A	G	F	G
D	D	G	X
F	D	X	A
D	F	F	D
A	G	X	A
F	G	X	G

A	D	E	H
F	G	G	A
G	X	D	D
X	A	D	F
F	D	F	D
X	A	G	A
X	G	G	F

FGXFXX GXADAG
GDDFGG ADFDAF

5 The message to be sent is then assembled by column to form the ciphertext above. Usually the messages to be sent were longer than this, and the key word or key phrase for transposition up to 24 letters long, which were changed daily, as was the order of the fractionation letters on the Polybius square.

The ADFGVX cipher

In early 1918, just before their final Spring Offensive on the Western Front, the German military added an extra letter, V, to the code, ADFGVX, making a 6x6 grid, which could incorporate all 26 letters of the alphabet, plus the numbers 1-10, shortening messages by avoiding having to spell out numbers.

French cryptanalyst Georges Painvin managed to crack the new cipher just as Ludendorff's offensive got under way, by concentrating on stereotypical words or phrases which might be found especially at the beginning of such messages, and comparing these to letters found at the column headings.

One problem with 'book cipher' keys was finding a canonical text, the most commonly used being the Bible.

Extreme keys

The problem of keys is that both the sender and the receiver need to have access to the key. Concealing the key from a third party is critical, and how to convey the key to the recipient even more so. This has become a major concern for modern digital cryptographers.

One answer is a 'running key,' today typically generated using random numbers (or massive prime numbers, *see page 274*). Another is an agreed codebook, but this is vulnerable to theft. A third is the virtually impregnable 'one-time pad,' in which each individual message has a unique key (*see page 83*). However, another solution might lie in both the sender and receiver having the same edition of a text. This might avoid the need for a shared codebook, whilst still providing a readily available key.

In earlier times, the idea of 'book ciphers' was popular, frequently used examples being pre-agreed passages from the Bible or a dictionary. Using such a text or look-up table of coordinates, the key could be easily accessed by the receiver, and applied to a *tabula recta*; the longer the piece of key text, the more laborious it would be to decrypt (every letter of the ciphertext could have its own key), and the key text could run across several de Vigenère or other tableaux (*see page 105*).

GRILLES

Girolamo Cardano

Renaissance Italy was fascinated by methods of secret writing. The physician, inventor, and mathematician Girolamo Cardano (1501-76) was only one of a number who published works on the subject. In *Ars Magna* (1545) he demonstrated various algebraic proofs; constantly short of money, he was an inveterate gambler and chess player, leading him to write about the laws of probability and techniques of cheating in games. His invention of the grille was to lead to a number of cryptographic techniques of increasing complexity.

The Italian mathematician Girolamo Cardano (1501-76) invented a system of secret writing straddling a number of cryptographic disciplines, which depended upon the use of a physical key – or grille – shared by sender and receiver. Although in essence quite a simple device, Cardano's grille can be made considerably more sophisticated, and varieties of the technique were used across the spectrum for simple correspondence to high-level military and spying purposes.

The Cardano grille

Cardano's basic idea was that a seemingly innocent message is written down, but embedded in it is a secret message. The secret message is revealed by placing a second sheet (a grille) in which holes have been cut (the key) over the first; this eliminates the unnecessary letters in the overall message to reveal the hidden message. The grille can be adapted to include groups of letters or whole words. The system is very vulnerable, partly because of the physical key, which might be stolen or lost, partly because the sender needs to smooth over any stilted writing in the original text.

The Trellis cipher

Sometimes called the 'checkerboard' or 'chessboard' cipher, this is a transposition technique that involves a variant on the grille key in the form of a standard checkerboard design. Take the plaintext 'I will be at the opera tonight, but will meet you for dinner later, if you like' adding three nulls ('x') at the end to make up the required 64 letters. If the plaintext is longer than 64 letters, rotate the checkerboard again and continue to make a message of 128 letters. The ciphertext can be unscrambled by reversing the process. One of its obvious limitations is the need for the message to contain at least 64 or 128 (or more) letters, although the ciphertext could be disguised by adding further pre-agreed nulls at various points which would be automatically discarded.

Starting with the checkerboard in one position, with a black square top left, write the message in the white spaces vertically.

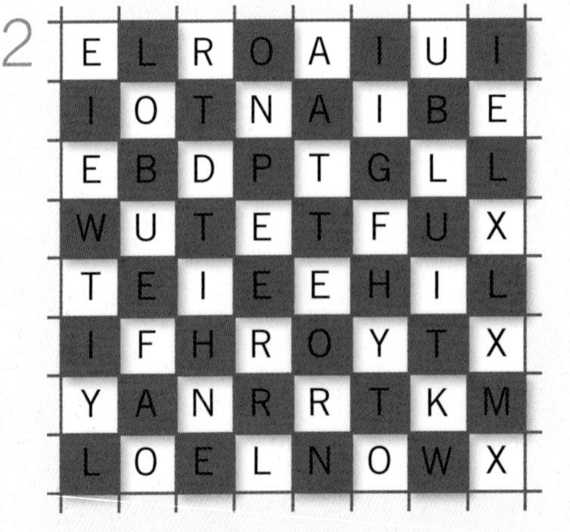

After 32 letters rotate the checkerboard pattern through 90 degrees and continue with the message in the same fashion. The three 'x's in the final column are nulls to make up the table.

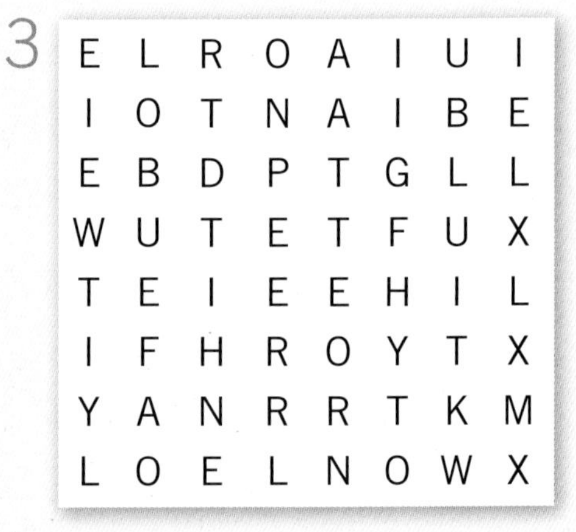

The letters should be read off in order left to right, row by row, producing the ciphertext: ELROAIUIIOTNAIBEEBDPTGLLWUTETFUXTE IEEHILIFHROYTXYANRRTKMLOELNOWX

The Turning grille

A more sophisticated variant on the Cardano grille was the 'Turning grille' developed by the retired Austrian cavalry officer Eduard Fleissner von Wostrowitz in 1880. A square 8x8 grid is made up of four quadrants of 16 spaces each. The grille would have a total of 16 holes, four in each quadrant, but when the grille is rotated through all four positions, the holes must eventually uncover all the squares 1-16 in each quadrant. As with the Trellis cipher, nulls may be used to fill up the 64 squares. The plaintext can be easily read if both parties have access to the pattern of holes on the grille, and the orientation and order in which each quadrant of the grille is to be used.

1

Using the same plaintext message as left, the first 16 letters are written by the sender using the grille.

2

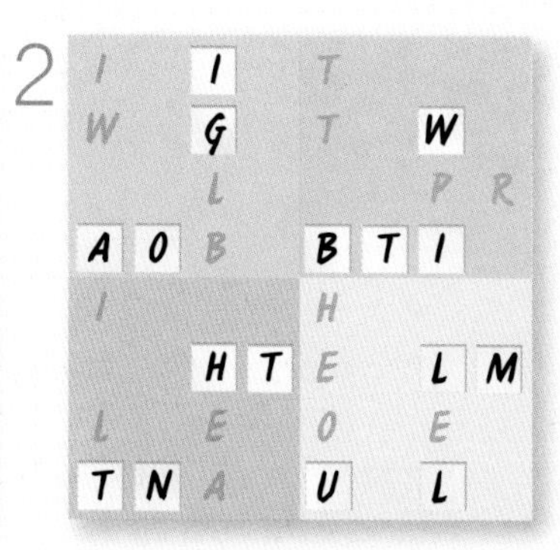

The grille is then turned, in this instance through 90 degrees counter-clockwise, and more of the plaintext transcribed.

3

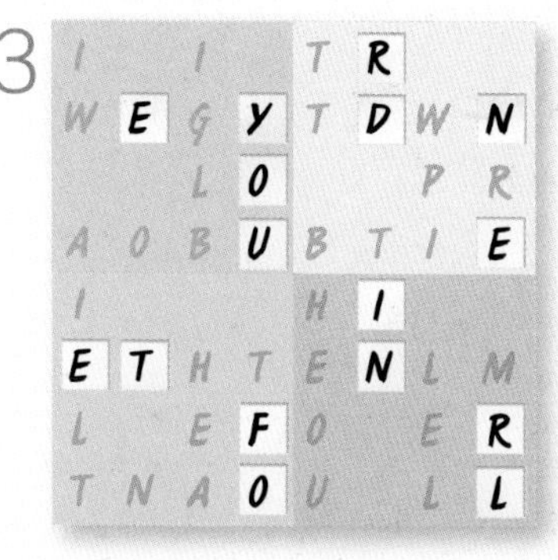

The process is repeated with another turn of the grille.

4

Finally, after another turn of the grille, the underlying grid is completed, producing the ciphertext (*below left*). This can be assembled by row or column.

5

I	T	I	Y	T	R	E	X
W	E	G	Y	T	D	W	N
A	E	L	O	U	L	P	R
A	O	B	U	B	T	I	E
I	R	F	O	H	I	X	X
E	T	H	T	E	N	L	M
L	I	E	F	O	I	E	R
T	N	A	O	U	K	L	L

The German army briefly used Fleissner grilles during World War I for encrypting and decrypting messages which were transmitted by telegraph or field telephone.

Variations on the Turning grille

There are numerous elaborations of the Fleissner idea – the grid can be of varying size (5x5 or 6x6 for example), and the holes in the grille do not have to be equally distributed across the four quadrants. The starting point and order in which the grille is rotated can be varied. The system was used briefly by the German army for encrypting field signals in 1916, where grids and grilles varied considerably, each having their own code name, but it was not deemed particularly secure and was abandoned after a few months.

SPIES AND BLACK CHAMBERS

The gathering of intelligence, its transmission, and its interpretation, is an important organ in the body politic of most states. To make such information available, networks of spies and secret messaging need to be established. The Old Testament makes mention of spies, and the Roman dictator Julius Caesar created a sophisticated intelligence system during his campaigns in Gaul. However, it was during the Renaissance, with competitive states being established all over Europe, that covert observation, messaging, and the interception of secret messages came into their own. The first *cabinet noir* (black chamber) was established in France by Henry IV in 1590.

Espionage in Renaissance Italy

By the beginning of the 15th century three city states had emerged in Italy as powerful mercantile centers: Florence, renowned for banking, and Genoa and Venice, both commanding trading networks spanning the Mediterranean and beyond. One of the keys to Venice's success was its location between eastern and western Europe. The other was its gathering of commercial and political intelligence. Marco Polo was only one of many emissaries the doge sent far abroad to garner trading information and contacts; Venetian diplomats sent coded reports almost daily to their home city from the courts of Europe and the East. The Vatican archives also contain thousands of pages of coded intelligence gathered by the papacy over the years. Most codes used at the time were monoalphabetic substitution ciphers using nomenclators (*see page 70*).

The Bacon cipher

The use of secret writing became widespread in the 16th century for both private correspondence and espionage. The English courtier Sir Francis Bacon (1561-1626) developed a simple coding system suitable for secret messages. It was based on a binary five-bit alphabet, to which both sender and receiver must have access:

A	aaaaa	**N**	abbaa
B	aaaab	**O**	abbab
C	aaaba	**P**	abbba
D	aaabb	**Q**	abbbb
E	aabaa	**R**	baaaa
F	aabab	**S**	baaab
G	aabba	**T**	baaba
H	aabbb	**U/V**	baabb
I/J	abaaa	**W**	babaa
K	abaab	**X**	babab
L	ababa	**Y**	babba
M	ababb	**Z**	babbb

Firstly, transform the plaintext into the five-bit alphabet equivalents:

B aaaab
E aabaa
W babaa
A aaaaa
R baaaa
E aabaa

Then choose a harmless message of the same length as the encoded text; Bacon recommended writing this using upper-case and lower-case letters to indicate which letters signified an 'a' and which a 'b,' but this is not particularly secure, as it looks odd and immediately raises suspicions. A better scheme is to nominate all the letters of the normal alphabet from A-M as indicating 'a,' and the rest of the alphabet N-Z as indicating 'b.' Then write an apparently harmless 30-letter message, such as: 'Did my father break Jade's magic doll' which the receiver will then break down into 'a's or 'b's to reveal the secret message.

Black chambers

By the 17th century most European powers had developed their own cryptography departments, which came to be known as *cabinets noirs*, or 'black chambers.' The most notorious was the Geheime Kabinets-Kanzlei in Vienna, which ran to a strict timetable. Each day, around 100 letters to or from foreign embassies were intercepted at the post office, delivered to the black chamber, the seals broken, the contents meticulously copied, the letters resealed, and returned to the post office within three hours for delivery. This cycle occurred three times daily, for incoming, outgoing, and domestic mailings. The copies were passed to teams of cryptanalysts, amassing huge quantities of intelligence. Unlike the secretive Venetian and Vatican systems, much of this was then sold to other states, providing a considerable income for the Hapsburg emperors.

"Gentlemen do not read each other's mail."

US SECRETARY OF STATE HENRY L. STIMSON, UPON WITHDRAWING STATE DEPT. FUNDING FOR MI-8, 1929.

In Venice, by 1542, three full-time cipher secretaries were employed in the Doges' Palace to develop new codes as old ones were cracked; these men enjoyed high status in the Republic, but the penalty for betraying the secrets of the codes was death.

Spooks

The services of freelance cryptographers and code masters, such as John Dee or Thomas Phelippes, were highly valued in Renaissance courts; both traveled widely working for many masters – usually the highest bidders. The growing sophistication of cipher messages by the end of the 19th century, along with technological advances in methods of encryption, transmission, and decryption, provoked the creation of larger 'black chambers,' often dedicated to specific areas of military intelligence or governmental departments.

Britain

Room 40 (NID25) was established in the Admiralty in 1914, and read over 15,000 intercepted German messages before being wound down in 1919; it was merged with British Army Intelligence (MI1b) to form the Government Code and Cypher School (GCCS), which relocated out of London to Bletchley Park during World War II (*see pages 118-121*). In 1946 it was renamed the Government Communications Headquarters (GCHQ) and is now based at Cheltenham.

USA

It was not until the end of World War I that Herbert O. Yardley founded the US Cipher Bureau (MI-8, often known as the American Black Chamber) in New York. Disguised as a commercial company, it created commercial codes for business, while focusing on diplomatic signals, notably those of the Japanese. After state funding was withdrawn in 1929, William Friedman helped to establish the Army Signals Intelligence Service (later known as SIGINT) in 1931, dedicated to the interception and decryption of potential enemy traffic – cracking the Japanese Purple code early in World War II. The governmental National Security Agency (NSA) was established in 1952. The FBI and CIA both maintained their own cryptographic units.

One-time pads

Developed after World War I, just as spying became virtually industrialized, one-time pads proved almost uncrackable, and were to be used widely among the espionage community on all fronts. The principle is quite simple: for every enciphered message that is sent, there is only one unique key. Invented by Major General Joseph Mauborgne, then head of the US Army's cryptographic research unit, pads of random letters were created, each to be used once only as a key for a de Vigenère tableau cipher (*see page 105*). And the principle works well unless the shared key is stolen or misplaced or unless, due to laziness, the same one-time pad key is used more than once, which gives the cryptanalyst the opportunity to compare the messages. The most notable example of this lapse was the Venona project intercepts during World War II (*see page 124*).

MECHANICAL DEVICES

In the world of secrecy, ever since the early modern era of the de Vigenère's square (*see page 104*), that involved considerable calculations to even encode a message let alone attempts to decrypt it, the idea of a mechanical device or machine that could perform these functions automatically haunted the imaginations of Enlightenment mathematicians. Although Babbage's engines never really saw the light of day during his lifetime, the mechanical logistics inherent in the Industrial Revolution saw a number of innovations that paved the way for the advances of the 20th century. Among them were systematize implements we take for granted today, such as the telephone, the keyboard, and the calculator.

Calculating engines
Gottfried Leibniz (1646-1716) the German father of the calculus had, by around 1700, posited in principle a mechanical 'Logick Mill' or 'Difference Engine'; employed by various German courts he, like his great rival Isaac Newton, could see the value of a machine capable of generating or decoding mathematically-based ciphers, but it was not until Charles Babbage designed such a machine for the accurate calculation of mathematical tables in the 19th century (*see page 268*) that such implements began to become a reality.

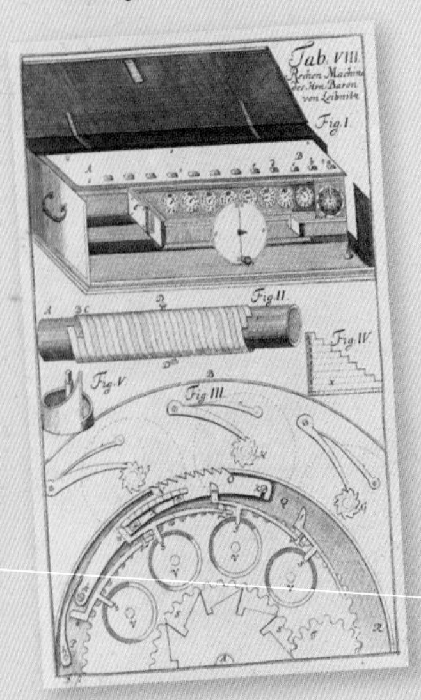

Leibniz drew up elaborate plans for his 'Logick Mill,' in effect the first computer.

Cipher disks

The Alberti-style cipher disk was mass-manufactured for use during the American Civil War (1861-65). Although simple in design (often an advantage among largely illiterate users), it could be used together with a preset daycode or a polyalphabetic algorithm to rapidly encipher urgent information from the front line or command headquarters. This could then be sent via Morse code across the newly-invented electric telegraph, or by semaphore or heliograph. The Union forces appear to have been able to decipher most of the Confederate signals, but there is little evidence that the reverse happened.

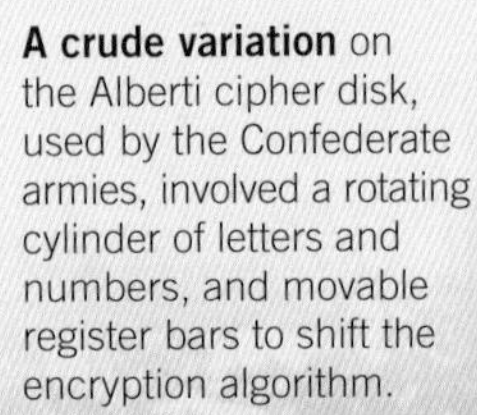

A crude variation on the Alberti cipher disk, used by the Confederate armies, involved a rotating cylinder of letters and numbers, and movable register bars to shift the encryption algorithm.

The Alberti cipher disk as used by Union forces allowed for more variations in the polyalphabetic cipher than Alberti's original design (*see page 72*), and included not only more numerals, but punctuation marks as well.

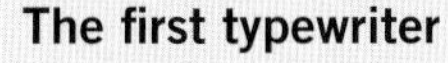

The first typewriter

A major invention in the history of communication was the typewriter. The first commercially available model was the Danish Hansen Writing Ball (1870, *left*), and by the end of the decade both Remington and Underwood were producing machines that more closely resemble modern models, although early machines had an alphabetic key arrangement. The invention in 1874 of the more efficient QWERTY keyboard layout for typing English increased typing speeds to a good average of 100 words per minute, and provided the model for most mechanical encryption machines and computer keyboards; (there are minor variants for other languages: QWERTZ in German, AZERTY in French, QZERTY in Italian). The typewriter transformed administrative, office, and commercial life, and produced millions of adept typists.

The Burroughs adding machine

Another 19th-century innovation that, like Babbage's Difference Engine, also relied on cogs and ratchets to produce accurate calculations, but for the first time incorporated a keyboard, was the mechanical adding machine – forerunner of the till and of the personal calculator – patented by the American William S. Burroughs in 1888. The first versions of the machine merely added up the sum total of a customer's basket of purchases, but soon it was possible to also input the cash tendered and the change required, in effect becoming an accounting machine, with a printout receipt. By the middle of the 20th century, the type or variety of goods being purchased could also be registered through the keyboard, anticipating the stock control innovation of bar codes (*see page 205*).

From telegraph to telephone

The invention of electric telegraphy and the binary Morse code system (*see pages 94-97*) inaugurated a series of new opportunities for encryption, and encouraged the invention of space-saving commercial telegraphic codes (*see page 204*). But it was the introduction of one-to-one wire telephones toward the end of the 19th century that saw a new and very pragmatic problem emerge: with a single telephone wire connecting each telephone exchange (and by 1900 most communities had them), how could multiple simultaneous calls be passed through a single exchange? The solution was a 'stepping' system. Most early telephones had rotary mechanical dials, which could then engage the rotation of mechanical cogs linked to the incoming call: with each number that was dialed, a different level of cog was engaged, making each dialed number unique, and guaranteed to reach the intended receiver. The system was developed on the principle of layered codes (*see page 109*), and remained in use until the last quarter of the 20th century.

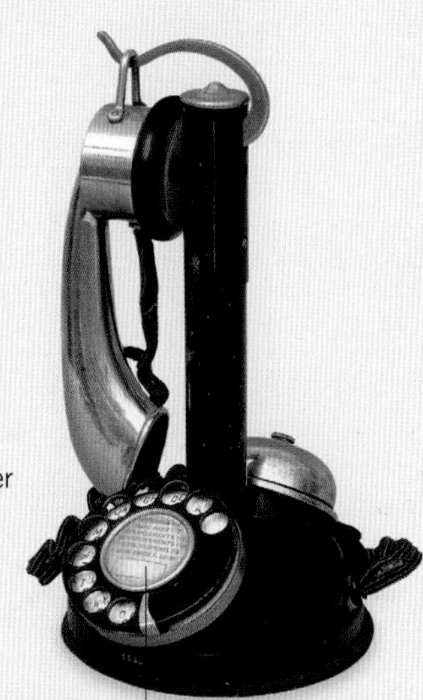

Number please
The circular dial on early telephones engaged a 'stepping cycle' ensuring that the electrical impulse reached the right destination, although this would normally be a local exchange.

The US Civil War was immediately distinguished as a 'modern' war by its reliance on railroads and the telegraph. Field telegraph wire systems were immediately established wherever a command center was set up, usually as 'spurs' from the lines that ran along railroad tracks.

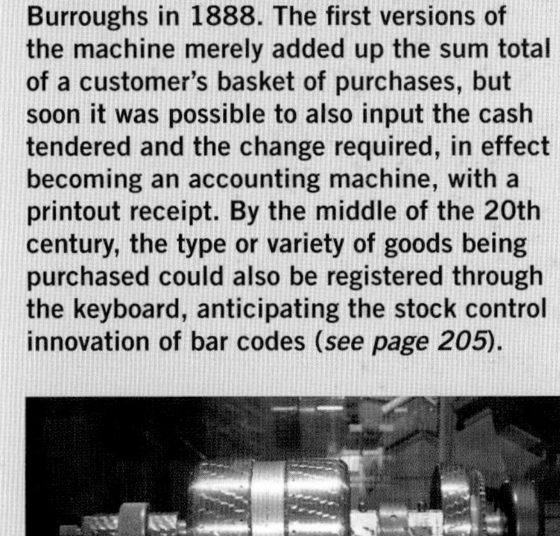

Mechanical encoding machines

The idea of linking the typewriter keyboard to a series of 'stepped' mechanical or electronic cycles in order to encrypt a message, the most famous example being the Enigma machine (*see page 116*), had its roots in the 19th century. The Hebern rotor engine *(above)* combined cog technology with electrical impulses and was one of the most successful machines, being sold to the US Navy in the interwar period.

Hidden in Plain Sight

Cryptic clues

Crossword puzzles became an enormously popular feature of daily newspapers in the 1920s. They fell into three types: those with general knowledge clues (Q: The capital of France (5); A: Paris); simple puzzles involving anagrams and puns (Q: Canine slain at a carve-up; A: Alsatian); and those with highly cryptic questions, notably in the London newspapers, *The Times* and *The Daily Telegraph*. It was widely rumored that, among the cryptic-style puzzles, selected questions, and even the grid of answers when the puzzles were successfully completed, were used for conveying coded messages to agents abroad.

The development of the newspaper industry from the late 18th century – the first form of mass media – provided a new opportunity and new challenges for those keen to send covert messages. Coded messages which could appear completely innocent could suddenly be broadcast far and wide. The personal, or 'agony' as they became known, columns were enthusiastically taken up by sweethearts as a means of sending coded messages of passion, bypassing the ever-present chaperone. From this, the idea of sending disguised messages through the press, whether to spies and agents abroad, or simply to anonymous readers, rapidly became popular.

> "Snow conditions bad. Advanced base abandoned. Awaiting improvement."
>
> JAMES MORRIS, MESSAGE TO *THE TIMES* ANNOUNCING THE CONQUEST OF EVEREST, 1953.

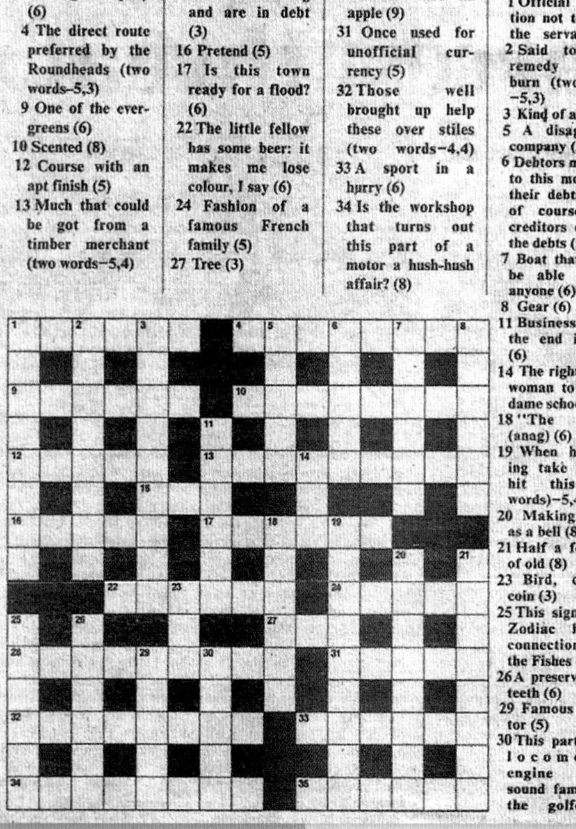

ACROSS

1 A stage company (6)
4 The direct route preferred by the Roundheads (two words–5,3)
9 One of the ever-greens (6)
10 Scented (8)
12 Course with an apt finish (5)
13 Much that could be got from a timber merchant (two words–5,4)
15 We have nothing and are in debt (3)
16 Pretend (5)
17 Is this town ready for a flood? (6)
22 The little fellow has some beer: it makes me lose colour, I say (6)
24 Fashion of a famous French family (5)
27 Tree (3)
28 One might of course use this tool to core an apple (9)
31 Once used for unofficial currency (5)
32 Those well brought up help these over stiles (two words–4,4)
33 A sport in a hurry (6)
34 Is the workshop that turns out this part of a motor a hush-hush affair? (8)
35 An illumination functioning (6)

DOWN

1 Official instruction not to forget the servants (8)
2 Said to be a remedy for a burn (two words –5,3)
3 Kind of alias (9)
5 A disagreeable company (5)
6 Debtors may have to this money for their debts unless of course their creditors do it to the debts (5)
7 Boat that should be able to suit anyone (6)
8 Gear (6)
11 Business with the end in sight (6)
14 The right sort of woman to start a dame school (3)
18 "The War" (anag) (6)
19 When hammering take care to hit this (two words)–5,4)
20 Making sound as a bell (8)
21 Half a fortnight of old (8)
23 Bird, dish of coin (3)
25 This sign of the Zodiac has no connection with the Fishes (6)
26 A preservative of teeth (6)
29 Famous sculptor (5)
30 This part of the locomotive engine would sound familiar to the golfer (5)

A cryptic crossword from *The Daily Telegraph* was used to assess lateral-thinking skills to help the war effort when recruiting potential intelligence personnel. A competition crossword was placed in the newspaper in 1940, and successful applicants were then invited to a crossword 'play-off,' the most successful being recruited to the staff at the intelligence and decrypting base at Bletchley Park (*see page 118*).

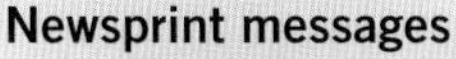

Newsprint messages

During World War I newspapers in both Britain and France were frequently used to send messages and coded information. The 'agony,' or personal, columns often contained cryptic statements, anagrams, or ciphers based on an agreed key. One of the most daring messages sent via newsprint took the form of a fashion sketch, where the arrangement of the dots on a woman's dress was designed to be read as a specific message concerning enemy troop dispositions. The signature on the sketch, Mary Helen Shaw, indicated their location – the Arras sector of the Western Front.

Sitting on top of the world

Getting a scoop in some far-flung place back to a newspaper's head office, often across public wire systems, could be a major challenge. The London *Times* reporter James Morris was covering the 1953 British and Commonwealth assault on Everest, and agreed a series of coded messages in advance. They were designed to confuse any other journalists who may have come across them. There were two key phrases, plus code names for the various members of the climbing party, to indicate who had made it to the summit.

Coded message	Meaning
Snow conditions bad	Everest climbed
Wind still troublesome	Attempt abandoned
South Col untenable	Band
Lhotse Face impossible	Bourdillon
Ridge camp untenable	Evans
Withdrawal to West Basin	Gregory
Advanced base abandoned	Hillary
Camp 5 abandoned	Hunt
Camp 6 abandoned	Lowe
Camp 7 abandoned	Noyce
Awaiting improvement	Tenzing
Further news follows	Ward

The triumphant message
Morris sent included the phrases 'Advanced base abandoned' and 'Awaiting improvement,' immediately telling *The Times* office in London that Edmund Hillary and Sherpa Tenzing had reached the top of the highest mountain in the world on May 29 – a story which broke on Queen Elizabeth II's coronation, June 2, 1953.

Lonely Hearts

The 'personal ad' columns of many newspapers today contain an abundance of coded messages, designed to send the right message to the right person. In the 'Lonely Hearts' columns a widely-recognized series of coded acronyms has developed, partly as a way of saving money as these advertisements are normally charged per letter.

A	Asian
B	Black
BI	Bisexual
C	Christian
D	Divorced
DDF	Disease/drug free
F	Female
FTA	Fun/travel/adventure
G	Gay
GSOH	Good sense of humor
H	Hispanic
HWP	Height/weight proportional
ISO	In search of
J	Jewish
LD	Light drinker
LDS	Latter Day Saints
LS	Light smoker
LTR	Long-term relationship
M	Male
MM	Marriage-minded
NA	Native American
NBM	Never been married
ND	Non-drinker
NS	Non-smoker
P	Professional
S	Single
SD	Social drinker
SI	Similar interests
SOH	Sense of humor
TLC	Tender loving care
W	White
W/	With
WI	Widowed
WLTM	Would like to meet
W/O	Without
YO	Years old

04

Sending messages as directly as possible beyond the range of the human voice has long been a necessity. From the earliest times, communication of this nature required the development of coded languages, whether aural or visual.

communicating at distance

The Industrial Revolution produced a number of technologies, most significantly the first telecommunications, that enabled the transmission of messages over distances and at speeds previously inconceivable; often regarded as the Second Industrial Revolution, these inventions led directly to the interconnected world of today. This world, based on binary and electronic systems, required a new breed of codes.

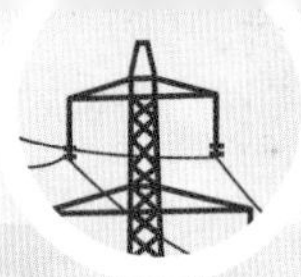

LONG-DISTANCE ALARMS

The ability to convey simple coded messages across considerable distances emerged amongst the earliest tribal and village communities, usually as a means of warning of a threat or as a call to arms. Considerable ingenuity gave rise to methods as various as the use of fire – creating smoke signals by day and beacon flares at night – drums, or heliography, reflecting the rays of the sun.

Plains Indians were noted for their use of smoke signals to convey warnings from one group to another.

Smoke signals

Although today primarily associated with Native North American peoples, the idea of using puffs of smoke, controlled by some sort of blanket to relay basic messages, was probably more widespread, and was certainly used in ancient China. Naturally limited by atmospheric conditions, not least the wind, nevertheless a simple message could be sent over distances of up to ten miles (16 km). The Native American system was sophisticated, and realizing that either friend or enemy might read the puffs, specified meanings were usually prearranged. Traditionally, and as used by Boy Scouts, one puff is usually a call for attention, two puffs means all is well, but three puffs is a signal that something is wrong (a basic code true also of whistle blasts or gunshots).

Drums along the river

Loud sounds, initially drumming but later bells or varieties of trumpet, provided an effective method of communication day or night, and in most weather conditions. Drumming was also linked to individual tribal rituals in many cultures, notably in Africa and North America, so localized meanings for different rhythms and tones evolved, meaning that there is no standard interpretation for drum messages. Early armies certainly used drums and other instruments as a means of terrifying their opponents as well as conveying messages on the battlefield.

Both beacons and church bells were effective ways of conveying a pre-agreed message rapidly from one point to another.

The alarming fire: beacons

In addition to their use from 2,000 years ago in lighthouses (*see page 166*) the lighting of fires on raised towers could relay an important pre-agreed message very swiftly, such as the English invasion alarm heralding the approaching Spanish Armada. Church bells were used to raise alarms and convey messages from one parish to another throughout the Christian world for over a millennium.

Heliography

The use of mirrors or polished metals and stones to reflect the sun's rays is once again very ancient, but limited by weather conditions and the need for pre-agreed meanings for the signals. Still, it has the advantage of being sent in a specific direction, thus making it a reasonably private, one-to-one messaging system, with a potential range of 50 miles (80 km). In the 19th century, mechanical heliographs were developed using double mirrors mounted on movable tripods, with a shutter system for interrupting the reflection. These could be precisely directed, and were used to send Morse code messages especially in remote territories, such as during the Boer War and the campaign against the Apache in the American Southwest. Many modern armies used heliographs as part of their signaling equipment until the second half of the 20th century.

Mechanical heliographs were used to convey messages in Morse code among many colonial armies in remote regions.

FLAG SIGNALS

The use of flags for signaling dates back to Han China some 2,000 years ago, where they were used to indicate different parts of an army on the battlefield. At much the same time, the Roman cavalry carried a square flag or *vexillum* (from which the term 'vexillology' for the study of flags originates). During the Middle Ages, battlefield flags bearing coats of arms identified the forces of particular kings or lords, and much of this heraldry persists in flag design today. The use of flags to send messages at sea and to identify the nationality or function of ships also dates back to the Middle Ages, but it was not until the Napoleonic Wars (1799-1815) that a detailed flag code was invented by the British to convey messages from shore-to-ship and ship-to-ship. The international marine flag code used today was established in 1932.

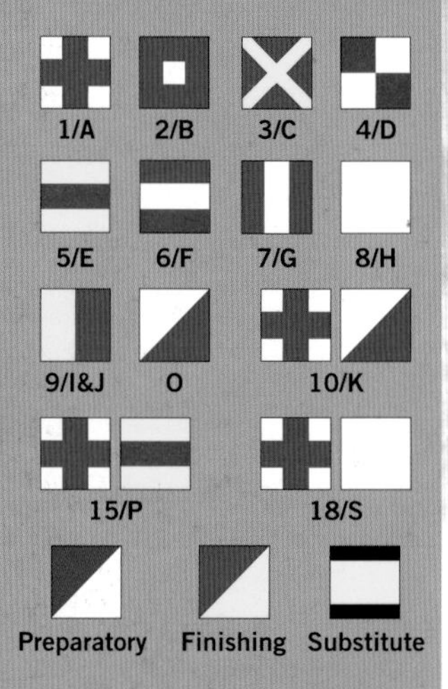

The first naval code

Derived from a system devised by Admiral Lord Howe in 1790, the flag code 'Telegraphic Signals of Marine Vocabulary' was introduced by Rear Admiral Sir Home Popham in 1800. It was based on a system of ten numerical flags, which could also be used as alphabetical letters. Combining numerical flags produced letters 'k' and above. The code included three additional flags: a substitute and two flags indicating the beginning and end of a signal. The flags were used in combination with frequently revised secret naval codebooks. To avoid these falling into the enemy's hands, the books were weighted with lead so they could be disposed of overboard in case of capture or sinking. These codebooks used combinations of the number flags to cover a wide range of words and phrases.

The naval flag code in action

Probably the most famous use of the naval flag system was the signal sent by Admiral Horatio Nelson to his fleet before the battle of Trafalgar in October 1805. It followed Popham's latest codebook, which contained numerical flag combinations covering more than 6,000 useful words and phrases. Each combination of flags was read from top to bottom, and raised on the mizzenmast in order. Nelson originally wanted to say 'confides' but it didn't appear in the Popham code, so the three-flag code for 'expects' was used. Only the last word of Nelson's famous signal had to be spelled out letter-by-letter. The substitute flag was used to represent 2 in the code for 'do,' and the 'u' in 'duty' is numbered 21, rather than 20, because in the early 19th-century English alphabet the letter 'v' occurred before 'u.'

"England expects that every man will do his duty."

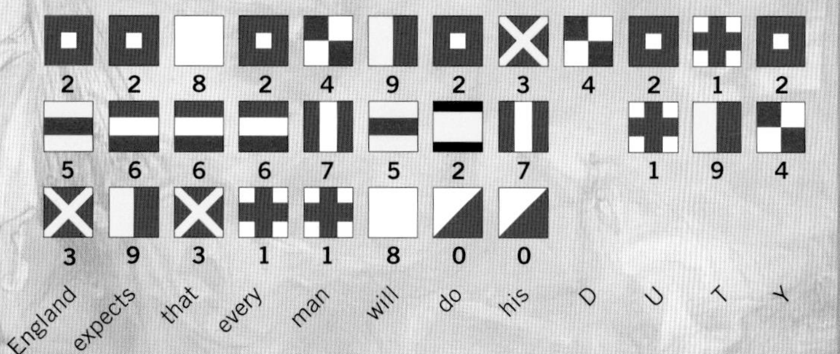

Close action

Nelson sent his final signal minutes later, before he was fatally wounded: "Engage the Enemy More Closely." It required only two flags, the 1 and the 6, making 16, the standard naval code at the time for this common battlefield command.

1

6

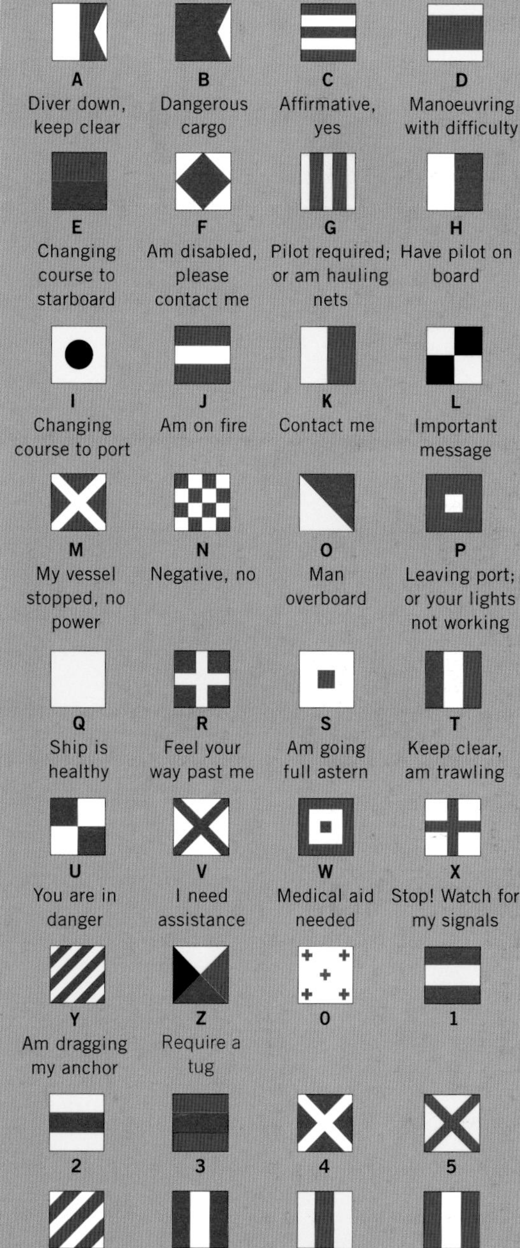

The international maritime code

The Commercial Code of Signals (later called the International Code of Signals) launched in 1857 consisted of 18 flags. Unlike the secret naval codes, this maritime code was designed to be memorized and did not require a detailed code book to be understood. Each flag represents a letter or number, but the letter flags, when raised individually, have specific meanings – usually a warning. Ships still use a modified version of this code today. For example, a three-letter flag combination identifies a ship's nationality.

SEMAPHORE AND THE TELEGRAPH

The challenge of sending detailed messages over long distances using various technologies was addressed by several scientists and technicians in the 18th century. Early experiments with electricity showed promise, but it was an ingenious French mechanical system called semaphore, involving coded messages that, by the turn of the century, provided the first rapid and reliable long-distance telegraph service.

The mechanical semaphore

With post-Revolutionary France under attack from all sides, the need for a fast and efficient method of communication became critical. In 1790, the Chappe brothers devised a system of towers with windmill-like arms which could relay messages by way of the position of the blades or shutters. After various tests, a message could be sent from Paris to Lille (143 miles; 230 km) via 15 towers in 32 minutes. The method was not cheap, involving the cost of tower building and staffing, and would soon be outflanked by electric telegraphy, but the Chappe brothers were not slow to realize that their system had commercial advantages for the French stock market too. It was rapidly imitated as an essential intelligence tool in England – linking the Admiralty in Whitehall to key naval bases around the country – and in Sweden (where it was used as a coastal defense system until the 1880s), Prussia, and Japan, where a new version of the encoding system had to be invented to accommodate the much longer Japanese syllabary alphabet.

One of the few remaining mechanical semaphore towers in France which, in good weather and across favorable terrain, could transmit messages letter-by-letter across a ten-mile (16 km) span to the next relay tower.

HELP! Another Beatles myth The photographer Robert Freeman conceived the Moptops as spelling out the title in semaphore on the cover of this seminal pop album. Unfortunately, graphically the image didn't work as a design, and the resulting image, even if reversed, means nothing. On the US release the images are even more jumbled up.

A / 1 B / 2 C / 3
D / 4 E / 5 F / 6
G / 7 H / 8 I / 9
J K / 0 L
M N O
P Q R
S T U
V W X
Y Z cancel
error

Semaphore using flags

The system of hand-held flag positioning was developed in the early 19th century, and grew out of the mechanical semaphore system (*above*) as a means of rapidly sending messages over short distances, both at sea and on land. The flags themselves are of less importance than the position in which they are held.

Electric telegraphy

Shortly after the introducton of semaphore systems, the first inventions using electric impulses were created. Experiments earlier in the 18th century had demonstrated that electrical impulses could be sent over long distances, but these had not been developed, as the duration and intensity of the impulse could not be controlled. The invention of the Voltaic pile allowed all-round scientist and inventor Samuel Thomas von Sömmering to build his 'electrochemical' telegraph system in Bavaria in 1811. This was followed up in Germany by the Russian-born Baron Schilling, with an electromagnetic telegraph system in 1832; it used a keyboard for transmitting individual signals, and a series of needle and flag pointers for the receiver to read. The idea was adapted by Carl Friedrich Gauss and Wilhelm Weber for regular use in 1833. Within ten years the 'needle' electric telegraph system was spreading throughout Europe, while the first commercial telegraph system in the US was set up in 1845 in Pennsylvania, along the Lancaster/Harrisburg railroad.

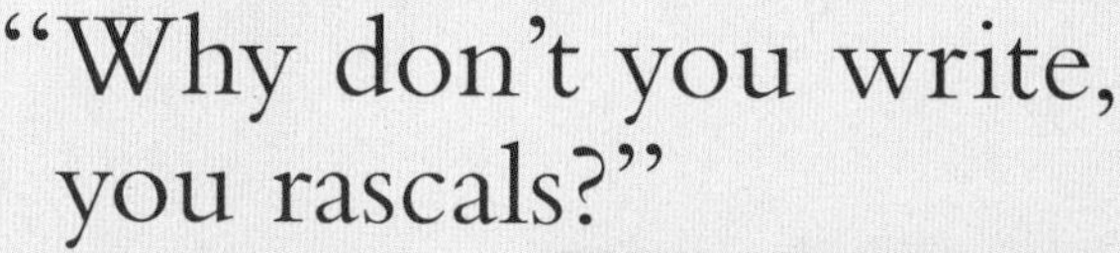

"Why don't you write, you rascals?"

THE FIRST US TELEGRAPH MESSAGE, RECEIVED JANUARY 8, 1846.

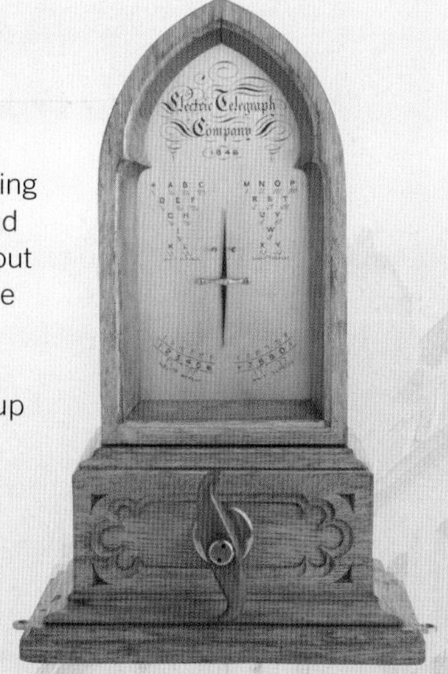

The needle telegraph receiver worked rather like a modern speedometer. As each electrical impulse was received, usually in the form of binary Morse code, this would send an impulse to the needle which would indicate the "letter" received – literally spelling out each word of the signal.

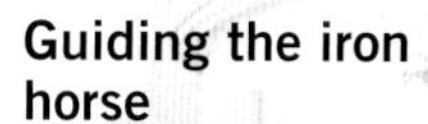

Guiding the iron horse

Another great 19th-century innovation, the railroad, was greatly enhanced by telegraphy. When railway networks began to grow in the 1830s, the need to convey information down the line (to avoid crashes, or to provide updated timetable information) became crucial. The first commercial needle telegraph system was introduced by Sir William Fothergill Cooke in the UK between London's Paddington station and West Drayton some 13 miles (21 km) away, and the system rapidly spread, with overhead telegraph lines shadowing the iron tracks.

Signaling systems also needed to be developed to convey information to train drivers, and here a combination of electric telegraphy linking signal boxes and an adapted semaphore system of mechanically-operated towers using shutters and paddles providing visual instructions for the drivers was developed.

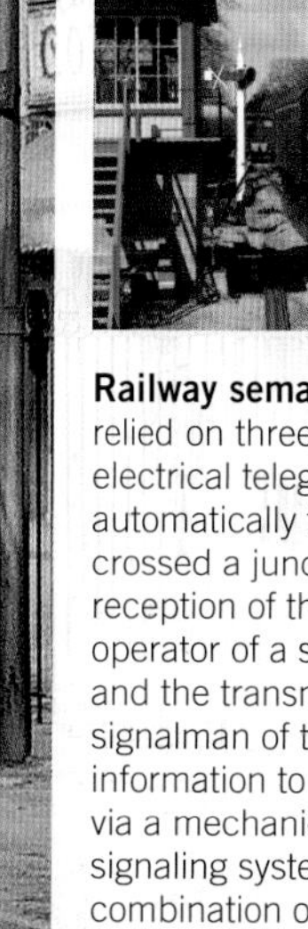

Railway semaphore systems relied on three elements: electrical telegraphy, sometimes automatically triggered as trains crossed a junction or points; the reception of these signals by the operator of a signal box (*above*); and the transmission by the signalman of the appropriate information to the train driver via a mechanical semaphore signaling system using a combination of colored blades.

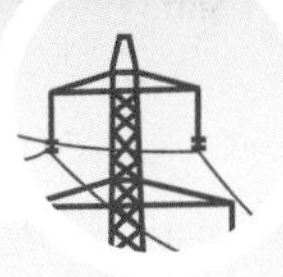

MORSE CODE

A new system

Samuel Finley Breese Morse (1791-1872) was born in Charlestown, Massachusetts. His interest in electricity dated from his years at Yale. From 1832-44, under the influence of the French physicist André-Marie Ampère, Morse worked with Leonard Gale, Alfred Vail, and Joseph Henry to perfect a new system of telegraphy. In addition to developing the transmitting technology, Morse also had to perfect a way of encoding letters and numbers into electric impulses. Although neither telegraphy nor the basic code were entirely original (and Morse had to defend his patent in court), with the backing of Congress his method rapidly acquired global recognition, and remained one of the standard systems for binary encoding until the 21st century. In 2008 it was abandoned as an official messaging system.

One of the challenges confronting the early industrial and imperial nations in the 19th century was efficient – and rapid – long-distance communication, whether on land or at sea. A simple letter to London from a colonial outpost in India could take anything up to eight weeks to reach its destination, and if a quick decision was needed, then a wait of up to four months could ensue between a question being dispatched and an answer being received. In the spirit of the times, the American artist and inventor Samuel F. B. Morse developed and patented a system of substituting a combination of dots and dashes for numbers and letters of the alphabet. Although designed to be used by the emerging telegraph, the system can be adapted for other long-distance communication systems, such as heliographs, foghorns, and flashlamps.

This early telegraph receiver was made mainly of brass, to avoid rusting or corrosion.

Remembering the Morse code
This alphabet was designed as an effective way of memorizing the dots and dashes which make up the Morse code.

A B C D E F G H I J K L M
N O P Q R S T U V W X Y Z

Save our souls This is the internationally recognized telegraph distress signal, which originated because the dot and dash combination of the letters in Morse was the simplest to remember. The acronym 'save our souls' was merely an afterthought. 'Mayday' became the international distress signal using radio, and is derived from the French *'m'aider.'*

Morse code in action

Morse's system was first used in warfare during the Crimean War (1853-56), when *The Times* newspaper's correspondent William Russell filed copy from the battlefront. It was later critical in the conduct of operations during the American Civil War (1860-65). There were other stories too:

Catching Crippen

The infamous American-born wife murderer Dr. Crippen was trapped by using Morse code. In 1910, having poisoned his wife and buried her in their basement, he and his mistress escaped from Britain on the liner *Montrose* bound for Quebec. Upon discovering Crippen's wife's remains, Inspector Dew of Scotland Yard sent out a general alert. One of few liners then to be fitted with Marconi radiotelegraphy, the captain contacted Dew at Scotland Yard: "Have strong suspicions that Crippen and accomplice are among saloon passengers." Dew set off in pursuit on a faster ship and Crippen was arrested before the *Montrose* docked. He was subsequently tried and hanged.

Iceberg alert

The *Titanic* disaster might have been averted if the wireless operator on the nearest vessel, only ten miles (16 km) away, had not retired for the night. A state-of-the-art vessel, the *Titanic* was equipped with the latest radio-telegraphy equipment. Her captain received advance warnings of drifting icebergs, but wanted the 'unsinkable' liner to break a trans-Atlantic crossing record, and ignored them. When disaster struck, her radio operator sent SOS signals, but only the following day did rescue ships arrive to save a mere 700 from the 2,200 on board.

Sending Morse code

Morse code is made up of a binary system of dots and dashes. The 'dot' is one contact, flash, or bleep, while the 'dash' is a similar signal of normally three times the length of a 'dot.' Gaps between the individual letters are usually signaled by a pause the equivalent of a 'dash,' and between words a pause of two dashes.

The Morse code alphabet

The International Morse Code system is shown below. There are some variants in the US system which are also shown.

A	.–	N	–.	1	.––––
B	–...	O	–––	2	..–––
C	–.–.	P	.––.	3	...––
D	–..	Q	––.–	4	....–
E	.	R	.–.	5	.....
F	..–.	S	...	6	–....
G	––.	T	–	7	––...
H	....	U	..–	8	–––..
I	..	V	...–	9	––––.
J	.–––	W	.––	0	–––––
K	–.–	X	–..–		
L	.–..	Y	–.––		
M	––	Z	––..		

The American Morse code differs in the following symbols:

C	.. .	1	.––.
F	.–.	2	..–..
J	–.–.	3	...–.
L	——	4	....–
O	. .	5	–––
P	.....	6	......
Q	..–.	7	––..
R	. ..	8	–....
X	.–..	9	–..–
Y	.. ..	0	———
Z	... .		

PERSON TO PERSON

The invention of the telephone (*see page 85*) redefined how people could communicate directly over long distances, and paved the way for the modern, interconnected planet familiar to many of us through the invention of the Internet. Indeed, in the last two decades the main thrust of technological development has been witnessed in the fields of information technology and personal communication, the keys being miniaturization of high-performance computer chips, and the integration of coding systems. It is now easier – and faster – to e-mail or text a message to someone across the office than it is to walk over and talk to them. And it takes no longer than sending the same message to somebody on the other side of the world.

The mobile revolution

Kentucky melon farmer Nathan B. Stubblefield (1860-1928) is frequently cited as the inventor of the mobile phone due to his demonstrations of radio technology to send voice communications via wireless telephony, gaining a patent in 1908. Unfortunately the world took another 70 years or so to catch up with his visionary ideas.

Introduced in the 1970s, the mobile phone is now the most widely spread consumer electronics item on the planet. There are now over three billion mobile phone subscribers worldwide. The BlackBerry extended the mobile handset, providing a miniature keyboard and screen for use with PC-style applications such as Microsoft Office. The integration of various developed technologies has now produced Apple's iPhone, combining phone, Internet, e-mail, music, camera, TV, and video playback in one slim electronic package. Touch screen interfaces optimize Internet, video, and TV viewing, while third-party systems can be easily linked in, such as iTunes for music downloads, and GoogleEarth for locational data.

Telephone numbers and directories

The problem of the unique telephone number was initially developed *ad hoc* by individual phone companies. In order to dial up an individual line, an individual number was required. The use of manned local exchanges initially solved the problem, as did the combination of numbers and letters on the telephone dial, allowing telephone numbers to be broken down into a mixture of geographic and numeric codes, especially within large cities like London or New York. KEN 162 might have stood for a number in Kensington, London. Now, regional and national codes have been added. Glenn Miller's famous song 'PEnnsylvania 65000' references what is said to be the oldest telephone number now in use in the city, that of the Pennsylvania Hotel, the PE now transformed into 73 which, with the national and area code added, is currently: 001-212-736-5000.

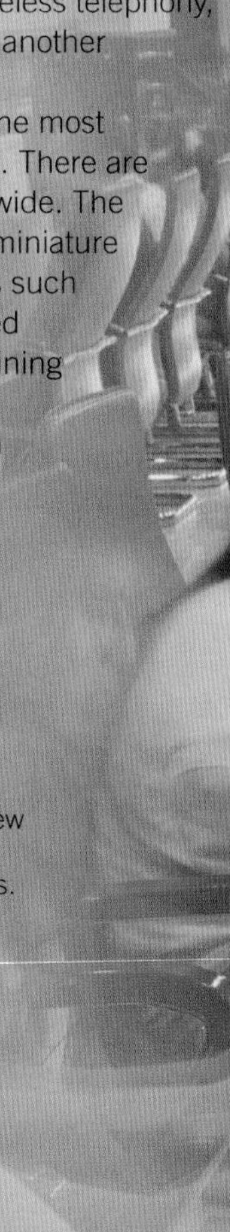

Introduced in 2007, Apple's iPhone set a new standard in handheld communication devices.

Screen
In order to maximize the screen size for reading e-mails and surfing the Internet, most of the functions are accessible through touch screen displays.

Slimline
Despite its multifunctionality, the iPhone is a masterpiece of miniaturization.

The Internet

The origins of the Internet can be traced back to data transmission research conducted at the United States Defense Advanced Research Projects Agency (DARPA) in the late 1960s. The key development was a new way of sending electronic information, known as 'packet switching.' Previously, remote electronic communication had been conducted through 'circuit switching,' functioning in much the same way as old-fashioned telephony systems: a circuit being used to connect two people could not be used by anyone else. By breaking up the information into small 'packets' before transmission, to be reassembled at the receiving end, a single link could be shared by more than two systems, as well as each being able to 'route' packets independently through the network. A common analogy is that of a single mailbox which can be used to post letters to more than one recipient. The first such network was known as ARPANET.

The Internet we use today did not really exist until the 1990s. Previously, due to cost, networks existed only in universities and government agencies. However, the introduction of e-mail and Web-browsing software in the early 1990s, making use of the public network, created a revolution: the term 'Internet' entered common vocabulary. For the Internet to be able to expand, a common 'language' for the exchange of information had to be established that had fixed rules, known as 'protocols.' The most important is the 'Internet protocol,' which deals with splitting data into packets, and the complex task of assigning each device connected to the Internet a unique 'address' (*right*). As the Internet became more mature, and as the bandwidth (or the number of packets that can be sent every second) increased, the number of applications that the network could be used for increased.

Though the Internet is essentially a public network, it is regulated to ensure its consistency. The Internet Corporation for Assigned Names and Numbers (ICANN) controls 'domain names,' which makes the location of a device easier to remember, while the World Wide Web Consortium is in charge of the design and maintenance of many standards that define the way Web pages work.

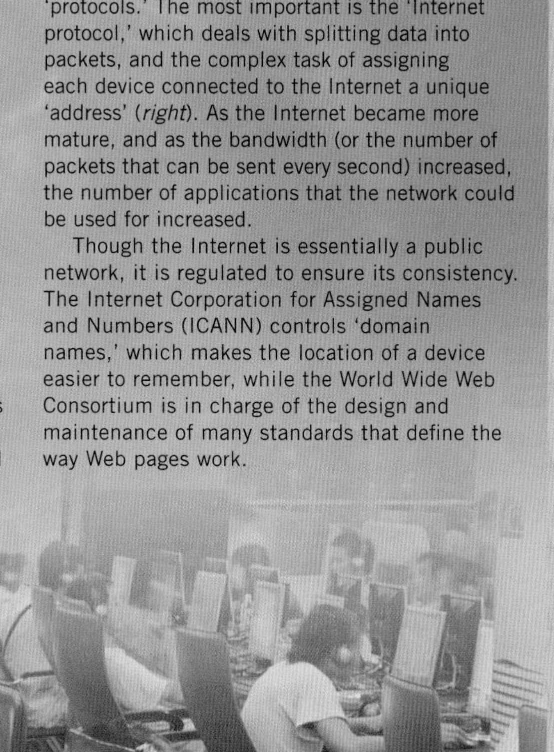

URLs and the Internet phone book

Resources on the Internet are each located using a coded URL (Uniform Resource Locator). What is a URL? A URL is an address to some kind of resource; a Web site, an image on your computer, a secure Web server, or one of many other types of resources. The most common URLs we see are Web site addresses, for example:

http://www.google.com A URL for the Google Web site. The first part, http, indicates that the resource is accessed via Hypertext Transfer Protocol, meaning it is an HTML Web site (*see page 272*).

file://c:/notes.txt The 'file' part of the URL indicates that the resource 'notes.txt' is stored on the local computer.

ftp://name:password@www.downloads.com A more complicated URL. It represents a connection to the Web site 'www.downloads.com' for File Transfer Protocol (FTP), which is a common way of moving files across the Internet. This URL also specifies a user name and password for security.

URLs normally contain an address, such as www.something.com, but how does this actually navigate to a Web site? When you enter an address such as www.google.com into your Web browser, the browser sends the address to a DNS (Domain Name System) server. A DNS server is like an Internet phone book – it turns names into numbers. These generally take the form of four one-to-three-digit numbers separated by dots (such as 206.34.2.100), which means that it is possible to identify 4,294,967,296 separate devices. For example, putting the address 'www.google.com' into a DNS server returns the value '74.125.45.99.' This is the Internet 'phone number' of the Google Web site – typing http://74.125.45.99 into a Web browser will take you direct to www.google.com.

In times of armed conflict, the need for secrecy, intelligence gathering, and secure channels of communication has always been of paramount importance. The activities of spies are described by the earliest writers such as Herodotus and Thucydides, and in the Old Testament.

codes of war

The significance of security and secrecy has become even greater in the modern world, demonstrated by the critical role played by radio traffic decryption in World War II. The sheer scale of modern military budgets has permitted considerable expenditure on research and development of new and ever-improving methods of encrypting and decrypting strategic information. However, one of the persistent challenges with the acquisition of secret intelligence is how to use it without alerting the enemy to its loss.

CLASSICAL CODES OF WAR

Sparta at war
The Spartans were noted in ancient Greece for their prowess in the arts of war. Warriors were trained to undergo extreme hardships from a young age to hone their combat and survival skills, and Bronze Age Spartan armor and weaponry was regarded as technologically advanced. It is no surprise then that one of the first examples of secret messaging, the scytale, was developed by the city state.

Diplomacy and warfare are the two main areas in which codes and ciphers first came to be developed and used. Considerable levels of ingenuity were used to ensure secrets remained secret. The tradition appears to have begun in the Greek world – unsurprising considering the Greek fascination with words and mathematics – a legacy that would be passed on to medieval scholars via the Arab world. The principles of the substitution cipher, as developed by Julius Caesar, dictated the basic form of cryptography for the next two millennia.

The Spartan scytale

The earliest cryptographic device we know of is the scytale, used in Sparta from the 7th century BC. It produced a simple form of transposition cipher. The scytale was a wooden rod, around which the the sender would wind a strip of parchment or leather, write the full message out in successive lines along the length of the rod, then unwrap the strip. The sender will have produced an apparently meaningless string of letters. The strip would then be secretly conveyed to the intended recipient – if written on leather, this might be reversed and worn by the messenger as a belt to conceal the letters. The recipient would then wrap the strip around their scytale (assuming it was of the same diameter and had the same number of faces as the sender's) to reveal the hidden message.

Encryption

In this example the scytale rod allows the sender to write five letters each in four rows. The plaintext 'Under siege send forces' would be written across the strip of leather:

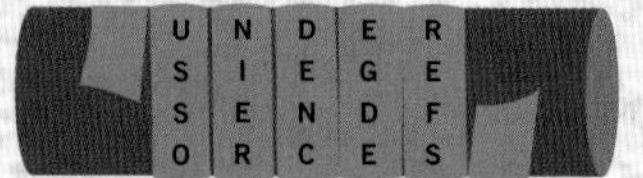

The plaintext written on the scytale

Unwound, the ciphertext on the leather strip would look like this:

USSONIERDENCEGDEREFS

Decryption

To understand the message, the leather strip would be wound around a scytale of the same dimensions, revealing the original message.

Gaius Julius Caesar

The Roman general and dictator Julius Caesar (100-44 BC) was a prodigious writer of histories and correspondence, using both Latin and Greek. He was also an enthusiastic user of secret writing, so much so that a treatise was written by the Roman grammarian Marcus Valerius Probus in the 1st century AD on Caesar's various techniques, unfortunately now lost. Caesar was an inveterate intriguer, and developed an extensive intelligence network whilst conquering Gaul. In his own account of the Gallic Wars, he describes how, upon finding that his fellow Roman leader Cicero is besieged, he sends him a message assuring support, but written in Greek "lest the letter being intercepted, our measures should be discovered by the enemy." Attached to a spear, the message was hurled into Cicero's camp, although not discovered for two days.

The Caesar Shift cipher

Probably the first substitution cipher to be described is that used by Julius Caesar in parts of his private correspondence to friends and colleagues in Rome while he was campaigning in Gaul. The code Caesar used to plan a political intrigue upon his return to Rome was described by Suetonius in his *Twelve Caesars* 150 years later. It involves a simple 'shift' of the alphabetical order (known as the algorithm); if the letter A is to be encrypted, and the shift used is four letters, then A appears encrypted as E, M as Q, and so on. The Caesar Shift cipher therefore has 25 potential ciphers, and in turn 25 separate keys depending on which shift is chosen. However, it is not very secure: cryptanalysts, if they suspect a simple Caesar Shift has been used, only have to check 25 potential keys.

REINFORCEMENTS
ON THE WAY

Plaintext message

VIMRJSVGIQIRXW
SR XLI AEC

Ciphertext after 4-letter shift

As with any substitution cipher, the ciphertext can be made more impenetrable using various further techniques, such as deleting the word spaces, or including other symbols or numbers for selected features of the plaintext. Other ways of making a substitution cipher more complicated are explored elsewhere (*see page 66*).

A 4-letter Caesar Shift

Plain alphabet	Cipher alphabet
A	E
B	F
C	G
D	H
E	I
F	J
G	K
H	L
I	M
J	N
K	O
L	P
M	Q
N	R
O	S
P	T
Q	U
R	V
S	W
T	X
U	Y
V	Z
W	A
X	B
Y	C
Z	D

THE 'INDECIPHERABLE' CODE

Blaise de Vigenère
The French career diplomat, Blaise de Vigenère (1523-96), developed an interest in cryptography whilst serving in Rome. Italy at the time was the centre of European cryptography. In his retirement he wrote several books on a variety of subjects, including *Traicté des Chiffres ou Secrètes Manières d'Escrire* (1586), which looked at various codes and ciphers, including the *tabula recta* of Bellaso, which was mistakenly attributed to de Vigenère in the 19th century. The latter did, however, produce a number of elaborations to Bellaso's basic idea, extending the tableau to 26 x 26, and inventing the autokey system for encryption (*see opposite*). Interestingly, de Vigenère's cipher system was not used widely until the 18th century.

Although various potential polyalphabetic enciphering systems had been identified, notably by the Italian Renaissance genius Alberti among others (*see page* 72), the formulation of a mathematical encryption system based on a *tabula recta* which used a progressive Caesar Shift for encryption, and which in turn depended on a key, was developed in 1553 by an Italian, Giovan Battista Bellaso, but popularized by a Frenchman, Blaise de Vigenère. It represented a turning point in the theory of cryptography, and provided a very secure (though cumbersome) system for sending secret messages, especially in times of crisis or war, which was used until the end of the 19th century. It was regarded as so secure that it was called *le chiffre indéchiffrable* – the indecipherable code.

The Thirty Years War (1618-48) inaugurated three centuries of intense and often bloody rivalry between Europe's leading states, which provoked a surge of interest in secure methods of encoding secret messages.

The de Vigenère tableau

The basic principle of the de Vigenère cipher was very simple. What it produced was complicated.

1 Draw up a 26 x 26 *tabula recta* formed of a series of single Caesar Shifts running through the alphabet. Across the top of the tableau is the plaintext alphabet, while down the side the rows may be numbered.

2 Unlike a single Caesar Shift ciphertext, the tableau could produce 26 potential ciphertexts. By shifting between the rows of ciphertext, a polyalphabetic system is established: if row 5 is selected the plaintext 'a' would be enciphered as 'F,' if row 22 is selected, 'a' would be enciphered as 'W.'

3 Now the cryptographer has to produce a guide to which of the 26 available ciphertexts are to be used and in which order, known as a 'key word,' or 'keytext.' This can take the form of a word, or a sequence of words, or a sequence of numbers.

4 Using a relatively short key word, such as 'ENEMY,' this word is repeated across the top of the tableau to form the keytext. This tells both the sender and the receiver in which order the letters of the plaintext have been encrypted. It does this by indicating which row, beginning with each successive letter of the key word, should be used, which is then matched to the plaintext alphabet at the top of each column.

A de Vigenère tableau, beginning with a single Caesar Shift.

	a	b	c	d	e	f	g	h	i	j	k	l	m	n	o	p	q	r	s	t	u	v	w	x	y	z
1	B	C	D	E	F	G	H	I	J	K	L	M	N	O	P	Q	R	S	T	U	V	W	X	Y	Z	A
2	C	D	E	F	G	H	I	J	K	L	M	N	O	P	Q	R	S	T	U	V	W	X	Y	Z	A	B
3	D	E	F	G	H	I	J	K	L	M	N	O	P	Q	R	S	T	U	V	W	X	Y	Z	A	B	C
4	E	F	G	H	I	J	K	L	M	N	O	P	Q	R	S	T	U	V	W	X	Y	Z	A	B	C	D
5	F	G	H	I	J	K	L	M	N	O	P	Q	R	S	T	U	V	W	X	Y	Z	A	B	C	D	E
6	G	H	I	J	K	L	M	N	O	P	Q	R	S	T	U	V	W	X	Y	Z	A	B	C	D	E	F
7	H	I	J	K	L	M	N	O	P	Q	R	S	T	U	V	W	X	Y	Z	A	B	C	D	E	F	G
8	I	J	K	L	M	N	O	P	Q	R	S	T	U	V	W	X	Y	Z	A	B	C	D	E	F	G	H
9	J	K	L	M	N	O	P	Q	R	S	T	U	V	W	X	Y	Z	A	B	C	D	E	F	G	H	I
10	K	L	M	N	O	P	Q	R	S	T	U	V	W	X	Y	Z	A	B	C	D	E	F	G	H	I	J
11	L	M	N	O	P	Q	R	S	T	U	V	W	X	Y	Z	A	B	C	D	E	F	G	H	I	J	K
12	M	N	O	P	Q	R	S	T	U	V	W	X	Y	Z	A	B	C	D	E	F	G	H	I	J	K	L
13	N	O	P	Q	R	S	T	U	V	W	X	Y	Z	A	B	C	D	E	F	G	H	I	J	K	L	M
14	O	P	Q	R	S	T	U	V	W	X	Y	Z	A	B	C	D	E	F	G	H	I	J	K	L	M	N
15	P	Q	R	S	T	U	V	W	X	Y	Z	A	B	C	D	E	F	G	H	I	J	K	L	M	N	O
16	Q	R	S	T	U	V	W	X	Y	Z	A	B	C	D	E	F	G	H	I	J	K	L	M	N	O	P
17	R	S	T	U	V	W	X	Y	Z	A	B	C	D	E	F	G	H	I	J	K	L	M	N	O	P	Q
18	S	T	U	V	W	X	Y	Z	A	B	C	D	E	F	G	H	I	J	K	L	M	N	O	P	Q	R
19	T	U	V	W	X	Y	Z	A	B	C	D	E	F	G	H	I	J	K	L	M	N	O	P	Q	R	S
20	U	V	W	X	Y	Z	A	B	C	D	E	F	G	H	I	J	K	L	M	N	O	P	Q	R	S	T
21	V	W	X	Y	Z	A	B	C	D	E	F	G	H	I	J	K	L	M	N	O	P	Q	R	S	T	U
22	W	X	Y	Z	A	B	C	D	E	F	G	H	I	J	K	L	M	N	O	P	Q	R	S	T	U	V
23	X	Y	Z	A	B	C	D	E	F	G	H	I	J	K	L	M	N	O	P	Q	R	S	T	U	V	W
24	Y	Z	A	B	C	D	E	F	G	H	I	J	K	L	M	N	O	P	Q	R	S	T	U	V	W	X
25	Z	A	B	C	D	E	F	G	H	I	J	K	L	M	N	O	P	Q	R	S	T	U	V	W	X	Y
26	A	B	C	D	E	F	G	H	I	J	K	L	M	N	O	P	Q	R	S	T	U	V	W	X	Y	Z

As E 'f' will be encrypted as 'K.'

As M 'f' will be encrypted as 'S.'

As N 'f' will be encrypted as 'T.'

As Y 'f' will be encrypted as 'E.'

5 Thus the plaintext message 'wait for men' combined with the key word 'ENEMY' would produce the following:

plaintext:	w	a	i	t	f	o	r	m	e	n
key word:	E	N	E	M	Y	E	N	E	M	Y
ciphertext:	B	O	N	G	E	T	F	R	R	M

6 To decipher the text, reverse the process: find the cipher letter in the row indicated by the key word.

Increasing complexity: keys

Once the principle of a *tabula recta* along the lines of de Vigenère's system had been established, there remained two further ways of enshrouding a message. The first was to shuffle the logical order of the *tabula recta*. The second was to introduce increasingly elaborate keys: a key word might be formed of a random assembly of letters or numbers; an autokey, in which a key word is simply added at the beginning of the plaintext producing a shift, the plaintext itself then forming the key, avoiding the vulnerable repetition of a key word; or, lastly, a running key – an assembly of letters or numbers as long as the plaintext – such as a book cipher (*see page 79*).

Computing power

Although claims have been made for Friedrich Kasiski's formula for decrypting a de Vigenère cipher, published in 1863, it was probably the father of the computer, Charles Babbage (*see page 109*), who first devised a method of cracking the system a decade earlier, some 300 years after it was invented. Fascinated by the mathematics underlying this variety of cryptography, and provoked by an English dentist who claimed to have developed a new enciphering system – but one which largely replicated de Vigenère's method – Babbage set about looking at the repetitive cycles which might appear in a typical de Vigenère ciphertext.

The key, as always, is the key word; Babbage's basic principle was to establish the length of the key word by searching for repetitions of strings of letters in the ciphertext. This is only possible with a ciphertext of some length – 30 letters or more. Various instances of repetition might enable the cryptanalyst to firstly establish the length of the key word, and also to identify instances where a short word, likely to appear several times in a long ciphertext, such as 'the,' might have been encrypted using the same letters of the key word. A secondary line of attack was to search for probable digraphs, the most common in English being: 'th,' 'he,' 'an,' 'in,' 'er,' 're,' and 'es.' Such a search might again reveal a pattern in the ciphertext, allowing the cryptanalyst to begin to identify the key word.

Although Babbage identified the principles for attacking the de Vigenère cipher, he realized that a machine like his Difference Engine (*see page 268*) would be needed to achieve the amount of analysis required – especially in a lengthy message.

THE GREAT CIPHER

Louis XIV (1638-1715) came to the throne in 1661, and reigned France for over five decades. He regarded himself, with good reason, as the most powerful man in Europe, and his construction of the Palace of Versailles symbolized his pre-eminence as the 'Sun King.' It was a time of rabid competition between the emerging European nation states and, behind the scenes of diplomacy, an intense atmosphere of secrecy was fostered. It is no surprise, then, that alongside his court, his palace, his navy, and his army, Louis should also invest in one of the most devious cipher systems yet invented. Other nations too were keen to develop secure codes.

Louis at war

The Sun King liked to present himself as a martial hero, and he was widely suspected of harboring grand military ambitions for European conquest. In fact, most of his warfaring was designed to strengthen France's northeastern borders, and closer analysis reveals his huge efforts to consolidate the French nation internally. He created a strong, centralized ministerial government and state bureaucracy, an important aspect of which relied upon the need for security in both civil and military matters. The Great Cipher in part provided this. His palace and gardens at Versailles (*right and below*) were an outward expression of Louis' diplomatic mindset, ornamented with intricate decorations and mirrors.

The Rossignols

Louis XIV's chief cryptographer was Antoine Rossignol (1600-82) who was succeeded by his son Bonaventure and grandson Antoine Bonaventure. Antoine came to prominence in 1626 under the Sun King's father, Louis XIII, when he decrypted a message revealing that the Huguenot city of Réalmont that the French army was besieging was on the point of collapse. The knowledge led to a bloodless victory. The young king and his advisor Cardinal Richelieu realized the value of such cryptanalytical gifts, and encouraged the Rossignols to develop an impregnable cipher, in which they succeeded, the secrets of its construction being lost and the many letters they produced resisting all attempts at decryption until the 19th century. It is said that the Rossignols were so highly valued as cryptographers and code breakers that their name has passed into the French language as slang for a lock-pick, although there is evidence that the term was used much earlier.

The Great Cipher, and how it worked

It would be a French military cryptanalyst, Étienne Bazeries (1846-1931), who finally unlocked the secrets of the sheets of numbers discovered in Louis XIV's correspondence by historian Victor Gendron in 1890.

Unlocking the puzzle

Although there were thousands of numbers written across the sheets, Bazeries counted only 587 different ones. He proceeded step by step, initially up several blind alleyways: he first assumed that the Rossignols had developed a substitution cipher involving a huge number of homophones (*see page 70*), whereby many numbers represented the same alphabetic letters. Months of analysis drew a blank. He then assumed that the numbers represented digraphs, or pairs of letters. Using frequency analysis he searched for the most frequent numbers in the ciphertext, which were 22, 42, 124, 125, 341. He then compared these to the most frequent digraphs in the French language: es, en, ou, de, nt. Again, laborious work led nowhere. His third assumption stemmed from his second one: maybe the numbers represented syllables rather than simple digraphs. Having attempted various permutations, he isolated a set of frequently recurring numbers – 124, 22, 125, 46, 345 – and guessed that, given the context of the letters, a phrase such as 'les ennemis' might occur quite often.

Bazeries constructed a table to match his findings, breaking the words into syllables:

les	en	ne	mi	s
124	22	125	46	345

Bazeries then applied this assumption across large areas of the ciphertext, which began to reveal parts of other words, and sometimes whole words, which in turn led him to identify other syllables represented by numbers.
The Rossignols had seeded their cipher with a number of traps, such as one number which simply deleted the previous number. Nevertheless, over the next three years Bazeries worked his way around these traps and deciphered the Rossignols' masterpiece.

Bazeries' scratch pad shows how he tentatively matched syllables and words with the original numbers in the Rossignols' coded messages.

The struggle for secrecy

While the Great Cipher of Louis XIV remained a closed secret until the end of the 19th century, the search for ever-more secure ciphers continued. The de Vigenère cipher remained seemingly impregnable, but was cumbersome to use, especially if messages need to be encrypted and decrypted rapidly. One system with its roots in classical Greece was seen as a potential source of new cryptographic ideas: the Polybius square (*see page 78*), which also influenced the development of the Playfair cipher (*see page 109*).

19th-century Innovations

Technical innovations

The invention of semaphore, the telegraph, and Morse code (*see pages 94, 96*), logical technical solutions to practical problems, in turn created both dead ends and opportunities for cryptographers. Morse code especially, in its reliance upon a system of dots and dashes, paved the way for a whole new area of thought about how coded messages might be enciphered in a purely binary way. On the other hand, simple but mass-manufactured cipher disks based on a simple system of polyalphabetic ciphers continued to be used during the American Civil War and later (*see page 84*).

For two hundred years, from the 17th century until the 19th century, it was assumed that the complexities of polyalphabetic substitution – especially those based on the de Vigenère system or Louis XIV's Great Cipher, with their reliance upon shared keys – and the sheer amount of time it would take to unravel such codes meant that military ciphers were regarded as relatively secure. However, the scientific, mathematical, and mechanical advances that flowed from the Enlightenment to produce the Industrial Revolution soon changed the landscape of cryptography and cryptanalysis and suggested new ways of approaching the problem of secure ciphers – and new ways of rapidly deciphering codes.

The Playfair cipher

Invented by Sir Charles Wheatstone, one of the early developers of the electric telegraph in Britain, and the politician Baron Lyon Playfair, this code relies on creating digraphs (pairs of letters), then substituting these with a pair of cipher letters.

1 Firstly, a keyword needs to be agreed, such as PLAYFAIR. Lay out a five-row by five-column alphabetical grid, using the keyword to begin with (removing any repeated letters), then fill in the grid left to right with the remaining letters of the alphabet in their usual order, combining **I** and **J** as a single letter.

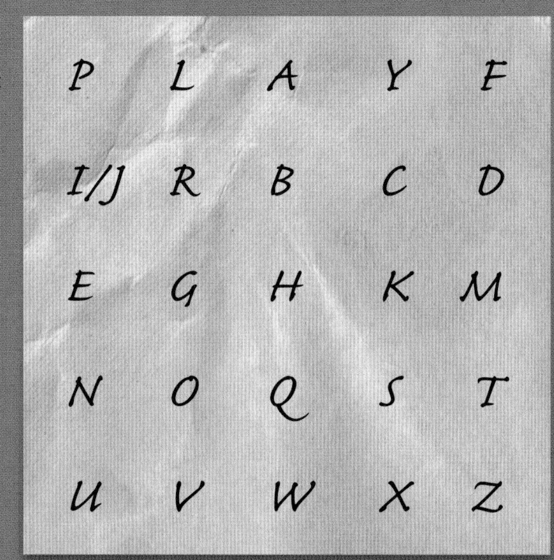

2 Break the plaintext message into pairs of letters (digraphs). Each digraph needs to be made up of different letters; if two letters occur together within a digraph, insert an x between them, and also add an x at the end if you are left with a single letter, for example – plaintext: **help I really need somebody** becomes: **he lp ir ea lx ly ne ed so me bo dy** Now, referring to the grid, the digraphs will occur in three different ways:

3 If in the same row, then the digraph is encrypted by the letter to its right (so **ly** becomes **AF**); if a letter occurs at the end of the row, it is replaced by the one at the beginning of the same row (so **me** becomes **EG**).

4 If in the same column, the letters are encrypted by the letter immediately below (so **ne** becomes **UN**); if one of the letters is at the bottom of a column, it is replaced by the one at the top of the same column.

5 If the letters of the digraph are neither in the same row nor the same column, then find the first letter on the grid, then look along that row until you find the column with the second letter in it; the letter at the intersection becomes its enciphered replacement, while enciphering the second letter, look along its row until you reach the column in which the first letter occurs, and the letter at the intersection becomes its enciphered replacement. Thus, **bo** becomes **RQ**.

6 Following this, the complete encryption for the message reads:

plaintext in digraphs:
he lp ir ea lx ly ne ed so me bo dy

ciphertext:
KG AL RB HP YV AF UN MI TQ EG RQ CF

7 The receiver of the coded message, knowing the keyword, simply reverses the process. It is hardly impregnable, as frequency analysis can be used to examine the ciphertext for recurring digraphs (as Charles Babbage proved, *right*), which can then be compared to the most common pairings of letters in English, which are: **th**, **he**, **an**, **in**, **er**, **re**, **es**.

The Babbage connection

The inventor Charles Babbage, in addition to inventing the prototype computer (*see page 268*), also seems to have cracked the de Vigenère code (*see page 104*). A serial non-completer of projects, he never published his solution to the de Vigenère problem, but in this instance it seems probable that he was encouraged not to do so. A mixer in the higher echelons of the British government (usually seeking funding for his ambitious projects), it seems likely that he revealed his solution, and was told to 'keep it under wraps.' Although Babbage's solution would inevitably have relied on massive comparative calculations – of the sort that required huge numbers of mechanical computations – and his Difference Engine was only built in prototype, it does not seem unlikely that elements of his mechanical invention could have been put to such a purpose. By the last half of the 19th century, with the 'Great Game' for control of Asia being played out from the Crimea to Tibet between Britain and Russia, and a growing unease about the militaristic intentions of Bismarck's Prussia, the British secret services saw intelligence gathering and decryption of potentially threatening signals as of the utmost importance.

Layered codes

In computer programming terms today, 'layered codes' are little different from those developed in the 19th century for use across telegraph wires, often for commercial reasons (*see page 204*). Layered codes combine a level of secretive encryption over other, often merely mechanical, encoding systems but produce a level of complexity that requires considerable ingenuity to unravel. The International Telegraph Convention regulations classified 'code telegrams' as 'those composed of words the context of which has no intelligible meaning' and 'cipher telegrams' as 'those containing series of groups of figures or letters having a secret meaning or words not to be found in a standard dictionary of the language.'

plaintext	LOADING SHIP TODAY NOON
ciphertext	NQCFKPI UJKR VQFCA PQQP
this is converted into 5-bit units	NQCFK PIUJK RVQFC APQQP
which is then converted into Morse code	-. --.- -.-. ..-. -.- .--. .. ..- .--- -.- .-. ...- --.- ..-. -.-. .- .--. --.- --.- .--.

An example of four-layered codes centers around Morse code: a secret message is encrypted using a simple Caesar Shift of say two letters; this is converted into 5-bit units, which in turn is encrypted in Morse code, a binary system which is then layered onto a physical telegraph wire.

Of course other layers of coded information – pre-agreed code words or phrases, or numeric codes – could be added to the pile of layers. In the heat of battle or commerce there was usually enough complexity to make the cost and time of decrypting unjustifiable.

MILITARY MAP CODES

Imperial rulers during the classical age recognized the necessity for order, hierarchy, and discipline to maintain effective fighting forces. Early Chinese and Roman armies wore uniforms and were divided into self-sufficient units (divisions, cohorts, or regiments) with different functions – foot soldiers, cavalry, engineers – controlled by various ranks of command and authority. Both carried flag-like emblems in the field for identification. During the Middle Ages, when armies were largely feudal, uniforms disappeared, but banners and colored surplices were widely used for identification in the field, and it was not until the emergence of large nation states in the 16th century, maintaining substantial standing armies, that the symbology of modern military organizations developed.

Visual drill
Jacob de Gheyn's *Exercise of Arms* (1607) provided an illustrated manual for weapons drill. The intention was to ensure that whole regiments could act effectively in unison when properly trained. The various stages involved in using a pike, lance, or firearm were illustrated and numbered, and being purely visual, they could be used by soldiers of any nation, irrespective of language or literacy.

Showing force

Military symbols identifying the types and sizes of military units, their disposition and deployment, and the location of features such as forts or trenches became increasingly necessary as armies grew in size and complexity and the physical scale of campaigns increased. This system also differentiates national armies by color.

Sizes of military unit

XXXXX name/number	XXXX name/number	XXX name/number	XX type	X type	III type	II type 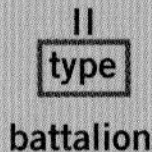
army group (name of commander to right of box)	**army/fleet/air force** (name of commander to right of box)	**corps** (number in roman numerals/name of commander to right of box)	**division** (nationality/number in arabic numerals outside the box)	**brigade** (nationality/number in arabic numerals outside the box)	**regiment**	**battalion**

Types of military unit

infantry

cavalry

artillery

armored

mechanized

airborne

airforce unit

naval troops

Military mapping

During the 18th century an internationally-recognized system for identifying forces and dispositions in the field was created. As warfare became more far-reaching and mobile, with huge armies deployed across vast areas, strategic maps needed to be swiftly and accurately updated, while campaign plans had to be distributed and understood, often by illiterate soldiers in the field, or by allied armies of different nationalities. At the battle of Waterloo in 1815, where English, Prussian, and Dutch troops massed against Napoleon's forces, clear communication of the various troop positions on the ground, and the orders they were to act upon, was vital. Here Napoleon's troops are shown in blue, his allied adversaries in red.

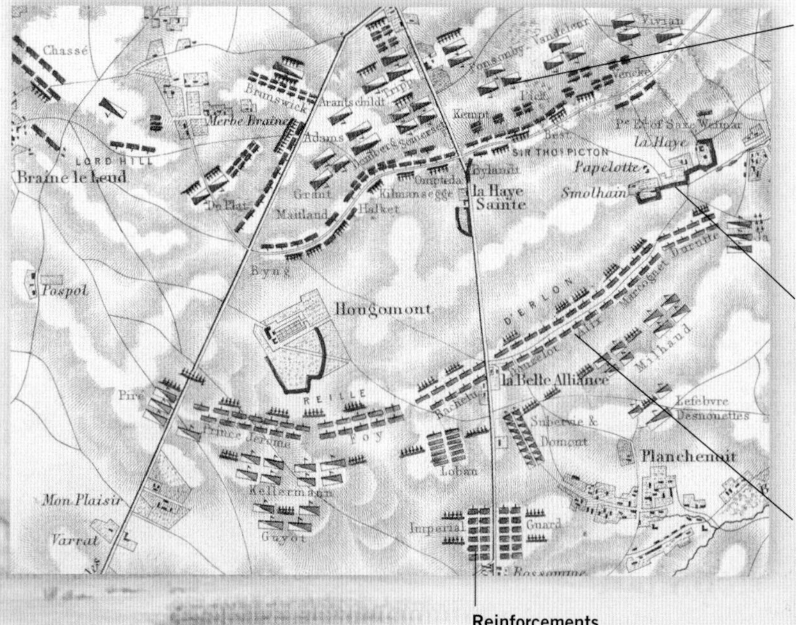

Cavalry units
Ranged behind the Allied infantry front line were a mass of cavalry detachments, ready to break through once the infantry had engaged. Modern cavalry units symbols have the diagonal reversed.

Defensive positions
The forward Allied emplacements around the hamlets at Hougomont, La Haye Sainte, and Papellotte proved to be focal points of the battle.

The French front line
The map shows huge numbers of French infantry units ready to move forward, supported by regiments of cavalry.

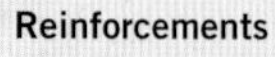

Reinforcements
The French Imperial Guard of infantry, flanked by cavalry moving up the main road from the south, joined the battle late in the day.

Examples of military units

XXXX
EIGHTH | MONTGOMERY

Eighth Army commanded by Montgomery

XXX
AFRIKA | ROMMEL

Afrika armored corps commanded by Rommel

III
fr

French artillery regiment

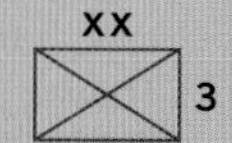

XX
3 mar

3rd Marine Division

Troop movement and location symbols

- troop movements/advances
- previous movements/advances
- troop withdrawals/retreats
- troop position
- previous troop position
- field works or trenches
- unoccupied fieldworks
- strongly defended position
- unoccupied defended position

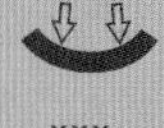

- troops in position, under attack
- boundary between units, with appropriate symbol showing size of unit
- fort or redoubt
- fortified area
- fuel pipeline
- minefield
- airborne landing

- naval base

FIELD SIGNALS

Drums and trumpets

From the earliest times, drums and horns or forms of trumpet were used in the battlefield, partly to intimidate the enemy, but more significantly as a means of communication. Massed drummers would beat out tattoos to coordinate marching and mass troop movements. Trumpeters, and later buglers, were used to send specific messages to troop deployments scattered across the battlefield.

The transformation of modern warfare from the mid-19th century, the result of new technologies such as railways and the telegraph, created a fast-moving and mobile battlefield. In turn, this produced a wide array of vital communications requirements across both short and long distances, and the need to develop new systems for transmitting messages rapidly and secretly, often from hazardous situations. By the time of the First World War, when telephonic voice communication, air power, and wireless radio telecommunications were emerging, the ability to send and receive information rapidly and efficiently had become of paramount importance both tactically and strategically.

In 1914, as the German army swept across the Low Countries and northern France towards Paris, they found the communications infrastructure of telegraph lines, railways, and telephone cables had been destroyed by the retreating French army. The invading forces had to rely on signaling systems such as semaphore and heliographs *(below)* as trenches were being dug along the Western Front. As a result, most German field signals were intercepted or read by the French and their allies.

> "Watchful for the Country."
>
> **US SIGNAL CORPS MOTTO.**

The Wig Wag code

The first dedicated signals corps was established by the US Army in 1860, shortly before the outbreak of the Civil War, under Major Albert J. Myer. The Confederate Army soon established its own. Although electric telegraphy was used, this was cumbersome in the field, and Myer adapted flag semaphore into a system using hand-held flags, or a torch or flashlamp at night, which replicated the binary Morse code (*see page 96*).

The Wig Wag or 'two element' code was eventually used by both sides during the war, and continued in use until recent times. The flags were large, ranging from six-foot (1.8 m) square to two-foot (0.6 m) square. Four-foot (1.2 m) square flags, mounted on 12-foot (3.6 m) poles, were most commonly used, with a black square on a white ground (red on white at sea).

The Union Signal Corps code

1 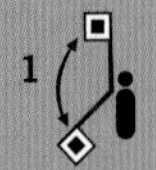
2 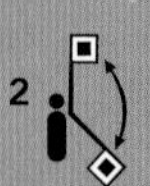
3 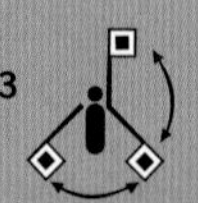
4 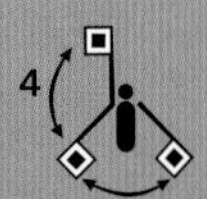
5 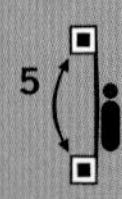

A flag waved continuously above the head from left to right attracted attention and indicated that signaling was about to start. For each number, the message sender always began and returned to holding the flag in the vertical position.

Alphabet

A	11	**Q**	2342
B	1423	**R**	142
C	234	**S**	143
D	111	**T**	1
E	23	**U**	223
F	1114	**V**	2311
G	1142	**W**	2234
H	231	**X**	1431
I	2	**Y**	222
J	2231	**Z**	1111
K	1434		
L	114		
M	2314		
N	22		
O	14		
P	2343		

Numbers

1	14223
2	23114
3	11431
4	11143
5	11114
6	23111
7	22311
8	22223
9	22342
0	11111

Special sequences of digits included:

5
End of word

55
End of sentence

555
End of message

11, 11, 11, 5
Understood

11, 11, 11, 555
Cease signaling

234, 234, 234, 5
Repeat

143434, 5
Error

It was only towards the end of World War I that the importance of a flexible, informed, and well-trained signals corps was recognized by the British Army, when the Royal Corps of Signals was established as a separate entity within the army. Hitherto each regiment maintained its own signals contingent alongside their regimental buglers, bands, and runners. However, World War I had proved that a signaler had to be able to receive and send messages by Morse code, semaphore, and heliograph, be proficient in operating radio, wireless telegraphy, and field telephones, and still be prepared, when all else failed, to run with a message across the trench lines.

'Roger, over, and out'

The unreliability of long-distance voice radio communications to ships and planes, often due to atmospheric interference on the airwaves, required a verification system when very specific instructions or details were being conveyed. Once again, it was the British who developed the first, and now largely standard, letter-by-letter confirmation code (still in use in civil aviation worldwide). This required the compilation of a lexicon of very distinct words, mainly of no more than two syllables, which were easily memorized:

A Alpha
B Bravo
C Charlie
D Delta
E Echo
F Foxtrot
G Golf
H Hotel
I India
J Juliet
K Kilo
L Lima
M Mike
N November
O Oscar
P Papa
Q Quebec
R Romeo
S Sierra
T Tango
U Uniform
V Victor
W Whisky
X X-ray
Y Yankee
Z Zebra

The system also required an indication of various parts of a message:

Roger Information received
Copy I understand what you said
Wilco Will comply, abbreviated
Over I have finished talking, am awaiting your reply
Out or **Clear** I have finished talking, no reply expected

'Roger, over and out' as a sign-off is a misleading movie cliché, as it makes no sense and is never used.

The Zimmermann Telegram

At the height of World War I, in early 1917, the Germans decided to restart their U-boat offensive against Allied shipping in order to force Britain and France to terms. However, the Germans understood that this action could provoke the United States into joining the Allied cause. In an effort to prevent this, German Foreign Secretary Arthur Zimmermann sent an encoded telegram for the Mexican premier, Venustiano Carranza, via the German Embassy in Washington D.C. He proposed an audacious plan to distract the Americans. Intercepting and decoding the telegram was a major coup for the British, as it drew the hitherto isolationist United States into the war against Germany.

The great plan
German Foreign Minister Arthur Zimmermann's scheme was to distract the United States by supporting Mexico in launching a cross-border offensive to recapture its 19th-century losses in the American Southwest. Further, the Mexicans would also invite Japan to launch an offensive against the United States across the Pacific.

The U-boat campaign against Allied shipping had been cut back in 1915 following the sinking of the British passenger liner *Lusitania*.

Pulling together a network of clues

The route of Zimmermann's telegram was his undoing: because the British had severed Germany's transatlantic cables at the beginning of the war, the telegram was sent to Washington via Sweden and Britain. The intercepted telegram was passed to Room 40, the British decoding office under the command of Captain William Reginald Hall, where the Presbyterian Reverend William Montgomery and Nigel de Grey, a peacetime publisher, were the first to attack it. The specific code used (Code 0075 in the German numbering system) had first been issued in July 1916. To add to the difficulty, it was enciphered. Room 40 had been working on the code through intercepts for about six months. However, they had some clues. Their Russian allies had captured a German codebook early in the war, having sunk the German light cruiser *Magdeburg* in the Baltic, and had shared its contents. Also, Wilhelm Wassmuss, a German agent planning to foment an anti-British rising among the Turkish tribes in Persia, had been briefly arrested in Behbahan in 1915, and his belongings had been returned to London. Room 40 now possessed the *Magdeburg* military codebook and Wassmuss's diplomatic codebook, which contained an earlier version (13040) of the cipher used in the Zimmermann telegram. But these provided only partial clues.

Wassmuss was the model for the German agent in John Buchan's spy novel *Greenmantle* (1916).

Deciphering the telegram

Arthur Zimmermann's telegram used a standard military numeric code, which relied on the receiver possessing the correct codebook for that day. Room 40 had to piece together scraps of information from the two captured codebooks.

To crack the code, Captain Hall's team in Room 40 concentrated on attempting to identify strings of consecutive numbers that appeared on the telegram (*right*).

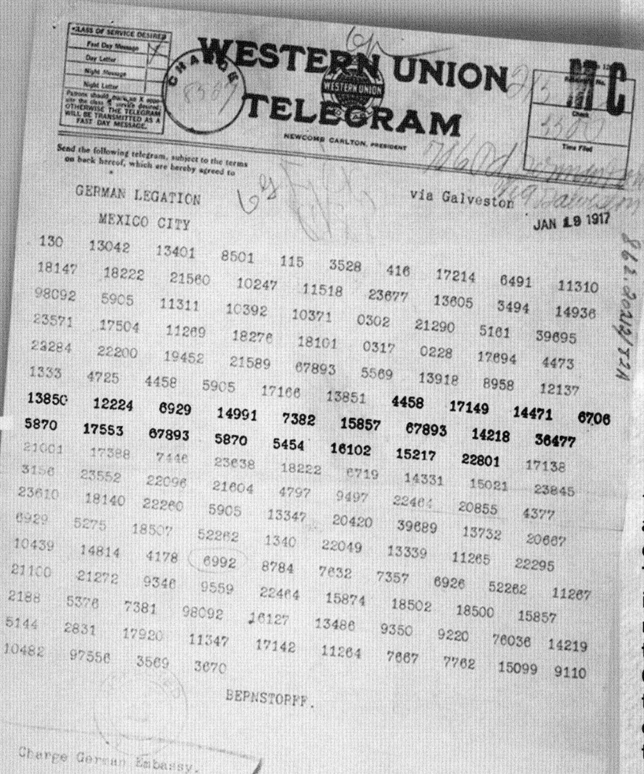

WESTERN UNION TELEGRAM

NEWCOMB CARLTON, PRESIDENT

Send the following telegram, subject to the terms on back hereof, which are hereby agreed to

GERMAN LEGATION via Galveston

MEXICO CITY JAN 19 1917

130 13042 13401 8501 115 3528 416 17214 6491 11310
18147 18222 21560 10247 11518 23677 13605 3494 14936
98092 5905 11311 10392 10371 0302 21290 5161 39695
23571 17504 11269 18276 18101 0317 0228 17694 4473
23284 22200 19452 21589 67893 5569 13918 8958 12137
1333 4725 4458 5905 17166 13851 **4458 17149 14471 6706**
13850 12224 6929 14991 7382 15857 67893 14218 36477
5870 17553 67893 5870 5454 16102 15217 22801 17138
21001 17388 7446 23638 18222 6719 14331 15021 23845
3156 23552 22096 21604 4797 9497 22464 20855 4377
23610 18140 22260 5905 13347 20420 39689 13732 20667
6929 5275 18507 52262 1340 22049 13339 11265 22295
10439 14814 4178 6992 8784 7632 7357 6926 52262 11267
21100 21272 9346 9559 22464 15874 18502 18500 15857
2188 5376 7381 98092 16127 13486 9350 9220 76036 14219
5144 2831 17920 11347 17142 11264 7667 7762 15099 9110
10482 97556 3569 3670

BERNSTORFF.

Charge German Embassy.

Code	Decrypt	Translation
4458	gemeinsam	solidarity
17149	Friedenschluß	make peace
14471	○	. (period)
6706	reichlich	generous
13850	finanziell	financial
12224	unterstützung	support
6929	und	and
14991	einverständnis	consent
7382	unsererseits	on our part
158(5)7	da/3	President?
67893	Mexico	Mexico
14218	in	in
36477	Texas	Texas
5870	ⓢ	, (comma)
17553	neu	New
67893	Mexico	Mexico
5870	ⓢ	, (comma)
5454	AR	Ar
16102	IZ	iz
15217	ON	on
22801	A	a

The scratch pad produced by de Grey and Montgomery was based on partial clues provided by the *Magdeburg* and Wassmuss codebooks. Much of their interpretation was based more on deduction and lateral thinking than cryptanalysis. Nevertheless, they decrypted enough (for example, 67893 denoted Mexico) to deduce that the telegram was of paramount importance.

TELEGRAM RECEIVED.

...tor 1-8-58
...ton, State Dept.

By Mark A. Ekhoff ...

Date Oct. 27, 1958

FROM 2nd from London # 5747.

"We intend to begin on the first of February unrestricted submarine warfare. We shall endeavor in spite of this to keep the United States of America neutral. In the event of this not succeeding, we make Mexico a proposal of alliance on the following basis: **make war together, make peace together, generous financial support and an understanding on our part that Mexico is to reconquer the lost territory in Texas, New Mexico, and Arizona.** The settlement in detail is left to you. You will inform the President of the above most secretly as soon as the outbreak of war with the United States of America is certain and add the suggestion that he should, on his own initiative, invite Japan to immediate adherence and at the same time mediate between Japan and ourselves. Please call the President's attention to the fact that the ruthless employment of our submarines now offers the prospect of compelling England in a few months to make peace." Signed, ZIMMERMANN.

The string of numbers Room 40 focused on revealed sufficient clues to enable the rest of the telegram to be unraveled and translated. It spelled out in detail Zimmermann's devious plan – a plan that was soon to backfire.

The decrypted telegram

The translation of the telegram presented a dilemma to the code breakers: how could they tell the Americans about it without revealing to the Germans that they had cracked their latest code? Captain Hall realized that the German embassy must have sent the telegram across public telegraph lines from Washington to Mexico, so a British agent in Mexico City was dispatched to steal a copy. To the delight of Room 40, it was encoded not in the newer Code 0075, but in the older 13040, so the Germans would assume that the British had learned of the plot using the stolen telegram and the captured codebook. Now the British could keep their secret and send a full version of the telegram to the Americans. Isolationist US President Woodrow Wilson received the decrypted telegram on February 25. It was published on March 1, and the US declared war on Germany on April 6, 1917.

The British satirical magazine *Punch* celebrated this with John Bull greeting Wilson: "Bravo Sir! Glad to have you on our side!"

ENIGMA: The 'unbreakable' system

Inventing Enigma
The Enigma machine was first patented in 1918 by Arthur Scherbius (1878-1929) for commercial use, but it soon attracted the attention of the German military. Over the next decade the encoding system was gradually made more sophisticated.

The German military recognized the need for a more secure enciphering system in 1923, after British official histories of World War I revealed that German messages had been read. They eventually acquired over 30,000 Enigma machines, with a more complex design than those available commercially. The Wehrmacht, Luftwaffe, and German Navy all issued separate daycode books throughout World War II. The beauty of the Enigma machine's mechanical enciphering system was that it was very fast and all but eliminated human error – the plaintext was typed in to produce the enciphered text, this was transmitted by radio, and the receiver merely typed in the coded message, and the machine produced the decoded plaintext. In addition, without access to the daycode settings it was almost impregnable.

Reflector
This did not rotate, thus ensuring that encrypted text was automatically sent back through the scrambler disks, mechanically producing the decrypted text as it was typed in.

Scrambler disks
Each contain the 26 letters of the alphabet, and were set in any start position from A-Z (determined by the daycode). They were geared to rotate cyclically. From 1938 the machines had five scrambler disks.

Plugboard
Originally, you could swap only six letters before the plaintext reached the scramblers, but in 1939, an enlarged plugboard increased this number to ten.

Keyboard
For typing in plaintext (or received encrypted text).

Lampboard
Shows the operator the encryption (or decryption) of each letter when it has been typed in.

Each disk has 26 contacts on each face (which correspond to letters of the alphabet) wired to 26 different contacts on its opposite face. Each numbered disk would be wired differently.

The portability of the Enigma machine was a huge advantage. One is seen here in use on General Heinz Guderian's half-track on the battlefield.

The daykey settings

Each month the German military would issue a new daycode book. This listed the individual settings operators were to use each day to set up all Enigma machines within each respective military unit. This ensured that the first message sent could be read by all members of the unit.

Setting up Enigma

Following the daycode setting, every morning the operators would: re-order the scrambler disks; adjust the scrambler orientation (which letter of the alphabet each scrambler should display at the start of the day); and change the plugboard settings. The systems combined meant a total of 10,000,000,000,000,000 calculations would have to be made to analyze the encryption.

Resetting the settings

During World War II, in order to increase the level of security, the Enigma operator would send an initial message, using the daycode settings, which would be a new setting for the scramblers. This would be repeated to ensure consistency. Thus, if the daykey required B–M–Q, a second signal might be preceded by a randomly chosen combination of three letters, for example, S–T–P–S–T–P, requiring the receiver to alter his scrambler settings accordingly.

Using the Enigma machine

Plaintext was typed in to produce the enciphered text, this was transmitted by radio, and the receiver merely typed in the coded message, and the machine produced the decoded plaintext.

Encryption

This schematic diagram follows the letter impulse for U, showing its passage to encryption as S. For the purposes of clarity, only four of the available switches on the plugboard have been set.

1 The operator types in plaintext, which is transmitted through the machine by electric current.

2 Letters that have been switched on the plugboard are first enciphered here. Remaining letters go straight to the first scrambler.

3 Passing beyond the plugboard, the letter impulses travel to and enter the first scrambler disk.

4 The arriving letter impulse passes through the disk to a different exit point, and thus a different letter entry point on the next scrambler. In addition, the first scrambler rotates by one notch with every letter that is typed.

5 The process is repeated at the second scrambler which turns one notch (or letter) once the first scrambler has completed its cycle of 26 letters.

6 The process is repeated at the third scrambler which turns one notch (or letter) once the second scrambler has completed its cycle of 26 letters.

7 Each letter impulse now reaches the reflector, which passes it back through the scrambler disks via a different path.

8 The impulse travels back through the plugboard and arrives at the lampboard, where the final encryption is displayed to the operator.

Reflector

Scrambler Disks

Entry Wheel

Q W E R T Z U I O
A S D F G H J K
P Y X C V B N M L
Lampboard

Q W E R T Z U I O
A S D F G H J K
P Y X C V B N M L
Keyboard

Q W E R T Z U I O
A S D F G H J K
P Y X C V B N M L
Plugboard

Decryption

Having set the machine using the same daycode settings as the encrypting operator, the receiving operator types in the received encrypted text. The letter impulses pass through the plugboard, the scramblers, and the reflector, and then returns through the system to be displayed, decrypted to plaintext on the lampboard.

WW II Codes and Code Breakers

The German Enigma machine was the most famous example of a number of electromechanical rotor enciphering devices that were used on all sides during World War II. The British relied on a similar machine called Type X (or Typex), while the Americans developed the more sophisticated SIGABA (or M-134-C) cipher machine. The Japanese cipher machine produced a code called PURPLE, which was cracked by June 1942. The ciphertexts produced by all these machines were deemed impregnable due to the sheer number of calculations involved to break a coding system that was being used daily, hourly, indeed every minute, to convey signals. But in this lay the key: it was human error, or laziness, that often proved to be the chink in the armor. The story of Enigma, and how it was cracked, is told on the following pages but here, as so often in modern warfare, code words came to play an important role. The intelligence received from the Enigma decrypts was in turn called ULTRA and, as with the Zimmermann telegram, the problem with the ULTRA intelligence was how much could be acted upon without giving away to the Germans that their secret messages were being decrypted and read.

Operational code names

A superficial level of security was imposed by most participants in WWII by assigning a code name to almost any operation or initiative.

***Adlertag* (Eagle Day)** Luftwaffe assault resulting in the Battle of Britain, 1940.

Attila Nazi occupation of France, 1940–1942 (later renamed *Anton*).

August Storm Soviet invasion of Manchuria, 1945.

Avalanche Allied landings at Salerno, 1943.

Avonmouth Allied expedition to Narvik, 1940.

Bagration Soviet offensive to liberate Byelorussia, 1944.

Barbarossa German invasion of the Soviet Union, 1941.

Cartwheel Allied combined operations in the Southwest Pacific, 1943.

Gomorrah RAF bomber raids on Hamburg, 1943.

Ichi-Go Japanese offensive in China, 1944.

I-Go Japanese counter-offensive in the Southwest Pacific, 1943.

Lightfoot Battle of El Alamein, 1942.

Market-Garden Allied airborne landings near Arnhem, 1944.

***Nordlicht* (Northern Lights)** German offensive against Leningrad, 1942.

Overlord Allied invasion of Normandy, 1944.

Steinbock German air raids against British cities, 1944.

Torch Allied coastal landings in French North Africa, 1942.

***Weiss* (White)** German invasion of Poland, 1939.

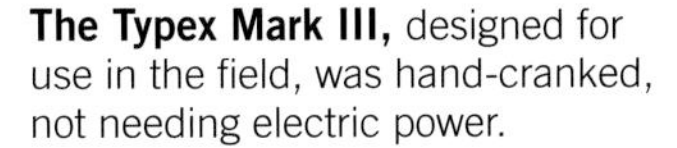
The Typex Mark III, designed for use in the field, was hand-cranked, not needing electric power.

The US SIGABA More sophisticated than Typex, from 1943 the machines shared messaging, forming the Combined Cipher Machine (CCM).

Bletchley Park

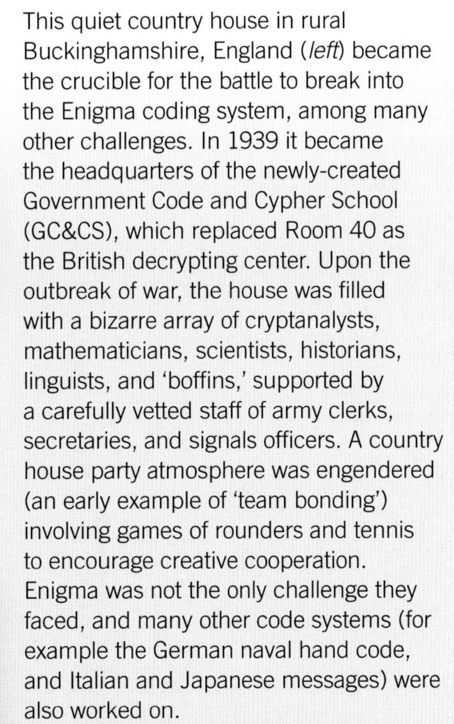

This quiet country house in rural Buckinghamshire, England (*left*) became the crucible for the battle to break into the Enigma coding system, among many other challenges. In 1939 it became the headquarters of the newly-created Government Code and Cypher School (GC&CS), which replaced Room 40 as the British decrypting center. Upon the outbreak of war, the house was filled with a bizarre array of cryptanalysts, mathematicians, scientists, historians, linguists, and 'boffins,' supported by a carefully vetted staff of army clerks, secretaries, and signals officers. A country house party atmosphere was engendered (an early example of 'team bonding') involving games of rounders and tennis to encourage creative cooperation. Enigma was not the only challenge they faced, and many other code systems (for example the German naval hand code, and Italian and Japanese messages) were also worked on.

Numerous huts were erected in the grounds (*above left*), each having a specific function on a 'need to know' basis, so that only a few senior staff had overall knowledge of what material had been intercepted and decrypted. Churchill was to refer to the staff as the "geese who laid golden eggs and never cackled."

The Manhattan Project

The single most important, and secret, undertaking of the war was the rapid development of a nuclear 'device.' Undertaken by the US with support from Britain, and a cosmopolitan roster of personnel under the control of the American physics genius J. Robert Oppenheimer, the project was given the code name 'Manhattan' – derived from the US Army Corps of Engineers, Manhattan Engineer District, New York City (MED), under whose auspices the project would be developed. Based at Los Alamos, in the inhospitable wastes of the Sangre de Cristo mountains in New Mexico, this top-secret undertaking was enshrouded in mysterious code names.

509th Composite Group Part of the 20th USAAF, using specially-adapted B-29 Superfortresses, designated as the delivery wing for the Manhattan Project.
Alberta The team who assembled the bombs at Tinian Island in the Pacific.
ALSOS Allied covert missions into occupied Europe to kidnap nuclear scientists and raw materials, such as uranium. There were three in all.
Bockscar The plane that delivered the Nagasaki atom bomb on August 9, 1945.
Box 1663 The Santa Fe postcode used by all involved on the project.
Enola Gay The plane that dropped the first atom bomb on Hiroshima on August 6, 1945, named after the pilot Col. Paul Tibbets' mother.
Fat Man The atom bomb detonated over Nagasaki on August 9, 1945.
Fission Coined by Otto Frisch for the splitting of atoms by bombarding them with neutrons.
Gadgets The general term used at Los Alamos for the nuclear devices they were developing.
Little Boy The atom bomb dropped on Hiroshima on August 6, 1945.
Site-Y The Los Alamos laboratory, also called 'The Hill' by locals.
Trinity The site of the first nuclear test, July 16, 1945.

Cracking Enigma

The German challenge

Upon the outbreak of World War II, Allied cryptographers were confronted by an awesome problem. The Enigma system (*see page 116*) had many variations. In addition to its existing complexity, in 1938 the Germans added a further two scrambler disks to many machines, and the plugboard was made more complex. There were also variants on the machines used by different parts of the German military, and each had different codebooks. The Afrika Corps used its own system, as did the *Kriegsmarine*, the German navy. It was the latter's Enigma signals (the Lorenz cipher) that were the most difficult to penetrate and the most vital for Bletchley Park to decrypt, as U-boat activity in the North Atlantic threatened to sever lifeline supplies from North America.

The German addition of two extra scrambler disks meant that their invasion of Poland in September 1939 was a surprise.

Since its introduction by the German military, it had been assumed by everyone that the Enigma system (*see page 116*) was unbreakable. Although versions of the commercial machine had been acquired by Germany's former adversaries, the workings of the military machine and the codebooks were unknown. But, in 1931 the French secret service bought copies of plans of the machine and daycode books from a disaffected German veteran, Hans-Thilo Schmidt, who continued to supply details of the daycode books for several years. The French made little of them. It was Poland that opened the door.

Marian Rejewski (1905-80), the man who cracked Enigma.

Poland fights back

In the 1930s, aware of German designs on their territory, the Polish cryptanalytical bureau, Biuro Szyfrów, prioritized breaking the Enigma coding system. An entente with France meant that much of the Enigma material was handed over to the Poles, who set about building replica machines. The Poles realized that Enigma was a mechanical system that required mathematical rather than linguistic skills to analyze. It was an inspired idea. Recent Polish history provided several mathematicians from the formerly German-occupied parts of Poland, who were familiar with the language. Among them was Marian Rejewski.

First letter	Fourth letter
A	U
B	M
C	X
D	N
E	C
F	B
G	V
H	Q
I	P
J	W
K	O
L	E
M	I
N	R
O	Z
P	T
Q	Y
R	L
S	A
T	K
U	S
V	J
W	D
X	H
Y	F
Z	G

1 The message key

Rejewski concentrated on the initial Enigma three-letter message key, sent twice at the beginning of each transmission. Realizing that with only three scrambler disks, every fourth letter must represent a different encryption of the first letter, he found a chink in Enigma's armor. He still had no idea of the daykey, but he started to look for links, or chains, of substitution. With access to enough messages in a day, he could build tables of relationships between the first and fourth, second and fifth, third and sixth letters of the message key.

2 Chains

By analyzing these tables he identified chains, that is, how many links there were before the first letter linked back to itself, in this instance A-U, U-S, S-A – three links (*left*). Rejewski realized that, while the plugboard settings were indefinable, the number of links in each chain was a reflection of the scrambler settings. Some chains were long, some short. Rejewski and his colleagues spent a year compiling tables of all the possible 105,456 scrambler settings, correlating them to the length of potential chains. Later, as the Germans changed their protocols, making his tables redundant, Rejewski developed electronic calculators called 'bombes' to recompile the tables.

3 Plugboard

The tables unlocked the scrambler settings, but not the plugboard settings. However, decrypting what they could using the scrambler setting tables, frequently a recognizable message might appear:

SONVOYC ON SOURCE

It is clear that the 's' and the 'c' might have been switched on the plugboard which, when adjusted, would read:

CONVOYS ON COURSE

Success

Rejewski's breakthrough enabled Poland to read Enigma signals for most of the 1930s. The addition of two further scramblers and an extended plugboard in 1938 set them back. A month before the German invasion of Poland in September 1939, the Poles managed to convey two replica Enigma machines, plans for the 'bombes,' and Rejewski's analysis to Britain.

Alan Turing at Bletchley Park

A gifted young mathematician at Cambridge, Alan Turing (1912-54) was among the mixed bag of recruits for the new British cryptanalysis center at Bletchley Park (*see page 118*). He had been working on binary mathematics and theoretically programmable computers and, confronted by what had been achieved in Poland (*left*), he set about designing an improved series of 'bombes' to analyze the newly increased scrambler settings of the Enigma machine. As the Enigma machine settings were altered at midnight every night, they had to work quickly. Nevertheless, the possible settings would be too numerous to work through in the time available without the aid of some further clues, some of which had been identified before Turing arrived in 1939.

Alan Turing, whose 'Turing machines' helped to unravel Enigma.

'Cillies' – Human error and laziness by some Enigma operators led them to use repeated message-key combinations instead of entirely random ones. Once identified these gave the cryptanalysts a useful clue, and signals from those operators were monitored.

Scrambler codes – The Germans assumed that ensuring no scrambler disk occupied the same position on consecutive days would make the system more secure. In fact, it made it weaker, as once one or two of the scrambler positions had been ascertained, it reduced the remaining potential combinations, while also reducing the possible combinations for the following day.

'Cribs' – Identifying known words in a message, a 'crib,' could help unravel the settings. Certain sorts of signal were predictable and formulaic, for example those from weather stations, often beginning with or containing the German word *wetter*. Such signals were monitored, and educated guesses made at identifying words of this sort. Another crib was to lay mines at a specific location, then try to find evidence of the known geographical coordinates in U-boat messages.

'Pinches' – The acquisition of German codebooks was a priority. During the Battle of the Atlantic both U-boats and weather ships were raided, codebooks captured, and the vessels sunk to avoid alerting the Germans to their loss.

Loops – Turing also worked on the problem of what might happen if the Germans stopped repeating the message key. He focused on the archive of decrypts, and began to detect a pattern of 'loops,' not dissimilar to Rejewski's 'chains,' which potentially revealed the scrambler settings if the plaintext was known or a 'crib' guessed. He had discovered another shortcut.

Many machines – If he organized enough 'bombes' working in sequence, each one imitating the action of a different scrambler disk, Turing reckoned he might stand a chance of churning through the 17,576 various possible settings in a short period, but he still required a mechanical shortcut. This he achieved by linking the sequenced machines together, and establishing circuits between them which revealed a matched loop by lighting a bulb on the circuit.

The plugboard problem – Like Rejewski, by setting aside the plugboard problem, Turing had minimized it. With an accurate 'crib' a decrypted word might appear with some odd letters in it which, when transposed, revealed the plugboard settings.

Jumbo and Colossus

Turing's plans for 'bombes' running in series, interconnected, and wired to reveal loops were approved, and £100,000 allocated to create them. Each 'bombe' consisted of 12 sets of replica scramblers, and the first, named 'Victory,' was in operation by March 1940. As the prototype was being tested and improved, the Germans changed their message key protocol, causing a decrypt blackout. An improved 'bombe' was in place by August and, by Spring 1942, 15 more 'bombes' were in place, running through cribs, scrambler settings, and message keys at an industrial rate. On a good day, the system could decipher all these within an hour, revealing the encryption behind the rest of the day's signals. By the end of the war, some 200 'bombes' were in operation. Nevertheless, the entire process still relied on accurate cribs, so human ingenuity still propelled the mechanical system.

Eventually the number of 'bombes' and the links between them created the world's first programmable computer, code-named 'Jumbo,' but known among the operators as 'Heath Robinson.' In 1942 Turing developed a further shortcut for decrypting the German naval Lorenz cipher used in an adapted Enigma machine, the *Geheimschreiber*, and passed on his ideas to Tommy Flowers and Max Newman, who went on to develop the Colossus computer, a more integrated programmable digital device, the true forebear of the modern computer.

In July 1942 Turing traveled to America to share his ideas with US cryptanalysts. Although the British shared their secrets with their Western allies, and Bletchley Park was involved in decrypting Italian and Japanese codes as well, the story of Enigma and its decryption was to remain secret until the 1970s.

Colossus in operation at Bletchley Park.

NAVAJO WINDTALKERS

With the Japanese attack on the US naval base at Pearl Harbor on December 7, 1941, the USA entered World War II. Although described as a 'surprise' attack, in fact the Americans had been decrypting coded Japanese messages for some years, and suspected something of the sort was imminent. In turn, the Japanese too had been reading US signals, allowing them to plan their campaign with clockwork efficiency. Within two months the Japanese had overrun key Allied bases in the Western Pacific, and invaded the Philippines, the East Indies, and Malaya, creating the largest battlefront and zone of occupation in world history. In preparation for the inevitable US counteroffensive across the Pacific, a top priority was ensuring their communication systems were secure.

Linguistic codes

The idea of using obscure languages as a form of code was not new. Julius Caesar had enciphered messages not in Latin, but in Greek, familiar to educated Romans, but not to their adversaries. During World War I, eight members of the Choctaw tribe communicated across trench telephones for the US Army's 36th Division in France. Early in World War II, the US Army used Basque speakers, although it was known that Basque missionaries had existed in the territories conquered by the Japanese. Also Basque speakers were in short supply. The British also experimented briefly with Welsh-speaking signals personnel. Native American languages had been considered before, but many of them had been studied by German anthropologists; for this reason the Navajo system was not used widely in the European theater (although 14 Comanche speakers were used during the D-Day operations in June 1944). The Japanese never cracked the Navajo spoken code system, and its central contribution to American success in the Western Pacific is beyond doubt.

Combat situations

One of the problems with the electromechanical US Army coding system SIGABA (or M-134-C, *see page 118*) was that, like the German Enigma system, it was relatively cumbersome, involving detailed encryption via a keyboard, checking daykey settings, and laborious decryption of the cipher message. It was secure and suitable for high-level strategic signals, but for rapid communication in the heat of battle this could hinder flexibility. In the Pacific theater, where immediate coordination between air, sea, and amphibious landing troops was essential, a more practical but still secure system was needed.

Creation of the Windtalkers

Early in 1942, a civilian engineer keen to help the war effort proposed the idea of using Navajo speakers to convey field messages by radio telecommunications. Philip Johnston had been raised in a Christian mission in a Navajo reservation, and understood the language of this Native American people. He proposed a trial to the US Marine Corps.

Several advantages, but several problems, confronted supporters of Johnston's scheme. The Navajo were not a particularly literate people, and a lack of federal funding had hindered education; finding and training Navajo operators might be difficult. On the other hand, Navajo was a largely oral tradition, with language, tales, and myths committed to memory. Also, one of the key advantages of Navajo was that intonation (as, oddly, in Japanese) could change the meaning of a word: for example, the word 'doo' with a high tone means 'and,' but with a low tone means 'not.' It was also rich in imagery, which could be readily adapted to military purposes.

Some 420 Navajo were trained in signals technology and incorporated into US Marine divisions during the war. The technique was also used in the Korean War (1950-53) and in the early years of the US engagement in Vietnam. The Windtalkers and their code remained an unacknowledged secret until declassified in 1968.

The Navajo Code

The Navajo language enjoyed a wealth of idiomatic forms and nuances, and lent itself to the creation of a suitably obscure vocabulary of terms, and one that could be memorized – avoiding the necessity for codebooks. An initial lexicon totaling 274 existing Navajo terms which stood for military expressions was rapidly drawn up (and a further 234 terms added later), as was a phonetic alphabetic vocabulary for spelling out military terminology and place names.

A Navajo lexicon

Fighter plane	Hummingbird
Observation plane	Owl
Torpedo plane	Swallow
Bomber	Buzzard
Dive-bomber	Chicken hawk
Bombs	Eggs
Amphibious vehicle	Frog
Battleship	Whale
Destroyer	Shark
Submarine	Iron fish
Grenade	Potato
Tank	Tortoise
Rolled hat	Australia
Bounded by water	Great Britain
Braided hair	China
Iron hat	Germany
Floating land	The Philippines

Navajo alphabet code

Letter	Word	Navajo
A	Ant	*Wol-la-chee*
B	Bear	*Shush*
C	Cat	*Moasi*
D	Deer	*Be*
E	Elk	*Dzeh*
F	Fox	*Ma-e*
G	Goat	*Klizzie*
H	Horse	*Lin*
I	Ice	*Tkin*
J	Jackass	*Tkele-cho-gi*
K	Kid	*Klizzie-yazzi*
L	Lamb	*Dibeh-yazzi*
M	Mouse	*Na-astso-si*
N	Nut	*Nesh-chee*
O	Owl	*Ne-as-jah*
P	Pig	*Bi-sodh*
Q	Quiver	*Ca-yeilth*
R	Rabbit	*Gah*
S	Sheep	*Dibeh*
T	Turkey	*Than-zie*
U	Ute	*No-ad-ih*
V	Victor	*A-keh-di-glini*
W	Weasel	*Gloe-ih*
X	Cross	*Al-an-as-dzoh*
Y	Yucca	*Tsah-as-zih*
Z	Zinc	*Besah-do-gliz*

Nevertheless, this alphabet was susceptible to frequency analysis, so two further Navajo words were added to the most common letters (e,t,a,o,i,n) to be used as substitutes, or homophones, on a cyclical basis, and one extra homophone word added to the next most common letters (s,h,r,d,l,u). This allowed a term such as Marianas to be spelt out using three separate homophones for the letter 'a.'

COLD WAR CODES

The freezing of relations between the USA and the USSR, the two superpowers that emerged from the ruins of World War II, led to an escalation of suspicion, secrecy, and nuclear arms stockpiling as mutual distrust developed into almost naked aggression. The activities of the secret services on all fronts cost many lives, but also captured the public imagination in the spy fictions of Graham Greene, Ian Fleming, Richard Condon, John le Carré, and many others. The era was filled with actual and fanciful acronyms, CIA, FBI, MI6, 007, SMERSH, and SPECTRE, but the code word that chilled all who saw it, the acronym MAD, stood for 'Mutually Assured Destruction,' the predicted outcome of a nuclear holocaust – regardless of who pressed the 'launch' button.

Ban the Bomb

One of the most famous symbols of the Cold War era was the 'Ban the Bomb' or 'Nuclear Disarmament' sign. Designed by British artist Gerald Holtom for the Direct Action Committee Against Nuclear WAR (DAC), it made its first appearance on the Aldermaston anti-nuclear march in April 1958, and was adopted by the Campaign for Nuclear Disarmament (CND). Holton claimed he was inspired by the pose of the doomed guerrilla in Goya's *Third of May 1808* (1814, *above*), but rationalized it by combining the semaphore signs for N (nuclear) and D (disarmament) within a circle.

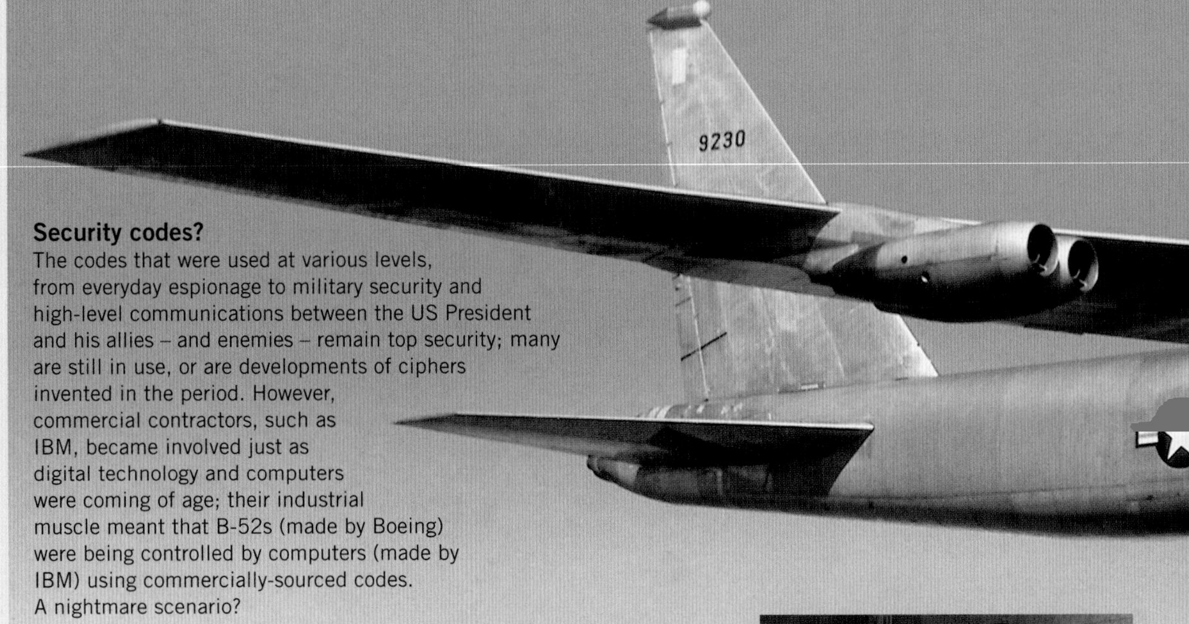

Security codes?

The codes that were used at various levels, from everyday espionage to military security and high-level communications between the US President and his allies – and enemies – remain top security; many are still in use, or are developments of ciphers invented in the period. However, commercial contractors, such as IBM, became involved just as digital technology and computers were coming of age; their industrial muscle meant that B-52s (made by Boeing) were being controlled by computers (made by IBM) using commercially-sourced codes. A nightmare scenario?

The Venona project

At an early stage of World War II the Western Allies began monitoring Soviet signals. Although encrypted and transmitted using the almost impregnable one-time pad system (*see page 83*), the signals were collated and distributed to a limited circle of US cryptanalysts. It was not until 1946 that any of the messages were decrypted, due to the Russians occasionally reusing the same one-time pads, although after 1945 this practice ceased. Nevertheless, a large amount of information was gathered from successful decrypts, giving a valuable insight into the workings of the Soviet military and intelligence, including information concerning Soviet collaborators in the West. Although this was so sensitive that the CIA and the White House were only fed partial briefs by the FBI, the intelligence named some 349 Americans; revealed that the Soviets had penetrated the Manhattan project (*see page 119*); contributed to the conviction of the spies Julius and Ethel Rosenberg; unmasked the UK Cambridge spies Donald Maclean and Guy Burgess just as they defected; and contributed to the cases against Alger Hiss and Harry Dexter White. The Venona project was wound down in 1980.

The Rosenbergs were tried, convicted, and controversially executed as spies based largely on Venona project evidence.

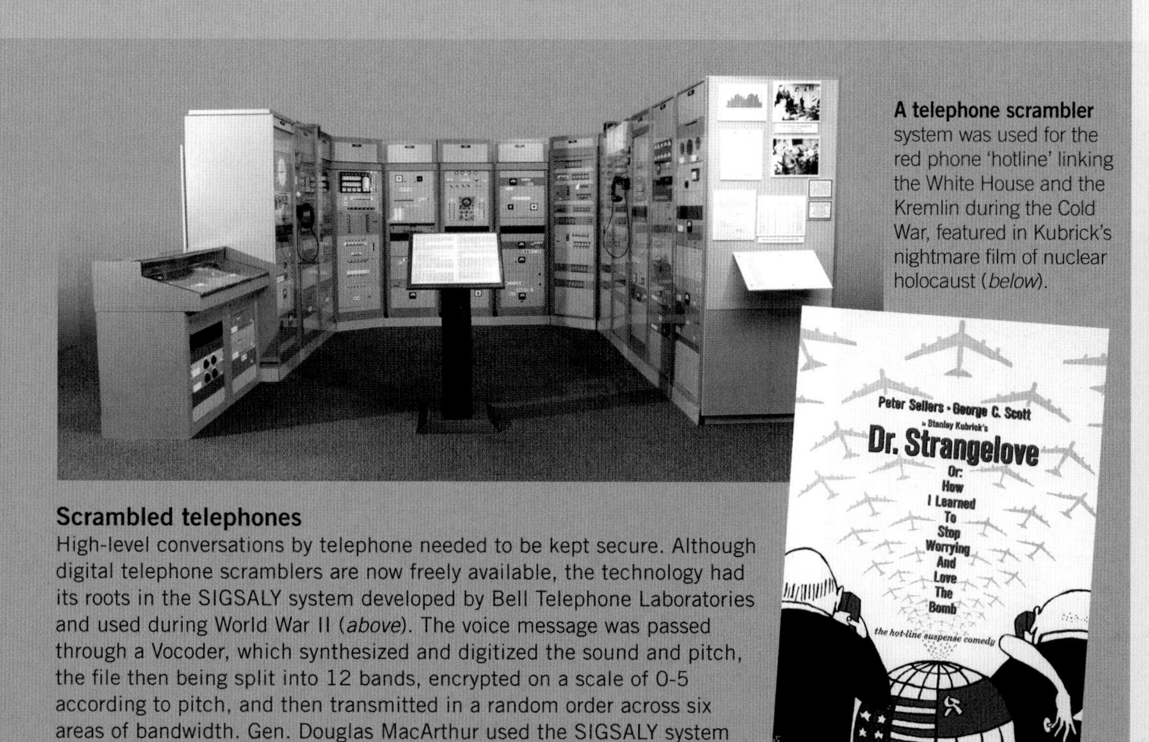

A telephone scrambler system was used for the red phone 'hotline' linking the White House and the Kremlin during the Cold War, featured in Kubrick's nightmare film of nuclear holocaust (*below*).

Scrambled telephones

High-level conversations by telephone needed to be kept secure. Although digital telephone scramblers are now freely available, the technology had its roots in the SIGSALY system developed by Bell Telephone Laboratories and used during World War II (*above*). The voice message was passed through a Vocoder, which synthesized and digitized the sound and pitch, the file then being split into 12 bands, encrypted on a scale of 0-5 according to pitch, and then transmitted in a random order across six areas of bandwidth. Gen. Douglas MacArthur used the SIGSALY system through much of the Pacific campaign, and over 3,000 of his phone calls were known to have been successfully scrambled.

The 'nuclear football'

A metallic black briefcase carried in a black bag, attached by a cord to the US President's military aide, could determine the fate of the world. It is known as the 'nuclear football.' It constitutes a mobile part of the US strategic defense system, contains a SATCOM radio and handset, a number of attack scenarios (the 'playbook'), and provisional plans in case of a national nuclear emergency. It accompanies the US President at all times, although the key to launching a nuclear strike, the Gold Code, which is revised daily, is not contained in the briefcase, but is always on the President's person (it is said that on one occasion a former Gold Code was found at a dry cleaner's in one of President Jimmy Carter's suits). In the event of an emergency or attack, the military aide and the President would open the briefcase, review the plans and, if necessary, use the radio handset to authorize a nuclear response using the Gold Code. The Russian premier is also accompanied by a similar briefcase.

The US Presidential military aide is in constant proximity to the President, and is constantly attached to the 'nuclear football.'

Surveillance field signals

Surveillance of possible spies became an almost obsessive activity during the Cold War. While telephone taps and undercover photography could assemble much evidence against suspects, tracking their actions often proved as useful. For agents in the field covertly tailing a suspect on the city streets, a code of body signals was devised, first used by the police and the FBI and then modified by the CIA. Of course, remembering not to do any of these things (if inappropriate) is important.

Watch out! Subject approaching Touch nose with hand or handkerchief.
Subject moving on, going further, or overtaking Stroke hair with hand, or raise hat briefly.
Subject standing still One hand against back, or on stomach.
Observing agent wishes to terminate observation because cover threatened Bend to retie shoelaces.
Subject returning Both hands on back, or on stomach.
Observing agent wishes to speak with team leader or other observing agents Open briefcase and examine contents.

06

Groups that find themselves, or place themselves, on the margins of society often develop secret communication systems to preserve their privacy.

codes of the underworld

The use of obscure symbols and private languages, argots, and cants are common among closed or criminal communities. Many of the latter have seeped into everyday speech, enriching the host language. Some remain impenetrable except to the initiated, and almost daily it seems that new cult groups invent fresh and bewildering languages and sign systems recognized only by those 'in the know.'

STREET SLANGS

Slang is any use of informal words or phrases that are not usual in either the speaker's language, or their particular dialect; specific slang is limited to certain groups or sub-cultures. An extension of slang is cryptolectic language, also known as 'argot' or 'cant' (*see also pages 132, 136*) These secret languages are primarily used to obscure and disguise the communication of a specific group or culture from society at large. In the 21st century, technology can easily be used to encrypt our secrets, but when speech was our primary means of communication (not everyone was able to read or write) secret languages were of greater importance. Concealment is not the only consequence of cants and argots: they also create a sense of unity among their speakers.

The language of the travelers

Shelta (also known as 'Sheldru,' 'Gammen,' or 'Cant') is thought to date from the 13th century; it is a cant based on both Irish and English with a mainly English-based structure (syntax), and shares some similarities with Roma or Romani (the language of the Romany gypsies). Like the Thieves' Cant, Shelta was (and is, as 86,000 people still speak Shelta globally) primarily used by travelers *(above)* to hide their meaning from non-travelers. Spoken Shelta can often be mistaken for garbled Irish, adding an extra dimension of secrecy to this cant.

Dorahoag	Twilight
Greetchyath	Sickness
Kawb	Cabbage
Myena	Yesterday
Sragaasta	Breakfast
Sreedug	Kingdom
Swurkin	Melody

Thieves' Cant

The 'idle poor' were a problem in European society in the 16th and 17th centuries: many became vagabonds and turned to crime. This criminal contingency created their own language in order to keep their less than legal activities hidden from the rest of society. In England around 10,000 of the country's four million population are thought to have spoken it. The language developed by this underground collective is often called 'Thieves' Cant'; Shakespeare put it in the mouths of his fools and lowlifes from *As You Like It*'s Touchstone to Autolycus of *The Winter's Tale*. Some of the vocabulary of these Elizabethan thieves lingers on in the language of modern criminal factions (*see page 134*).

Lurid accounts of highway robbery and other foul deeds *(left)* were related in the *Newgate Calendar*, published monthly in the 18th and early 19th centuries.

You've got to pick a pocket or two

Each criminal group developed its variation of Thieves' Cant, such as pickpockets.

Bung The targeted purse.
Cuttle-bung A pickpocket's knife (for cutting purses).
Drawing Taking the purse.
Figging Pickpocketing.
Foin A pickpocket.
Nip A cutpurse (someone who steals by cutting the strings of the victim's purse).
Shells The money in the purse.
Smoking Spying on the victim.
Snap A pickpocket's accomplice.
Stale An accomplice who distracts the victim.
Striking The act of pickpocketing.

Stand and deliver

Highway robbery too had its own terminology.

High-lawyer A highwayman.
Martin The highwayman's target.
Oak The accomplice who stands on watch.
Scrippet An accomplice who sets the watch.
Stooping When the victim yields to the highwayman.

The argot of Klezmer

Klezmer-loshn is Yiddish for 'musician's tongue.' It is based on the language of the Ashkenazi Jews of central and eastern Europe. Klezmer-loshn was spoken by traveling Klezmer musicians *(above)* and is known as a 'professional argot' – the secret language of a particular profession. Klezmer music, the music of the Ashkenazim, dates back to around the 15th century.

Geshvin Quickly
Katerukhe Cap
Klive Beautiful
Shekhte Woman
Shtetl Village
Tirn Chat
Yold Husband
Zikres Eyes

From Samurai to Yakuza

The *Bushido* code: the Seven Virtues

Courage

Right Conduct

Mercy

Loyalty

Honor

Respect

信

Honesty

These virtues were the keystones of the *samurai* warrior code, (and essentially the same as the US Army 'Core Values' adopted in the mid-1990s). Of many Japanese works on the subject, the best known in the West is the *Bushido Shoshinshu – Code of the Samurai* written by Taira Shigesuke, a *samurai* and military strategist of the early 18th century. It remains an excellent guide to the mindset of modern, and particularly corporate, Japan, and especially the deep-rooted concepts of *giri* – obligation which can extend as far as blood vengeance – and *ninjo* – the ability to feel compassion.

The ideas behind the concept of the *samurai* go back at least 1,000 years in Japan and are based on Confucian ethics, modified for a predominantly martial world. *Bushido* – the Way of the Warrior – was the code by which, ideally, the *samurai* lived and died. The *samurai* formed a powerful and prestigious section of Japanese society for centuries. But, from around 1600, the reforms of the Tokugawa shogunate reduced the opportunities for battle; peace and prosperity led to the rise of merchant classes, and the warriors found themselves increasingly marginalized. Finally, the Meiji reforms of 1868 swept away the feudal world. Many *samurai* were deeply resentful at what they felt was a betrayal of their way of life and the true nature of Japan. Nevertheless, the *samurai* provided a model for several more recent Japanese organizations and institutions, not least the notorious *yakuza*.

The *samurai* were the military elite, retainers of a feudal lord or *daimyo*.

Mon crests

From the 12th century in feudal Japan, identifying crests – *mon or kamon* – were used on the battlefield, on armor, banners, and personal possessions of all kinds. Unlike complicated Western heraldry, each *mon* was generally a single boldly-stylized symbol within a circle; color was irrelevant. The motif might be military, such as arrows, or an animal, such as the butterfly of the Taira clan, but plant motifs were the most common. The eldest son generally inherited his father's *mon*, while younger sons would use a slightly modified variant, so that there are an estimated 10,000 designs registered today. The only crests that were absolutely inviolable were those of the Emperor and his chief advisor. After the Muromachi period (c.1336-1573), *mon* became increasingly common across the social scale and the new merchant class adopted them as advertising logos, which persist today.

A *samurai* helmet displaying the *mon* of the wearer's clan.

Traditional *mon* crests Certain *mon* were reserved for the most powerful in the land.

Commercial *mon* logos Many modern Japanese companies still use *mon* as their logo.

The inviolable crest of the Emperor

Crest of the Prime Minister

Crest of the Tokugawa shoguns

Crest of the Taira clan

Yamaha

Mitsubishi

Toyota

Benihana

The *samurai* legacy

After the modernizing reforms of the 1860s, various organizations invoked the *samurai* past (*below*), among them Genyosha, or Dark Ocean Society, founded in 1881, which aimed to unite hundreds of secret societies, each with their own covert recognition codes. Highly successful and violent, they turned Japan's first election of 1892 into a bloodbath and, in 1895, assassinated the Korean queen, triggering the Japanese invasion that lasted 50 years. The successor to Genyosha was the Kokuryu-kai, or Black Dragon Society, founded in 1901. It promoted Japanese expansion into Asia, and was responsible for acts of violence against student and labor unions, politicians perceived as left-wing, and the democratic process in general. For muscle, they linked up with the gamblers and gangsters of the *yakuza*, which became one of the world's leading crime syndicates. Not traditionally politicized, the *yakuza* also romanticized the *samurai* past, which lent glamor to their occupations of extortion, rackets, prostitution, and people-trafficking.

Right-wing revivalists reveled in the dress and customs of medieval *samurai*.

The *yakuza*

The *yakuza* claim to have an inviolable code of honor (like the Italian Mafia), derived from the *bushido*. Within each *gumi* or gang, loyalties are extremely rigorous, hierarchies – as elsewhere in Japanese society – rigid, and feudal rituals are still observed. The *yakuza* are not, however, a secret society but an accepted part of the Japanese political and business scene – so much so that some headquarters have a plate on the door like any other company. *Yakuza* are easily recognizable, even without the *mon* lapel pins proclaiming their clan affiliation; the clothes, the large cars with darkened windows, the swagger – these are codes for gangsters almost anywhere, but especially in a country where even heads of major corporations are physically self-effacing.

Yakuza traditions

Yakuza are also famous for their spectacular full-body tattoos – *horimono*. These were always associated with the 'floating world,' marking out those living on the margins of society. To be tattooed is a sign of group solidarity and of physical courage, and a declaration of having chosen the dark side. The *samurai* who had disobeyed or failed his lord atoned by *seppuku* – ritual suicide by disembowelment. The modern *yakuza* atones for his offense by cutting off one joint of his finger – *yubitsume*. Initiation rituals and rituals marking agreements are also of great importance, with a certain number of cups of *sake* – an appropriate offering to the Shinto gods revered by the *yakuza* – being formally exchanged. Blood brotherhood rituals involving exchanging blood are now being phased out because of the threat of HIV.

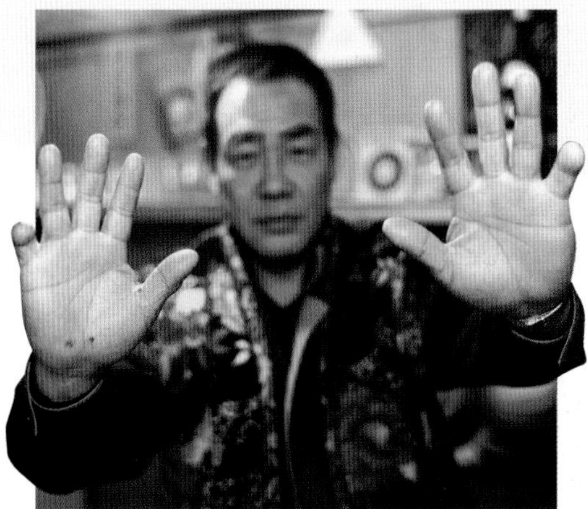

To atone for an offense, a *yakuza* will remove a finger joint, formally presenting it to his *oyabun* ('father').

Each tattoo is individually designed, the motifs including references to the owner's gang and *mon*, and represents hundreds of hours of work. Public baths often have a 'No Tattoos' sign, to the mystification of tourists.

Cockney Rhyming Slang

To be a true Cockney, it is said, you need to have been born within earshot of 'Bow Bells,' that is, the chimes of St. Mary-le-Bow, Cheapside, in the City of London (*left*). The Cockney East Enders have traditionally been the people that kept London, the world's first metropolis, supplied with its daily needs and worked to keep it vibrant as a capital of world commerce. Much to the frustration of foreign visitors, the English language is peculiar in being rich in idiomatic forms, localized dialects, slang, irregular constructions and bewildering relationships between spelling and pronunciation. Cockney rhyming slang is a pronounced example of this; developed in the fertile soil of London's highly cosmopolitan eastern districts during the Industrial Revolution, it has become a widely recognized and used form of verbal shorthand.

Roots of rhyming slang

London in the 19th century was supplied by three huge markets: Billingsgate (for fish), Covent Garden (*below*, vegetables, fruit, and flowers), and Smithfield (meat). These clustered around the great London prisons, Newgate and Bridewell in the City, and the Borough and the Clink on the South Bank of the Thames. Rhyming slang was probably used in all these locations, as well as in the burgeoning London docklands. Often described as a coded language which evolved in the criminal underworld as a means of disguising conversations in the pubs, dives, and coffeehouses of London's East End from police ears, in fact there seems a simpler explanation: the costermongers, butchers, fishwives, stevedores, and porters – not to mention the prisoners – simply didn't want their supervisors to understand what they were saying.

The porters at London's markets were famed for their ability to balance loads on their heads.

How rhyming slang works

A constantly evolving language, with new forms appearing almost daily, the principle of rhyming slang is simple: to adopt a common pair of associated words (or an expression or, more recently, a celebrity name) which rhymes with the intended word. Much has to do with the context in which the slang expression is used; for example: 'Would you Adam 'n Eve it? 'E's gone and changed his barnet' means 'Would you believe it. He has gone and changed his hairstyle.' To add to the confusion, Cockneys drop their 'aitches,' mangle proper grammar, and often the rhyming part of the expression is elided as in 'Let's have a butcher's' translates as 'butcher's hook' translates as 'look'; in addition, often the rhyme key will also refer to yet another slang expression, such as 'all time loser' translates as 'boozer' translates as either a drunkard or a pub.

"The *trouble* bought me a new *whistle* last week."

"'Ave yer got a *titfer* to go with it?"

"You'll 'ave to get yer *barnet* sorted out."

Examples of rhyming slang and their meanings

Adam and Eve	Believe
Airs and Graces	Braces or Faces
All Time Loser	Boozer (pub or drunkard)
Apples (& Pears)	Stairs
Barnet (Fair)	Hair
Boat (Race)	Face
Boracic (Lint)	Skint (broke, penniless)
Bottle of Sauce	Horse
Brass Tacks	Facts
Bread (& Honey)	Money
Bubble (& Squeak)	Greek
Butcher's (Hook)	Look
Chalk Farms	Arms
Chew the Fat	Chat (talk)
China (Plate)	Mate (friend)
Chocolate (Fudge)	Judge
Cream Crackered	Knackered (worn out)
Dickory (Hickory Dickory Dock)	Clock
Dog (& Bone)	Telephone
Down the Drains	Brains
Duchess (of Fife)	Wife
Duke (of Kent)	Rent
Dustbin (Lid)	Kid, child
Frog (and Toad)	Road
Frying Pan	Old man (husband)
Garden Gate	Date
Greengages	Wages
Ham and Eggs	Legs
Hampsteads (Heath)	Teeth
Ice-Cream (Freezer)	Geezer (man)
Iron (Tank)	Bank
Jack-and-Jill	Bill (check, account)
Jack (Tar)	Bar
Jam (Jar)	Car
Joanna	Piano
Linen (Draper)	Newspaper
Loaf (of Bread)	Head
Loop (the Loop)	Soup
Lump (of Lead)	Head
Mickey (Mouse)	House
Mince Pies	Eyes
Mother (Hubbard)	Cupboard
Mother's Ruin	Gin
Mutt and Jeff	Deaf
North and South	Mouth
Ones and Twos	Shoes
Oxford (Scholar)	Dollar
Peas in the Pot	Hot
Pig (Pig's Ear)	Beer
Plates (of Meat)	Feet
Porkies (Pies)	Lies
Pork Pies	Eyes
Potatoes (Taters, in the Mold)	Cold
Rabbit (& Pork)	Talk
Scotches (Scotch Eggs)	Legs
Sighs and Tears	Ears
Skin (& Blister)	Sister
Syrup (of Figs)	Wig
Tea Leaf	Thief
Teapot (Lid)	Kid
Tit for Tat (Titfer)	Hat
Tommy (Tucker)	Supper
Trouble (& Strife)	Wife
Turtle Doves	Gloves
Two and Eight	State (anguish)
Whistle (& Flute)	Suit
Wooden Plank	Yank

The Mob

Organized crime is by its very nature a dark mirror of the society it feeds off. And like most societies, it has its own codes of behavior and codes of communication. Among the first 'organized' criminals were probably the gangs of bandits who would haunt remote sections of highways and byways, intent on robbing or kidnapping travelers, and the notorious pirate rovers of the high seas. Europe had also seen loosely connected bands of vagabonds since the early Middle Ages (*see page 128*), but it was only with the burgeoning growth of city-dwelling in the industrial 19th century that international syndicates of highly-organized criminal activity began to develop.

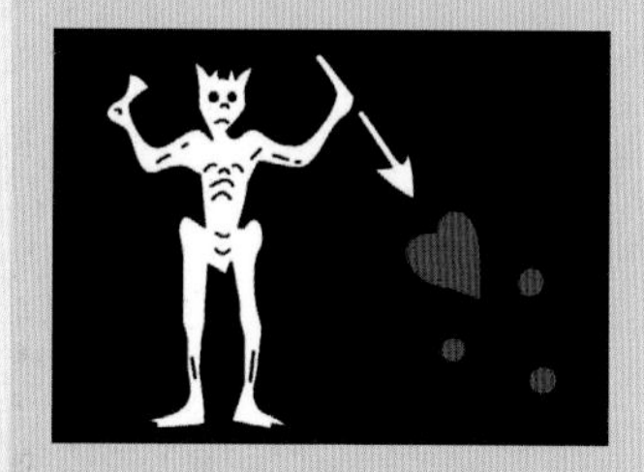

Blackbeard's flag was a variation on the skull and crossbones design widely used by pirates.

Pirates

The pirates, rovers, buccaneers, and freebooters who plundered the Spanish Main and other early commercial maritime routes were largely made up of escaped slaves, indentured laborers, or convict transportees, led by opportunists often in the employ of state governments in Europe. Despite their nefarious backgrounds, the rovers were bound by a code of honor and conduct embedded in loyalty to the group, and in a pre-agreed dividend of whatever booty they accrued – the origin of the term 'shares' on the modern stock market.

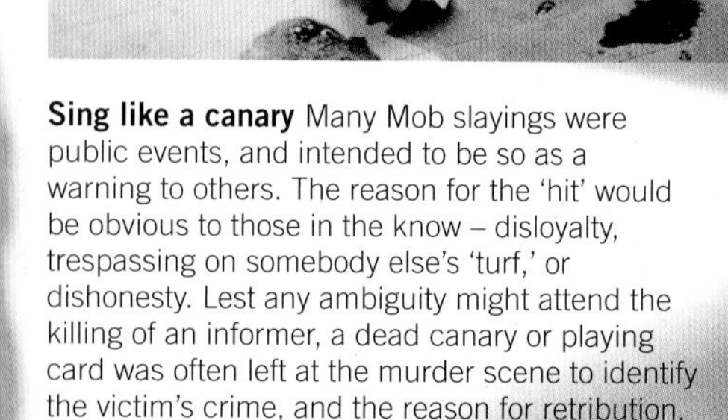

Sing like a canary Many Mob slayings were public events, and intended to be so as a warning to others. The reason for the 'hit' would be obvious to those in the know – disloyalty, trespassing on somebody else's 'turf,' or dishonesty. Lest any ambiguity might attend the killing of an informer, a dead canary or playing card was often left at the murder scene to identify the victim's crime, and the reason for retribution.

Codes of honor

Many of the communities who passed through US immigration portals such as New York's Ellis Island spawned criminal cliques in the New World. The Irish, Polish, Russian, and Jewish communities held among them at least a few who would seize the 'main chance' to earn a 'fast buck' at the expense of their fellows. Most notorious and successful were those from southern Italy and Sicily, whose traditions, codes of behavior, and language contributed to a rich polyglot of underworld idioms which are almost taken for granted in modern America.

Cosa nostra Literally, 'our thing,' or 'the thing we have between us,' a term adopted to play down in public speech (or in the presence of listening devices) organized criminal activity.
Omertà Meaning 'manliness' in Sicilian, the generally understood code of silence if questioned, punishable by death if broken; from the 1960s, law enforcement deals brokered with the police and district attorneys, involving plea bargaining and witness protection programs, have seen a number of high-profile instances of 'supergrasses' breaking the code, such as Joe Valachi and Henry King.
Onore Honor, whereby even the smallest slights to a man or a member of his family, require vengeance, the reaping of which creates 'respect.'
Big House Prison, originally used for New York's Sing Sing.
Canary Someone who 'sings' to the authorities.
Caper A moneymaking plan involving criminal activity.
Consigliere A negotiating middleman or adviser to a 'family' or 'boss.'
Contract A prearranged assassination, with a guaranteed fee.
Don The most senior figure, or head, of a 'family.' Also a 'boss.'
Family By no means bound genetically, the house or clan of a particular Mafia gang.
G-man Government-man, law enforcement officer. Legendarily coined by George 'Machine-Gun' Kelly upon his surrender to FBI officers in 1937: "Don't shoot, G-men."
Grift A scam or caper, usually involving trickery or fraud, often at card games.
Hit Murder, usually under 'contract.'
Made To be 'made' is to be formally accepted into a Mafia family structure, beginning at the lowest level, the *uomini d'onore.*
Scam A 'caper' involving deceit, trickery.
Stoolie/stool pigeon A 'canary,' someone who 'spills the beans' to the authorities.
Turf Area under control of a given gang.
Uomini d'onore Men of honor, the lowest level in a Mafia family.
Vig The sum required to buy into a gambling game or a caper.
To whack To kill.
Wiseguy A 'made' person.

British gangland

The British organized crime underworld developed between the world wars, mainly around sports involving gambling, such as horse racing and boxing. The black market during World War II saw it grow exponentially, as did tight restrictions on prostitution, and gambling and casino licenses. The notorious 1960s London gangs of the Kray brothers (*above*) and Charles Richardson brought extortion, protection, prostitution, illegal gambling, narcotics, and outright robbery into their fields of activity. By no means exclusively London-based, British criminal argot nevertheless drew heavily upon both cant and Cockney rhyming slang.

A long one/ a grand	£1,000
A monkey	£500
A ton	£100
A pony	£25
Cock-and-hen (ten)	£10
Beehive (five)	£5
Half-a-bar	Ten shillings
Blag	A tall story
Boiler	An older woman
Broads	Playing cards
Carpet	Year in prison
Cat's-meat gaff	Hospital
Do bird	Go to prison
Dot-and-dash	Cash
Drum	Room/flat
Flash/front	Nerve or face
Form	Criminal record
Gaff/crib	Home
Have it away	Steal (or sexual intercourse)
John (Bull)	Pull, or arrest
Kettle	Wristwatch
Kick	Pocket
Kite	Check
Knock	Credit
To lamp	To look
Lifters	Hands
Manor	Neighborhood
Minted	Rich
Mob-handed	A gang of three or more
To moisher	To wander
Morrie	An OK person
Nishte	Nothing
Nosh	Food or to eat
The old	Money owed
Old Bill/ Uncle Bill	Police
Punter	A gambler, or one with money to invest
Rabbit	To talk, chat
Readies	Cash
Screw	Prison warder
Shickered	Broke
Six-and-eight	Straight
Skint	Broke
Slush	Counterfeit
Snout	Tobacco
Spieler	Illegal gambling den
Stay shtum	Keep quiet
Stubs	Teeth
Sus/suss	To suspect/ understand
Tealeaf	Thief
Tomfoolery	Jewelry
To top	Kill
Twirl	A key

RAMBLERS' SIGN LANGUAGE

Hobo lingo

Like many 'gentlemen of the road,' hoboes developed a rough-and-ready camaraderie, which was enshrined in a code of honor and a secret language of verbal expressions which had been conceived, not unlike Cockney rhyming slang (*see page 132*) to confuse the cops and railroad 'bulls.'

Accommodation car The caboose of a train.
Angelina Young inexperienced kid.
Banjo A small portable frying pan.
Barnacle A person who sticks to one job.
Big house Prison.
Bone polisher A mean dog.
Buck A Catholic priest good for a dollar.
Bull A railroad officer.
Cannonball A fast train.
Catch the westbound To die.
Chuck a dummy Pretend to faint.
Cover with the moon Sleep out in the open.
Cow crate A railroad stock car.
Crumbs Lice.
Doggin' it Travelling by Greyhound bus.
Easy mark A person or place providing food or shelter.
Honey dipping Working in a sewer.
Hot A fugitive hobo.
Hot shot Express freight train.
Jungle A hobo camp or meeting place.
Knowledge bus A school bus used for shelter.
On the fly Jumping a moving train.
Spear biscuits Look for food in garbage cans.
Yegg A traveling professional thief.

From the time the railroads opened up the American West, the iron tracks stretching into the distance offered hope and opportunity to displaced, or misplaced, workers. From the closing years of the 19th century America saw not only a massive wave of immigrants from Europe and Asia (many of whom helped to build the railroads and highways) but a huge internal migration of seasonal pieceworkers, moving around the country by whatever means were available to find a grubstake or meal ticket.

In times of economic hardship, and especially during the Great Depression, hopping a train ride on an empty boxcar to a new and distant future remained a basic option – whether to follow an oil 'rush' or simply to escape a rural backwater and head for the bright lights of the city. And with the development of the national highways as part of Roosevelt's 'New Deal' in the 1930s, even more people were hitting the road. These were the 'ramblers' or 'hoboes,' among whom a distinct subculture was to evolve.

> "Now I been here an' I been there,
> Rambled aroun' most everywhere."
>
> WOODY GUTHRIE, *BOUND FOR GLORY* (1943).

Hobo chalk marks

Just how the hobo chalk mark system developed nobody knows, but as a means of providing vital information (frequently a matter of life or death – or prison) of often considerable complexity, it remains unparalleled. The marks were left on boxcars, signposts, town signs, mailboxes, and fence posts.

1 Main street good for begging.
2 Rock pile in connection with jail.
3 Saloons in town.
4 Prohibition (dry) town.
5 Police hostile to tramps.
6 Leaving railroad for highway.
7 Railroad police not hostile.
8 Railroad police hostile.
9 Town is hostile. Get out quick.
10 Church or religious people.
11 Good people live here.
12 Cranky woman or bad dog.
13 OK here. Black neighborhood.
14 Cooties in jail.
15 Good clean jail.
16 Jail good but prisoners starve.
17 Jail filthy.
18 Waiting for person named.
19 Circle town.
20 Jail good for night's lodging.
21 Police are hostile. Look out!
22 Police not hostile to tramps.
23 People do not give.
24 Bad man lives here.
25 City police are in plain clothes.
26 Policewoman lives here.
27 Danger!
28 Woman living alone.
29 Two women here. Tell good story.
30 Danger! Brutal man.
31 Get car fare here.
32 A crime has been done here.
33 A fence lives here.
34 Dog in the garden.
35 May sleep in the hayloft.
36 May get money here.
37 Nothing doing here.
38 OK here. Good chance for food.
39 Poor people.
40 May sleep here.

Cops and Codes

The art of criminal detection is one in which the gathering, organization, and interpretation of clues, the search for patterns, and the reliance upon databanks has long been recognized as of paramount importance – and one which has developed into a science not dissimilar to that of the code breaker: put together the right way, the available evidence can provide an accurate interpretation of the crime, and who committed it. Today, the interaction between traditional police groundwork (investigation, canvassing, interviewing) and forensic science, most notably DNA sampling (*see page 174*), is the primary tool in most cases where human intervention at a crime scene is detectable.

Identifying criminal types

The first attempts to apply genetic methods and classification to crime-solving emerged with the pseudo-science of phrenology, in early 19th-century Germany. This claimed that the external shape and measurement of the skull provided information about personality, intelligence, and moral faculties – and thus could be used to identify criminal types. These dubious ideas were developed by the French criminologist Alphonse Bertillon (1853-1914) who formalized the science into anthropometry, whereby criminals were carefully measured and photographed, a forerunner of modern 'mugshot' archives. Bertillon also specialized in another proto-genetic technique – handwriting analysis. This assumes that handwriting provides a unique signature. Unfortunately, handwriting can be forged, faked, or misidentified: the case of Alfred Dreyfus was a tragic miscarriage of justice, his conviction for treason in 1894 being based on a faulty handwriting identification by Bertillon.

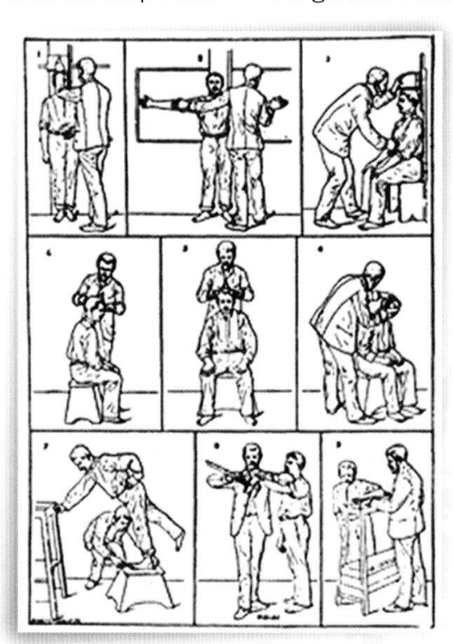

A page from Bertillon's book on anthropometry, the scientific measuring of criminal suspects.

Fingerprinting

The first credible precursor of DNA profiling, the scientific analysis of fingerprint patterns began in the 19th century as a result of early research into the genetic code. Although London's Metropolitan Police rejected fingerprinting as a method of criminal identification in 1886, in 1892 an Argentine police officer, Juan Vucetich (1858-1925), proved the guilt of a murderess from bloody handprints left at the crime scene. The first fingerprint bureau opened in Calcutta in India in 1897, where a classification system devised by Sir Edward Richard Henry (1850-1931), and his assistants Azizul Haque and Hemchandra Bose, was developed. The Henry system was adopted by Scotland Yard and by the New York Civil Service Commission in 1901. Within a decade it was internationally recognized as an essential tool in criminal detection and identification.

Modern computerized automated fingerprint identification systems (AFIS) use computers to scan the ridges, loops, whorls, arches, and bifurcations, identify key patterns, and compare them with prints held in the database.

The Black Hand

La Mano Nera **was a gang of Italian American immigrant extortionists – precursors of the American** ***Cosa Nostra*** **– who flourished in the early years of the 20th century. They would send threatening letters to their victims demanding money or death, identified by a coded message – the imprint of a hand in black ink. In New York alone 424 cases were reported in 1908, and in Chicago there were over 100 Black Hand killings between 1910 and 1914, plus 55 bombings of properties. Highly successful, their business model was transformed into bootlegging with the coming of Prohibition in 1920. As fingerprinting techniques developed, a crudely drawn hand was inscribed on the letters – the origin of our term 'blackmail.'**

Ignazio Saietta (1877-1947), also known as 'Lupo the Wolf,' was a leading Black Hander, and specialist in torture, which he carried out at the 'Murder Stable' at 323 East 107th Street in New York. He is thought to have murdered some 60 people, but was only imprisoned twice, once for counterfeiting, and once for racketeering.

The Unabomber cipher

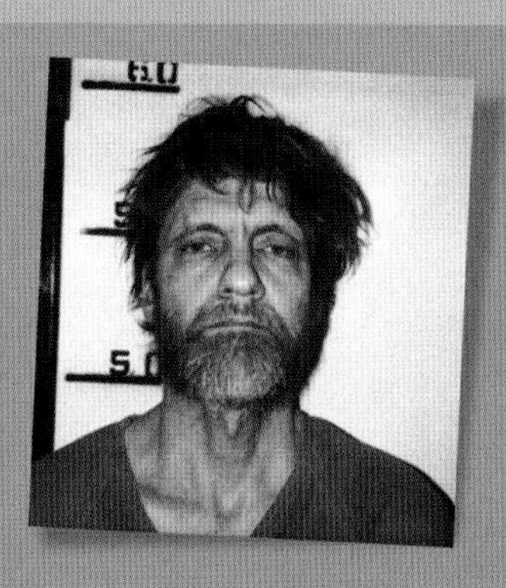

One of the most intriguing criminals of recent years left the police with an interesting challenge. Theodore Kaczynski (b.1942), the notorious lone anti-technology terrorist, was arrested by the FBI in a backwoods cabin in Montana in 1996, and was given a life sentence, without parole. During a campaign lasting from the late 1970s to the mid-1990s, he mailed or planted bombs (including one on a plane, that failed) targeting selected technology-based victims. He killed three people, and injured a further 23. His nickname 'Unabomber' was derived from the FBI's code name for him prior to identification, 'University and airline bomber.' The police failed to identify Kaczynski until he was informed on by his brother. He left a detailed account of his campaign – in code.

A mathematical prodigy, Kaczynski had studied at Harvard and taught at Berkeley before retreating to live a basic life, writing his manifesto *Industrial Society and its Future*, and launching his anarchic campaign. Among the many papers found at his hideaway were sheets of densely-packed numbers, commas, and spaces. These proved impenetrable to FBI and National Security Agency cryptanalysts until a notebook was found which provided two keys to the cipher (*left and below*). It was not until 2006 that the authorities announced that the code had been unraveled. The cipher was not intended as a confession, but a justification for his actions.

"It would not surprise me if this was the most complex cipher the FBI has seen since World War II."

BRUCE SCHNEIER, CRYPTOGRAPHY EXPERT.

The Unabomber's cipher was embedded in sheets of densely transcribed numbers found at his cabin. These meant little to investigators until the Unabomber's decoding notebook was found, which included an elaborate key.

The 'unscrambling sequence' described a series of steps or 'phases' indicating how the numbers should be read, added, subtracted, or multiplied, then paired to produce a new set of numbers.

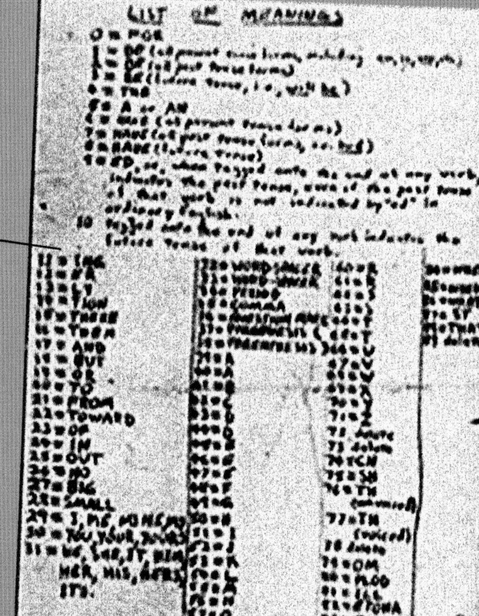

List of meanings
The key to the cipher was organized like a dictionary; each number produced by following the Unabomber's 'phases' was identified as having a particular meaning.

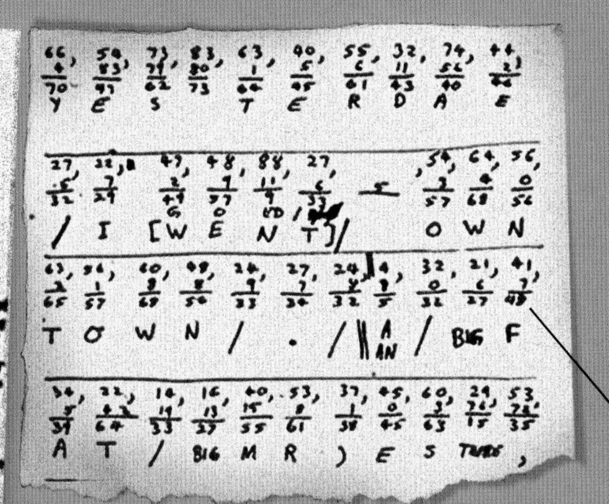

Decryption sheet
As the mathematical calculations were paired up with their meanings, so a detailed account of the Unabomber's activities slowly emerged.

The notebook went on to provide a 'list of meanings' (*left*) which linked the numbers to alphabetical letters, common combinations of letters, short words, and word-breaks. Only with the aid of this was decryption able to commence (*above*).

The Zodiac Mystery

Despite the popular belief that serial killers tend to play cat-and-mouse games with the police, this largely remains the stuff of crime fiction and slasher movies. Very few serial killers want to be caught. While Thomas Harris's Hannibal Lecter might crave public infamy, in reality most try to conceal their tracks. Although the first modern serial killer, Jack the Ripper, did taunt the police with notes and newspaper cuttings, often identifying his victims and supplying appalling details of his crimes, few others have done this. One major exception was the self-named Zodiac killer.

The verified murders

A picnic area on Lake Berryessa was the site of the attack on Bryan Hartnell and Cecilia Shepard on September 27, 1969. This artist's impression of the attacker is based on a description by Hartnell, who survived. Although carrying a firearm, the Zodiac used a plastic clothesline to bind them, then stabbed both victims. He inscribed the cross-and-circle symbol on Hartnell's car using a felt pen, and added "Vallejo/12-20-68/7-4-69/Sept 27-69-6:30/ by knife".

Despite the Zodiac's later claims, there remain only five official Zodiac killings. The first occurred on December 20, 1968, when lovers David Arthur Faraday and Betty Lou Jensen were shot on Lake Herman Road, Benicia, California.

On July 4, 1969 another couple were attacked and shot, at Blue Rock Springs Golf Course outside Vallejo; Darlene Elizabeth Ferrin died, but Michael Renault Mageau survived. This was followed by the Lake Berryessa attack as described above.

Finally, Paul Lee Stine, a cab driver, was shot dead by his passenger on October 11, 1969 at Presidio Heights, San Francisco.

"Dear Editor, I am the killer"

The Zodiac killer stalked the parks and lovers' lanes of the San Francisco Bay and Valley areas, killing five and injuring two in three attacks at remote places between December 1968 and 1969 (although some think he may have struck as early as 1966, and continued until 1974, or later; if all claims – including his – are counted, the body count could be nearer 40). He taunted the authorities with a series of letters and cards, four of which included encoded messages (*see page 142*). The first, and longest, message was sent in three parts to local newspapers – the *Vallejo Times-Herald*, the *San Francisco Chronicle* and the *San Francisco Examiner* – respectively, each received on July 31, 1969. Each coded message was accompanied by a scrawled cover note providing crime scene details that had not been made public by the police. The Zodiac demanded that they and the almost identical cover notes (which claimed credit for fatal attacks at Lake Herman Road and Blue Rock Springs) be published. As a result of this, there was considerable public interest. The police commissioned forensic tests and handwriting analyses in addition to sending the coded notes to cryptanalysts, but few solid clues emerged. However, by August 8, high school teacher Donald Harden and his wife Bettye, readers from Salinas, had cracked the majority of the coded message.

Dear Editor
I am the killer of the 2 teenagers
last christmass at Lake Herman &
the girl last 4th of July. To prove
this I shall state some facts which
only I & the police know.
Christmass
1 brand name of ammo - Super X
2 10 shots fired
3 Boy was on his back with feet to car
4 Girl was lyeing on right side
feet to west
4th of July
1 girl was wearing patterned pants
2 boy was also shot in knee
3 ammo was made by Western

Here is a cipher or that is part
of one. The other 2 parts are
being mailed to the Vallejo Times &
S.F. Chronicle
I want you to print this cipher on the front page by
Fry afternoon Aug 1-69. If you

The cover note sent to the *San Francisco Examiner*, accompanying one third of the first coded message. Each cover note revealed unpublished details of the Zodiac's attacks.

The Harden decrypt

The first Zodiac coded message comprised 408 characters organized in 24 rows each of 17 letters or symbols. It was written out on a single sheet of paper, which was then cut into three. This was an idiosyncratic substitution cipher cryptogram, which only partially followed systematic logic, and included misspellings (which may have been intentional). The Hardens assumed words such as 'killing' and 'fun' might appear somewhere, and that the killer 'had an ego,' and that the word 'I' would recur. Frequency analysis also revealed that the Zodiac was using homophones (meaning that certain letters in the plaintext were represented by two or more letters or symbols).

The key words or phrases identified by the Hardens are highlighted in red; having isolated these, the Hardens could begin to flesh out the rest of the decrypt. Some of the misleading homophones are picked out in blue. Interestingly, the key word/letter 'I' was one of them, being represented in turn by a triangle, P, U, reverse K, triangle. In contrast, 'K' is always represented by /.

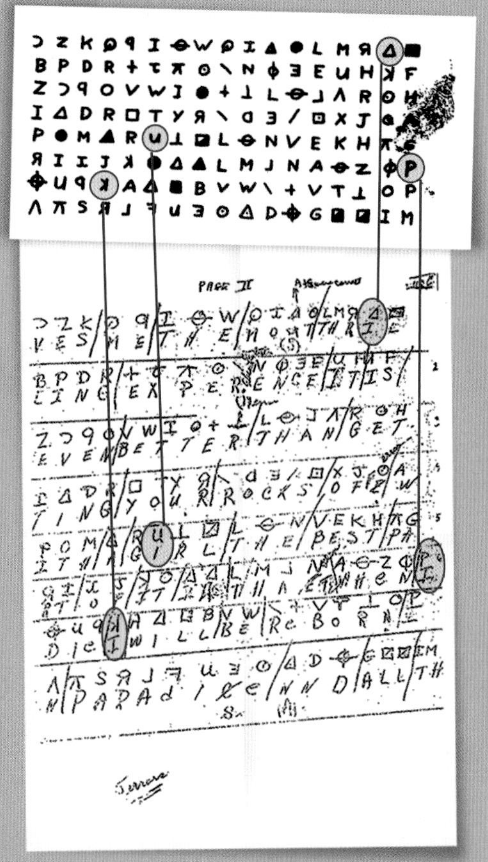

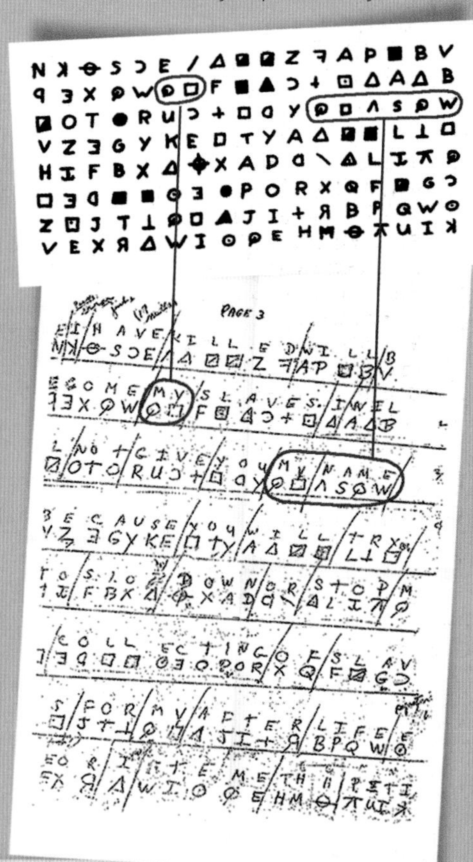

Donald Harden and his wife Bettye looked for predictable words in the cipher.

The Harden decrypt remains convincing and, although the meaning of the final 18 letters of the message remains unclear, it reveals inherent inconsistencies and misspellings (possibly intentional, as with the cover notes, to give an impression of illiteracy) which provide a chilling insight into the disorganized state of mind of the Zodiac. But the story didn't stop there. While the Hardens had provided a tantalizing glimpse of the inner workings of the Zodiac's mind, his subsequent ciphers and other chilling messages (*see page 142*) proved impregnable, and continue to fascinate cryptanalysts and conspiracy theorists alike.

The Hardens' decrypt reads:

"I LIKE KILLING PEOPLE BECAUSE IT IS SO MUCH FUN IT IS MORE FUN THAN KILLING WILD GAME IN THE FORREST BECAUSE MAN IS THE MOST DANGEROUES ANAMAL OF ALL TO KILL SOMETHING GIVES ME THE MOST THRILLING EXPERENCE IT IS EVEN BETTER THAN GETTING YOUR ROCKS OFF WITH A GIRL THE BEST PART OF IT IS THAT WHEN I DIE I WILL BE REBORN IN PARADICE AND ALL THE I HAVE KILLED WILL BECOME MY SLAVES I WILL NOT GIVE YOU MY NAME BECAUSE YOU WILL TRY TO SLOW DOWN OR STOP MY COLLECTING OF SLAVES FOR MY AFTER LIFE EBEORIETEMETHHPITI"

The Zodiac Legacy

The Zodiac's signature symbol was the most consistent coded image the killer used. Referencing alchemical and necromantic imagery, it also chillingly echoes a telescopic sight.

Despite the Hardens' breakthrough in deciphering the Zodiac's first anonymous cryptogram (*see page 141*), the killings continued, as did his taunts to the authorities (the 'blue pigs' or 'blue meanies' as he called them). Targeting mainly the *San Francisco Chronicle* or its staff, his subsequent 15 or so letters and cards built up a picture of an obsessive not only interested in killing, but in attracting attention by revealing details of his crimes, and ever more monstrous schemes. These mailings included a further three cipher messages which have remained unsolved.

The later letters

The inclusion of scorecards in his later letters, comparing his claimed body count (ultimately 37) versus the SFPD's success rate (0), reflects a detection success rate as true today as it was over 30 years ago. Several suspects were investigated, but only one remains a strong contender. The fact is that the self-named Zodiac appears to have been active for about two years, and his crimes and subsequent codes remain unsolved.

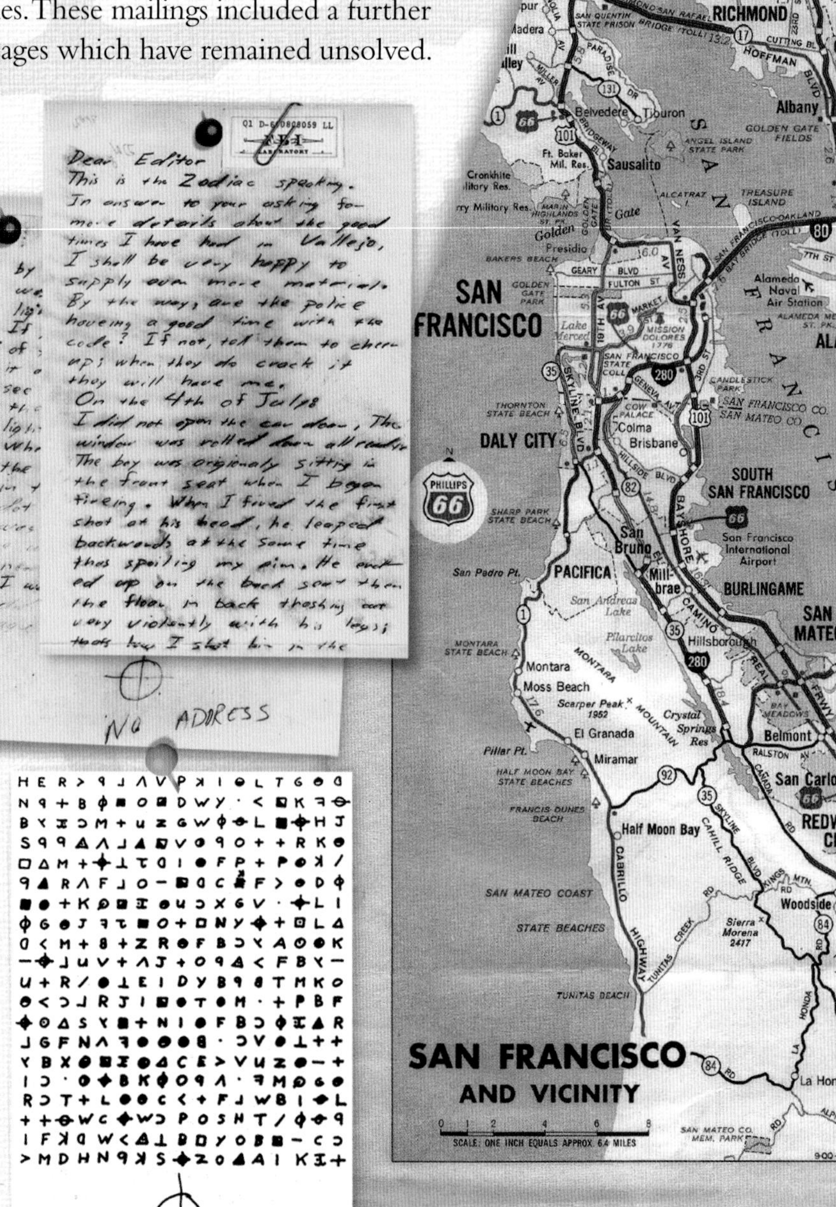

Dear Editor
This is the Zodiac speaking.
In answer to your asking for more details about the good times I have had in Vallejo, I shall be very happy to supply even more material.
By the way, are the police haveing a good time with the code? If not, tell them to cheer up; when they do crack it they will have me.
On the 4th of July
I did not open the car door, The window was rolled down all ready The boy was origionaly sitting in the front seat when I began fireing. When I fired the first shot at his head, he leaped backwards at the same time thus spoiling my aim. He ended up on the back seat then the floor in back thashing out very violently with his legs; thats how I shot him in the

NO ADRESS

"This is the Zodiac speaking." The killer revealed his *nom de guerre* in a letter postmarked August 4, 1969 to the *Vallejo Times-Herald*, and for the first time signed the letter with his characteristic cross-and-circle mark.

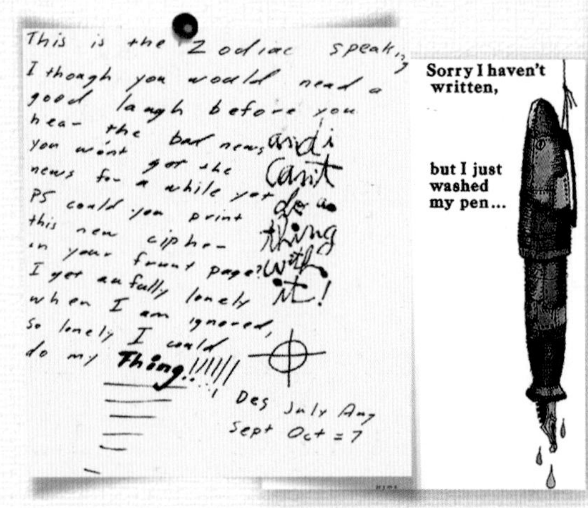

This is the Zodiac speaking
I though you would need a good laugh before you hear the bad news you won't get the news for a while yet
PS could you print this new cipher in your front page? I get awfully lonely when I am ignored, so lonely I could do my Thing!!!!!!
Des July Aug Sept Oct = 7

"Sorry I haven't written." Mailed to the *San Francisco Chronicle* on November 8, 1969, this cheap but sinister novelty card included a 340-character cipher. Superficially similar to his first coded message, the Hardens' decrypting method failed to crack it, and the meaning remains a mystery.

This is the Zodiac speaking
By the way have you cracked the last cipher I sent you?
My name is —

AEN⊕8K8M8⅃NAM

I am mildly cerous as to how much money you have on my head now. I hope you do not think that I was the one who wiped out that blue meannie with a bomb at the cop station. Even though I talked about killing school children with one. It just wouldnt doo to move in on someone elses teritory. But there is more glory in killing a cop than a cid because a cop can shoot back. I have killed ten people to date. It would have been a lot more except that my bus bomb was a dud. I was swamped out by the rain we had a while back.

The new bomb is set up like this

Sun light in early morning

Bus

String of Bombs

Sun

Timer

Car Bat

A&B are photo electric swiches when sun beam is broken A closes circut
" B opens "
which maks B the cloudy day dis con-ect so the bomb wont go off by accid.

PS I hope you have fun trying to figgure out who I killed

⊕=10 SFPD=0

This is the Zodiac speaking

I have become very upset with the people of San Fran Bay Area. They have not complied with my wishes for them to wear some nice ⊕ buttons. I promiced to punish them if they did not comply, by anilating a full School Buss. But now school is out for the sammer, so I punished them in an another way. I shot a man sitting in a parked car with a .38.

⊕-12 SFPD-0

The Map coupled with this code will tell you where the bomb is set. You have untill next Fall to dig it up. ⊕

"My name is ... " The *San Francisco Chronicle* had received a letter postmarked November 9, 1969 describing in detail a plan to bomb a school bus in the Bay Area. Such an attack never materialized (although it later inspired the plot of the 1971 Clint Eastwood film *Dirty Harry*). But some five months later, on April 20, 1970, a further threatened bomb attack was sent to the newspaper, at the end of which the Zodiac included for the first time a scorecard (Zodiac = 10; SFPD = 0). However, included in the letter was an even more explosive piece of information: the Zodiac revealed his name, again in cipher, and again it remains unsolved.

The last cipher The fashion for wearing symbolic badges or buttons temporarily distracted the Zodiac: he realized that his symbol would work just as well as a 'Smiley' or 'Ban the Bomb' logo, and he recognized the horrific celebrity he had acquired. A letter mailed to the *Chronicle* on June 26, 1970 suggests a new fashion in Zodiac buttons. It also included a map (possibly the site of a threatened bomb), a further scorecard, and the fourth and last cipher message which, like the previous two, has never been decrypted.

The Zodiac drops from view Further letters were received by the *Chronicle* postmarked July 24 and July 26, detailing more crimes, but with no further ciphers. *Chronicle* reporter Paul Avery took delivery of an ominous Halloween card mailed on October 27, 1970, but thereafter the Zodiac would seem to have stopped his activities. Two later letters are often included in the Zodiac canon, one postmarked March 13, 1971 to the *Los Angeles Times* threatening a renewed murder campaign targeting LA policemen, and four years later a letter extolling the 'satirical' qualities of 1974 movie *The Exorcist* was received by the *San Francisco Chronicle* postmarked January 29, 1974, but both seem more likely to be 'copycat' mailings and remain unconvincing.

Arthur Leigh Allen in 1969; he remains the most viable suspect in the Zodiac case.

Under suspicion

Popular speculation produced hundreds of possible perpetrators, but only one primary suspect emerged. Arthur Leigh Allen (1933-92) was a loner, who lived at home with his parents, and worked at various elementary schools, among other jobs. Police were alerted by an acquaintance of Allen's in 1971, based on bizarre and incriminating claims by Allen. He was interviewed several times, and the evidence accumulated: forensic techniques were still limited and, despite undoubted similarities, a Department of Justice analysis report in 1971 ruled out any connection between Allen's handwriting and that of the Zodiac. Nevertheless, he behaved erratically and drank heavily; he was known to humorously misspell words and phrases; he possessed guns, and bloodstained knives were found in his car (which he claimed he used for killing chickens); and he admitted reading Richard Connell's 1924 short mystery story *The Most Dangerous Game*, which appears to be referenced in the first coded message. Also, he owned a Zodiac watch, a present from his mother in 1967. Associates and friends provided further intriguing circumstantial evidence. Further, Allen was convicted of child-molesting in 1974. Investigations continued until Allen's death almost two decades later, but the police failed to establish any concrete links.

EMPLOYEE'S WITHHOLDING EXEMPTION CERTIFICATE

Print full name Arthur Leigh Allen — Social Security Account Number 572-44-8882

Print home address 32 Fresno St. City Vallejo State Calif.

HOW TO CLAIM YOUR WITHHOLDING EXEMPTIONS

(Date) Sept. 8, 1966 (Signed) Arthur Leigh Allen

The sample of Allen's handwriting used for analysis. The results were negative.

Graffiti

Used as a means of communicating anonymous political and social messages and comments since classical times, graffiti have a long if largely maligned pedigree. Toward the end of the 20th century, with the invention of aerosol spray paints, and urban utilitarian architecture creating vast acreages of inviting blank space, it enjoyed a reinvigoration as a controversial popular art form. Linked to the emergence of hip-hop culture, modern graffiti originated in New York, but has become a global phenomenon, with distinctive national schools. Various youth movements use graffiti to establish their territory – what may seem to be an unsightly daub in fact carries a lot of information.

Art in identity

Most graffiti art takes place on the fringes of the law. Thus, graffiti artists assume pseudonyms, often painted in abstract ways. The principal characteristic of the modern art of graffiti is elaborately disguised overlapping lettering, often on a gigantic scale, realized in bright, contrasting colors. This style of concealed writing echoes the psychedelic poster and album sleeve art which developed in the 1960s, but draws on the argot of rap and hip-hop, effectively forming a double-level code.

Pieces are elaborate and labor-intensive, often involving several members of a graffiti crew. These complex compositions frequently combine dubs with abstract and figurative elements, and may well imitate conventional fine art in being signed.

Dubs
The most intriguing aspect of graffiti art is the artful embellishment and distortion of letters and words to a point of sheer abstraction to the untrained eye.

Tags
These swift and ubiquitous squiggles are used to designate the territory of a graffiti crew. Tags generally take the form of a short acronym, and are rapidly applied.

Spelling
The style of spelling used for both dubs and tags started in the 1970s, and influenced the compressed spelling widely used in mobile phone texting.

Modern graffiti has lost none of its original political intent, with famous artists such as the pseudonymous Banksy creating powerful anti-establishment images which are increasingly prized in the high art market. Carefully composed, and swiftly executed, Banksy's work is very carefully sited and targeted. Here a stenciled Guardsman is in turn painting the Anarchy logo.

YOUTH CODES

Codified ways of establishing group identity among young people have always existed. Since World War II, many youth fashions have emerged and diversified, from the Bobby-soxers of the 1940s to the Beats of the '50s, the Hippies of the '60s, the Punks of the '70s, to the Grunge, Goth, and hip-hop clan followers of recent years. Young people use codes to establish their identities and their alignment with various groups, youth movements, and even criminal gangs, often in the form of dress codes, language, and symbols such as graffiti.

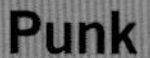

Punk

The 1970s music bands The Sex Pistols and The Ramones, the fashion designer Vivienne Westwood, and writers such as Richard Hell all played a large part in defining this influential counter-culture movement. Punk turned chaos and anarchy into a form of expression through extreme hairstyles, modified clothing, tattoos, and piercings.

Goth The black clothes, boots, and white make-up of Goth(ic) conceals a more complex and cultured response to the anarchy and nihilism of Punk. Much music, literature, and film inspire Goth culture. Various sects include Romantic or Aristocratic Goths, who wear Victorian-inspired clothing, whilst Cybergoths draw inspiration from Manga and Cyberpunk, sporting futuristic items of clothing.

EMO

Emo This recent and widespread movement evolved from Goth culture, adopting much of its music and literature, but with less socially alienating forms of dress and behavior.

Backslang, Pig Latin, and Double Dutch

In addition to the sudden popularity of variants of the Pigpen cipher (*see page 60*) in the playground, the 19th century witnessed a new verbal code that became popular among young people. Backslang simply involved saying a word backwards; it developed in British butcher's and grocer's shops as a means of disguising what was ordered from the store from the customer: for example 'yob' for 'boy' (an expression which has passed into the English language in its own right). A similar system evolved in France, known as *Verlan*.

Another easily disguised speech is Pig Latin, in which the syllable sound at the beginning of a word is placed at the end, followed by 'ay': for example, 'Pig Latin' becomes 'Igpay Atinlay.'

The interpolation of a meaningless sound before each vowel disguised it further, but with this the reversal of the word became redundant. Try interpolating the sound 'ayg' before every vowel and you can quickly scramble a sentence like: 'Two pounds of rice, please' into 'Taygoo paygounds aygof raygice, playgeese,' which most people will not immediately understand unless they know the key.

Other verbal codes include Double Dutch (or Tutnese), in which all consonants are given a syllabic value (B=Bub, C=Cash, D=Dud, F=Fuf and so on): 'Double Dutch' becomes 'Dudbubublul Dudtutcashlul'; unfortunately, this means it can take a long time to say a simple sentence. An American version is called Yuckish, or Yukkish.

HIP-HOP

Hip-hop

Hip-hop is a style of music that grew up in the USA in the 1970s out of Jamaican rhyming slang and various other influences, but rapidly came to encompass an entire lifestyle, largely but not exclusively in the 'projects' and gang cultures of disaffected Black youth. Hip-hop has now spread around the globe, with many different incarnations – Spanish hip-hop often incorporates flamenco music, while Japanese hip-hop visionary DJ Krush regularly includes traditional Japanese singing and music in his albums. Graffiti crews (*see page 144*) are only one facet of the hip-hop movement, which incorporates various other art forms: rap music, 'turntablism' music, breakdancing, and an entire dress and speech code; often the slang used by rappers is almost unintelligible to anyone unfamiliar with the subculture, as are their hand signs.

HOODIE

Hoodies Dress among the young can carry far more meaning than simple fashion sense. Choice of trainer brand can be critical to acquire 'street cred.' So can hooded tops (hoodies), now notorious in the UK, being associated (along with Burberry check) with petty crime.

Hip-hop codes

Rapping is the defining aspect of hip-hop, and it is thus natural that an extensive street language of terms has been developed, many scatological, sexual, or drug-related.

187	Murder (from the Californian penal code)
850	Prison (850 Bryant is the address of San Francisco county jail)
All gravity/ gravy	All good
Base	Weak
Bing	Prison
Biter	Rapper who steals another's lyrics
Blood	Friend, relative, fellow gang member
Boo	Lover
Boofer/duck	Ugly woman
Cabbage	Money
Faded	Drunk
Ghost	To leave
Grill	Face
Hood	Neighborhood
Jawzin'	Lying
Out the pockets	Getting out of hand
Piece/ heater/gat	Gun
Pulling licks	Robbery
Snake	Feeling stupid
Whip	Car
Wolfin'	Lying

Gangsta street signing

An important means of communication among street gangs is covert hand signaling. These are thought to have derived from similar signs used among Chinese Triad immigrants to the West Coast. Disguised numbers are significant as are signs showing allegiance to a particular gang or 'crip.'

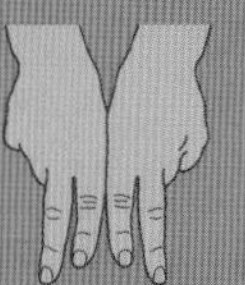

Mafia crips

Latin kingz

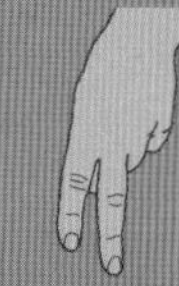

Hoover crip

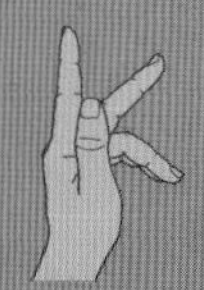

Killers

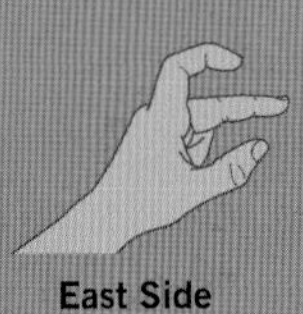

East Side

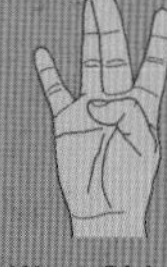

West Side

Youth cults in Japan

The *bosozoku* impacted on the urban scene in Japan in the 1980s, characterized by their customized small motorbikes (they have to be ridden with legs outspread). Although *bosozoku* members are very frequently school failures, they give their gangs exotic and bizarre names such as 'Don Quixote' or 'Tarantula,' and use right-wing slogans and symbols, such as the swastika – intended to shock rather than display political affiliation. The younger *yankees* name is derived from the brash US-occupation soldiers in Japan. Typically high-school students, or dropouts, they are in revolt against a highly disciplined society. Recognizable by their long permed hair, often dyed blond, their taste is for black and white or primary colors, *aloha* shirts, satin, and glitter; the men also wear pin heels. Like the *bosozoku*, it is a leisure time activity and most *yankees* will join mainstream society before they reach 20, but a number will go on to join the *bosozoku* and eventually the *yakuza*.

Lolitas A movement influenced by Romantic Goths, they wear clothes inspired by Victorian children's dress and rococo styles, obeying complex behavioral codes of extreme aestheticism.

Digital Subversion

let's warchalk!

KEY	SYMBOL
OPEN NODE	ssid)(bandwidth
CLOSED NODE	ssid O
WEP NODE	ssid access contact (W) bandwidth

blackbeltjones.com/warchalking

Wi-fi warchalking

In some cities, strange chalk circles and arcs on walls and pavements may be seen representing locations where it is possible to hack into company wireless (wi-fi) networks to give the hacker free access to the Internet, known as 'warchalking,' 'streetwarring,' or 'wardriving.' Often this is harmless, but the widespread use of wireless terminals by restaurants and some major chain stores to process customer payment transactions led, in 2008, to the revelation that an estimated 100 million credit card holders' details (the Track 2 data embedded on the magnetic strip) had been accessed by potentially criminal elements who had simply hacked into the wireless systems. The details were then auctioned online, and often paid for using a cyber-currency known as e-Gold.

The explosion of new communications technology in the last three decades has transformed the way in which terrorists and criminals operate, and has provided new challenges for the authorities set up to deal with them. Digital technology has opened many new avenues and opportunities for a range of activities, from data gathering and surveillance, to identity theft, fraud, or even detonating a bomb. Telecommunications also effortlessly cross political and physical boundaries. While cell phone use and Internet traffic are technically traceable and can be monitored, the sheer volume of traffic on the airwaves makes this an almost insurmountable problem for the security services.

Talking terrorism

Terrorists have always been confronted by the problem of keeping their communications secret. Russian anarchist cells in the 19th century (*left*) were known to use a version of the Polybius square code (*see page 78*) to convey secret messages. In more recent years many terrorist groups, notably the IRA and ETA, have used telephones and cell phones to ring dedicated security force phone line numbers (or newsapapers and broadcasters) using pre-agreed passwords to provide security forces with prior warnings about bomb attacks.

One of the many disturbing post-9/11 revelations was the extensive use of cell phone technology in organizing, coordinating, implementing, and executing the attacks. Pay-as-you go cell phones, which are disposable and largely untraceable, were used extensively (as they are by many criminal gangs also), and similar evidence was found in the wake of the Madrid and London bomb outrages. In addition, a signal sent to a cell phone merely by ringing it can be used to trigger an explosive device. E-mail and the Internet have become key channels of subversive communication, especially with the advent of encryption software such as PGP (*see page 274*) capable of resisting attacks by security services.

A breakdown in the agreed telephone code words system, and failure to monitor cell phone messaging, led to catastrophic civilian deaths in the Omagh bombing, Northern Ireland, 1998.

Surveillance

Digital technology has transformed our capacity to observe and monitor our fellow citizens. In addition to ubiquitous CCTV coverage of our streets, our movements can be traced through our cell phones, and our online activities through our Internet service provider (ISP) accounts, and 'cookies' transferred from servers to Web browsers. On limited networks we are likely to be monitored at work: 45% of US companies track the computer content, e-mails, and time spent at the keyboard of their employees. Remote-sensed satellite data can now track and analyze the shadows of people walking on the Earth's surface and will potentially be able to match this to a human motion 'fingerprint.' Meanwhile, huge databases are being compiled from a range of sources – utility bills, store loyalty cards, RFID tags on items we purchase, credit card use, medical records, and so on, making us all tiny but evaluable points in a massive coded matrix of data.

England has the highest *per capita* CCTV coverage of any country in the world. London's Metropolitan Police Special Operations room (*above*) opened in 2007, and can provide continual surveillance of any major public event or incident in the city.

Language under attack

Since the economy and convenience of texting on mobile phones (*see page 98*) has increased dramatically, a new type of digital slang has appeared. In early text-messaging systems there was a strict limit to the number of characters that could be sent (often 160 or fewer). As a consequence of this, many abbreviations started to come into common usage, often incorporating the phonetic sounds of number symbols.

lol Laugh out loud.
b4 Before.
l8r Later.
btw By the way.

A unique 'language' called Leet (or l33t or even 1337) has also been in use for years on Internet forums. One eccentric property of Leet is that it has no fixed grammar – in fact, if there is one rule it is that there are no fixed rules. Text in the language is almost deliberately obtuse and can often only be deciphered by one experienced in the language. Online gaming has expanded the language significantly – when playing fast-paced games, if one wants to communicate with the group it must be as quick as possible – if you're typing you're not playing!

sry m8 Sorry, mate.
np No problem.
gs, gg all Good shot, good game everyone!
noob Newbie: someone new to the game.

Growing concern over the impact of these new shorthand languages on literacy and education among the young erupted into a public debate among English academics in 2008, with even some Oxbridge voices recommending a more relaxed attitude toward 'correct' or 'traditional' spelling and punctuation.

07

The need to find ways of describing the often intangible functions of the natural world has provoked the invention of an enormous range of methods to define the seemingly indefinable.

encoding the world

Since the ancient times, mathematicians and scientists have provided the keys that unlocked these problems, creating frameworks that have been used to describe the abstract workings of concepts such as time, physics, mechanics, chemistry, biology, cartography, and sound.

DESCRIBING TIME

Long before the invention of writing, humans learned to mark time by the movement of heavenly bodies. The basic unit of every calendar system is the day, although not all begin their day at the same time. Today most calendars use a seven-day week, but in the past other periods were used. Months are measured by the phases of the moon, most calendars beginning their month with the first sighting of the moon's crescent; others, like the Hindu and Chinese calendars, begin the month with the full moon. Fixing the beginning of the year varies widely, from observation of the equinoxes or solstices, or the maximal northern or southern rising or setting of the sun or some other heavenly body.

Recording time

Various means of establishing diurnal time were tried before the invention of the mechanical clock, including the sundial (*above*), and the water clock. Early timepieces were driven by a clockwork system dependent upon coiled springs or carefully measured weights and pendulums.

Many cultures independently calculated their months and annual calendars using either a lunar cycle, a solar cycle, or a lunisolar system, where months are linked to the lunar cycle, but fitted into a solar year. Each system required the intercalation of extra days or months at regular intervals to resolve the problem of gearing days to lunations and solar years.

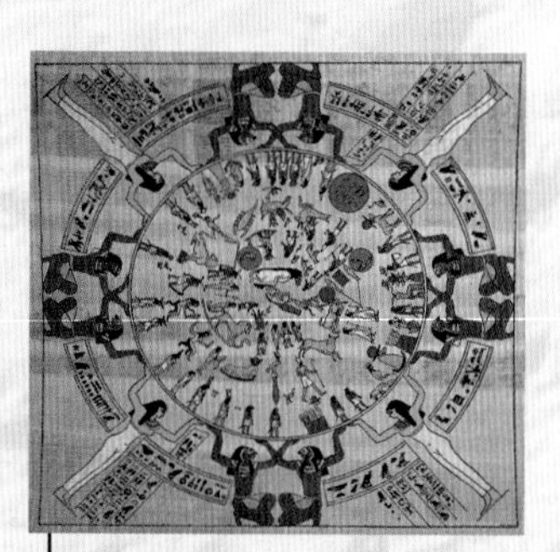

c.2500 BC Egypt
Solar. 12 months of 30 days, plus five extra days. Day starts at sunrise. Epoch: February 18, 746 BC. Parchment zodiac calendars *(above)* were produced.

A model of a Chinese water clock.

c.1300 BC China
Lunisolar. Based on astronomical observation. Days begin at midnight. Month begins with the new moon in Beijing. Years have 12 or 13 months, of 29 or 30 days, grouped in 60-year cycles. Epoch: March 8, 2637 BC.

c.500 BC India
Solar. 12 months of 29-32 days. Day starts at sunrise, divided into 30 *muhurtas* of 48 minutes. Epochs determined by regnal years, but also by the death of Buddha (c.544 BC) or by the death of the founder of Jainism, Mahavira (c.538 BC).

3000 BC | 1000 BC | 500 BC

c.3000 BC Europe
Solar. First stone circles, such as Stonehenge *(above)*, and alignments with position of sun at sunrise, sunset, and solstices.

c.1500 BC Babylonia
Lunar. 12 months, alternately of 29 and 30 days. Day begins at sunset. A 13th month intercalated whenever necessary to bring back in line with seasons. Under the Achaemenids, months were intercalated at fixed intervals within a 19-year period (the Metonic cycle) still followed by the Hebrew calendar. The seven-day week was unknown, but the Babylonian month names are still used in Arabic and Hebrew. Epoch unknown.

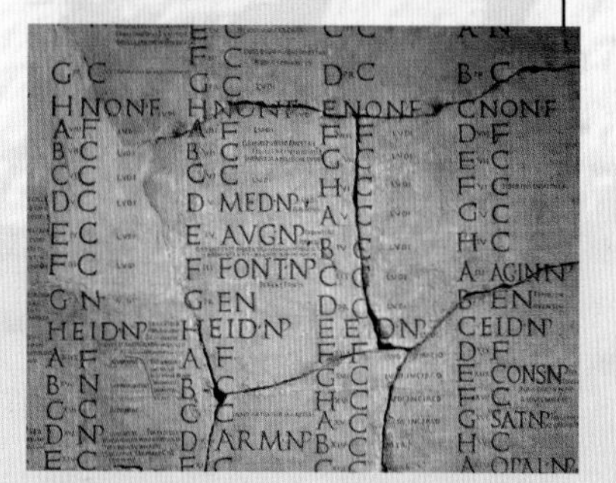

713 BC Rome
Lunisolar. In 713, two months, January and Feburary, added to the old Roman ten-month lunar calendar. 355 days; a month intercalated periodically to align the calendar with the solar year. Under the Republic, epochs were calculated by consulships; later an epoch from the founding of the city in 753 was used.

c.500 BC Greece
Lunisolar. There were many regional calendars in Greece. Athens used a lunisolar calendar, a solar civil calendar with ten arbitrary months, and an agricultural calendar based on the constellations. The lunisolar calendar had 12 months of 29 or 30 days with an extra month intercalated every three years. Months were divided into three phases of ten days. The civil calendar had 365 or 366 days and ten months, six of 37 days and four of 36.

c.250 BC Maya The epoch of the 'Long Count' calendar corresponds to September 8, 3114 BC. At the end of each 'Long Count' cycle, the earth is destroyed and then recreated. The present 'Long Count' cycle ends on December 21, 2012. The *Haab* is a 365-day year with no intercalation, and 18 months of 20 days plus five extra days at the end of the final month. The *Tzolkin*, or 'Calendar Round,' is an ever-repeating cycle of 52 solar years of 365 days, and seems to have been used throughout Mesoamerica. The Madrid Codex contains some 250 almanacs, dealing with rain ceremonies, sowing times, new years, sacrifice of captives, hunting, and even bee-keeping.

Life and death
The god of Death, with his head thrown back and holding maize seeds in his hands. These will be brought to life by the rain god, Chac.

Madrid Codex
This page (29 of 56) shows two agricultural almanacs.

c.250 BC Maya
Three interlocking calendars, the 'Long Count,' the *Haab*, and the *Tzolkin* (*see above*).

AD 532 Rome Era
Counted from birth of Christ (*Anno Domini*).

AD 1789 France
Revolutionary calendar (*see page 197*).

AD 0 — AD 1000

45 BC Rome
Solar. Julius Caesar reformed the Roman calendar. Year consisted of 365 days with a leap year every four years. Day began at midnight. The 12 months of the early Roman calendar were preserved. The average length of a year was 365.25 days, so eventually the Julian dates would regress through the seasons.

AD 359 Hebrew
Lunisolar. Epoch: September 7, 3760 BC.

c.250 BC Celts
Lunisolar. 12 months of 29-30 days, with one month intercalated every 2.5 years; months divided into two fortnights of 14 or 15 days.

AD 1753 Harrison's chronometer
The first truly accurate timepiece *(right)* was invented over a period of 40 years by Englishman John Harrison in response to a competition sponsored by the Admiralty and the British government. The precise measurement of longitude relative to the Greenwich meridian could only be achieved using a reliable weatherproof timepiece; the invention saved thousands of ships and sailors.

Epochs and modern calendars

Every calendar has a starting point – a specific day and year, known as an 'epoch.' The date chosen may be an historical event, a legendary date in the past, or simply arbitrary. It is almost never the date the calendar was actually adopted; it is a hypothetical date from which to begin counting the days. For cyclical systems, like the Chinese, the first day of any cycle can be chosen, although for longer counts an epoch can be used.

AD 2008 (Common Era) beginning January 1, corresponds to:

Muslim	1428 AH, next year began on January 10, 2008.
Julian	AD 2007, next year began on January 14, 2008.
Chinese	4644 AC or 4704 AC or Cycle 78 or 77, year 24 (Ding Hai), next year began on February 7, 2008.
Hindu	1929 (Saka as standardized by the Indian government), next year began on March 21, 2008.
Iranian	1386, next year began on March 21, 2008.
Ethiopian	2000, next year began on September 11, 2008.
Coptic	1724, next year began on September 9, 2008.
Jewish	5768 AM, next year began at sunset on September 29, 2008.
Japanese	2668 or Heisei 20, next year began on January 1, 2009.
Buddhist	2552, next year began on January 1, 2009.

The Gregorian calendar

As the Roman Julian calendar regressed with the seasons, in 1582 Pope Gregory XIII commissioned the first major calendar reform since antiquity. This calendar is the one most widely used today.

Establishing Easter

One of the mysteries of everyday life is how the date of the major Christian festival of Easter is calculated. It is, in fact, quite a simple formula, but one which has its roots in pre-Christian astronomical calculations. Easter is established by taking the date of the vernal (Spring) equinox, establishing when the next full moon will occur after this, Easter Day being the following Sunday.

DESCRIBING FORM

Geometry is the study and codification of shapes, areas, volumes, and angles – a way of defining the physical world around us. Mathematics has developed alongside geometry since 3000 BC – the first mathematics was geometry, finding ways to describe the lengths and sizes of things and the relationships between them. The ancient Greeks were particularly fascinated with geometry, believing that shapes and special numbers such as the 'golden ratio' and pi had an ethereal reality of their own. The 17th-century French polymath René Descartes later revolutionized the field with the concept of 'Cartesian coordinates' by which the position of a point may be described in terms of its relationship to lines or surfaces. Modern geometry is crucial in the fields of relativity, symmetry and asymmetry, and quantum mechanics. The evolution of geometry is closely tied in with the evolution of algebra (*see page 158*).

Euclid

Although by no means the 'father' of geometry, the Alexandrian Greek Euclid (fl. c.300 BC) produced history's most influential textbook on geometry and mathematics. In *Elements* he produced a summary of what had been established by Greek thinkers before him, and extended their investigations in a variety of directions, including investigations of factors, perspective, and optics. Only in the 19th century would mathematicians move beyond the boundaries of Euclid's ideas.

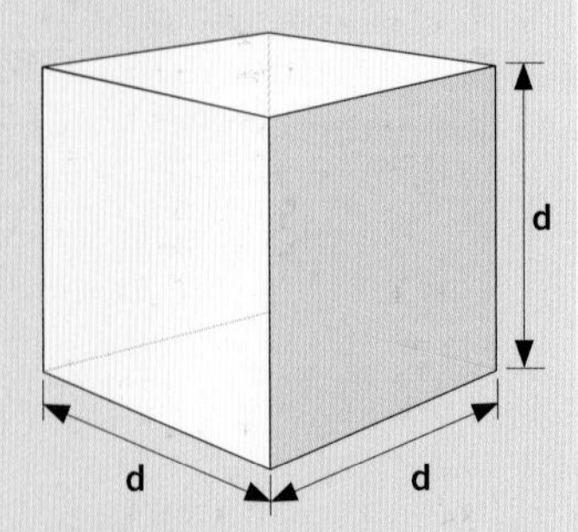

The square and the cube

The square: a simple symmetric geometric form, comprising four sides of equal length, which meet at right angles. To calculate its area you simply multiply the length of two sides:
Area = d x d

The cube: this is a three-dimensional shape. Some simple relations can be formulated:
Surface area = 6 x d^2
Volume = d x d x d, or d^3

Geometry

The definition and investigation of the properties of two-dimensional and three-dimensional forms by the ancient Greeks produced several fundamental principles of enduring validity which still govern many aspects of mathematical thought today. Many significant formulas were established by Pythagoras and his followers (*see page 158*). The most elegant of the ancient Greek geometric proofs demonstrate the properties of squares, cubes, triangles, and circles, and how they can be expressed algebraically.

The golden ratio

The discovery by the Pythagoreans of the golden ratio (or mean) provided a fundamental harmonic proportion which not only reflects the diatonic scale of musical harmony, but also came to be seen as an aesthetically pleasing proportion in architecture and pictorial composition. The golden ratio is based on the relation between a square, a circle, and a rectangle:

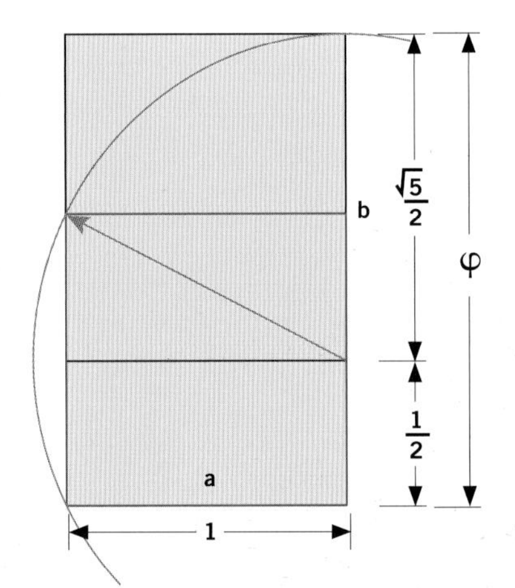

Draw a square, then take a line from the midpoint of one side to an opposite corner; use this measurement as the radius of a circle to then construct a rectangle. The proportion is defined by the Greek letter phi (φ) and may be expressed in various ways:

a + b is to a as a is to b
or
$\frac{a+b}{a} = \frac{a}{b} = \varphi$

The Gehry Pavilion at the Serpentine Gallery in Hyde Park, London openend in July 2008. It is a fascinating demonstration of the architectonics of geometry.

The Pythagorean theorem

This critical principle, combining ideas about squares and triangles, paved the way for over 300 further proofs concerning the properties of triangles, and linked directly to the field of trigonometry – the calculation of a variety of unknown numbers using functions derived from the functions of two known measurements.

> "The square of the hypotenuse of a right triangle is equal to the sum of the squares on the other two sides."
>
> PYTHAGORAS, QUOTED BY EUCLID, c.300 BC.

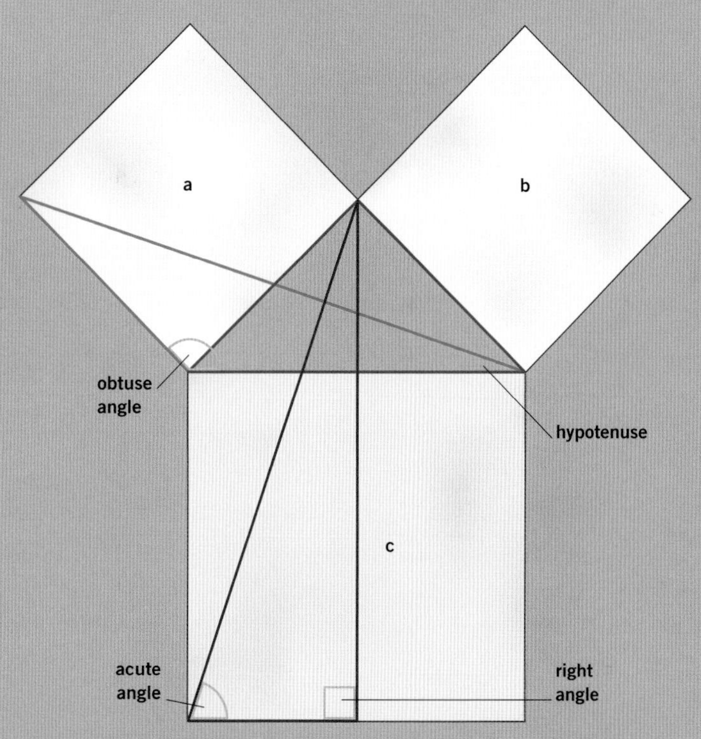

Triangles The theorem examines the properties of triangles, which have always been of great importance in the field of geometry and its relation to algebra. The Greeks particularly were fascinated by triangles. They defined four main types:

Right-angled triangle
One of the angles in this triangle is a right angle, which is 90 degrees. This triangle is the most important triangle in the field of trigonometry.

Scalene triangle
The sides are all of a different length and none of the angles between the sides are the same – it lacks symmetry.

Isosceles triangle
In an isosceles triangle two of the sides are the same length, which means that two of the angles in the triangle are the same.

Equilateral triangle
Every edge of the triangle is the same length, and every angle in the triangle is the same. This is the most symmetric triangle.

Pythagoras also demonstrated that if the lengths of two sides of a right-angled triangle are known, then the third may be calculated.

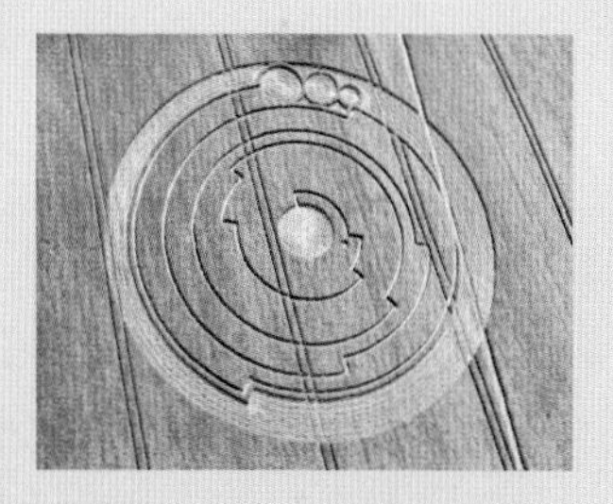

The transcendental properties of pi (ϖ)

A good example of why the ancient Greeks were fascinated by the mystical properties of numbers is pi. What is pi? It is a transcendental number (which means it goes on forever, it can never be written down completely) that is intimately linked to circles. It was first properly investigated by Archimedes of Syracuse (287–212 BC). Pi is a function of the two immediately measurable properties of a circle – its diameter and radius. It can be defined in various ways, in addition to its geometric demonstration in a recent crop circle (*above*).

The actual value of pi can never be written down completely, but here it is to 50 decimal places:

= 3.14159265358979323 846264338327950288419 716939937510 ...

If this looks large, think again: the latest supercomputers have calculated pi to 1.24 trillion digits!

Pi's function in the geometry and algebra of the circle can be demonstrated using the following algebraic formulas: Circumference of the circle = ϖ x d (diameter). Area of the circle = ϖ x $(d \div 2)^2$

Demonstrating pi When a circle's diameter is 1 its circumference is about 3.14 (pi).

FORCE AND MOTION

Eureka!

Once again, the first stepping stones to understanding how the universe works, and the forces which make it work, were laid down by the ancient Greeks. While the Pythagoreans contemplated harmony, proportion, and astronomy, Archimedes of Syracuse (c.287-c.212 BC) concentrated on more pragmatic problems, inventing siege engines, levers, a ray gun (probably using parabolic mirrors to concentrate the rays of the sun on the target), and methods of raising water using a spiral, known as Archimedes' screw; he realized that the accurate measurement of the volume of an irregular body could be achieved by displacement, familiar to students today through the story of making his discovery in his bath. Concepts like mass (density) were also identified, mass equaling weight divided by volume.

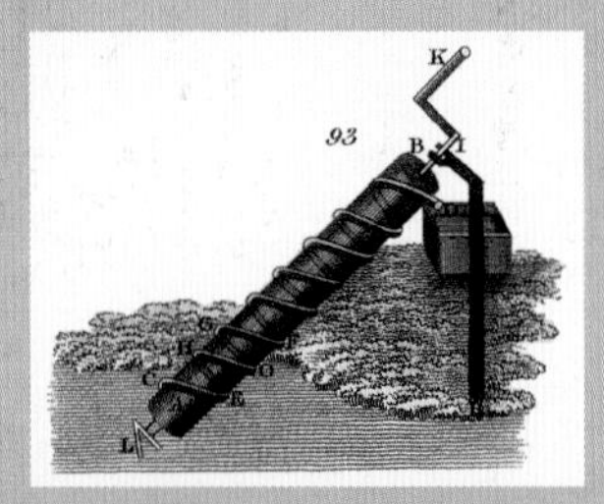

Still in use today in some traditional parts of the world to raise water for irrigation, Archimedes' screw uses the principle of the spiral.

How long is a meter? How heavy is a kilogram? How quickly does a ball drop to the ground? All of these problems are intimately tied in with forces, motions, and measurements. How do we standardize measurements? Also, how do we know how things like speed, distance, and time all relate to each other? This field of physics is mechanics – the study of forces, motions, and measurements, and the formulation of mathematical principles to define them. In order to decode these mysteries and explain how they work, a very precise language of measurements had to be established, a language that is still evolving today.

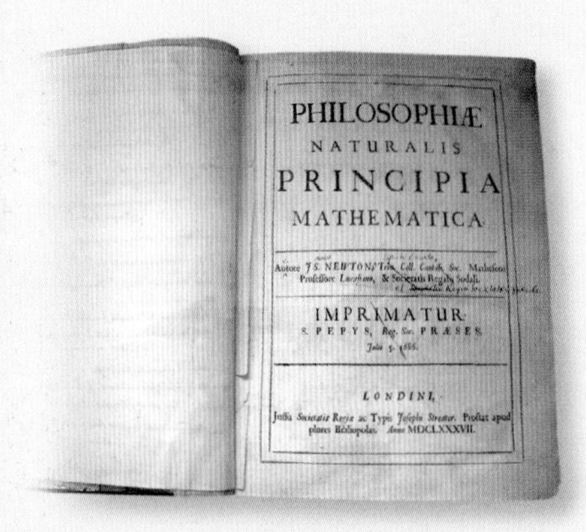

PHILOSOPHIÆ
NATURALIS
PRINCIPIA
MATHEMATICA

Autore JS. NEWTON, Trin. Coll. Cantab. Soc. Matheseos
Professore Lucasiano, & Societatis Regalis Sodali.

IMPRIMATUR
S. PEPYS, Reg. Soc. PRÆSES.
Julii 5. 1686.

LONDINI,
Jussu Societatis Regiæ ac Typis Josephi Streater. Prostat apud
plures Bibliopolas. Anno MDCLXXXVII.

Newton's famous work drew on some concepts investigated by earlier scientists such as Galileo and Kepler.

Newton and the *Principia Mathematica*

The pioneer of scientific descriptions of gravity, optics, and light was Isaac Newton. He published the most famous and important work in the field of forces and motion in 1687 called *Philosophiæ Naturalis Principia Mathematica* (The Mathematical Principles of Natural Philosophy). In this book he formulated equations describing how gravity affects masses, and the Laws of Motion.

First Law – Inertia

An object will stay at rest or moving at a constant velocity unless acted on by an external force. This means that if you throw a tennis ball, it should keep on flying in the direction you threw it indefinitely – unless an external force acts on it. If you throw a tennis ball on Earth, then the force of gravity pulls it to the ground, and the force of air resistance slows it down. If you threw a tennis ball in space it would fly in the direction you threw it in forever, never slowing down or changing direction.

Second Law – Resultant force, $F = ma$

The force on an object is equal to the product of its mass and its acceleration. This law is more easily expressed by saying that if you provide a force on an object, then it will cause it to accelerate – but the more massive the object the less the acceleration. Kicking a football causes it to accelerate away rather quickly, but kicking a cannonball with the same force causes it to accelerate away much more slowly – this is because the cannonball has more mass.

Third Law – Reciprocal actions

'Every action has an equal and opposite reaction.' This is the most famous and quoted of Newton's laws, but it has a more subtle meaning than the others. If you exert a force on an object, it exerts the same force on you. When you push against a brick wall, the brick wall is pushing back with the same force. If it wasn't, you would push through the wall.

These three laws and the equations of gravity in the *Principia* make up the earliest systems of the mechanics of motion, and together they make up Newtonian Mechanics. It would be many years before different laws of motion (such as Einstein's laws, which produce Relativistic Mechanics) would be developed.

Into the mystic

Albert Einstein (1879-1955) is the most famous scientist in modern history, noted principally for his formulation of the General Theory of Relativity, a groundbreaking work which states that the only constant in the universe is the speed of light relative to an observer – which gives rise to a range of uncomfortable ideas, such as time passing at different rates for people moving at different speeds.

In later life, Einstein would discover a phenomenon known as the 'Photoelectric Effect' which began the revolution that would become Quantum Mechanics. It shows a principle known in physics as Mass-Energy Equivalence. It tells us that anything that has mass contains an enormous amount of energy – even when it is completely stationary. Newton's Laws tell us that when an object is moving it has kinetic energy, or when it is held up high it has potential energy (which is released when it drops and falls to the ground). Einstein tells us that each tiny bit of mass also contains a vast amount of energy.

E – Energy
This is the energy that is present in the object. It is measured in joules.

c^2 – Speed of light squared
'c' is the speed of light; it is a fundamental constant with the value 300,000,000 meters per second – so when squared it is an enormous value.

m – Mass
This is the mass of the object, measured in kilograms.

$$E = mc^2$$

Energy
The c^2 in the equation means that for a small amount of mass we get a truly vast amount of energy. One gram of mass, which is about the same as a dollar bill, contains about 22 kilotons of energy. This is about the same energy as that of the atomic bomb dropped on Nagasaki. This energy is very stable in the form of mass however; it is generally only released in powerful nuclear reactions.

Complex equations are not unique to the modern world. The Incas developed a calculator in the form of knotted multicolored strings, called a *quipu*. It seems to have been used for keeping records, mathematical calculations, and sending secret messages.

Unusual units of measurement

There have been many interesting and unusual ways used in the past to define length, mass, time, and other critical measurements. Most commonly used units of measurement, such as the second and the meter, today have specific scientific definitions, which mean that should every clock and meter stick in existence be destroyed, they could be reproduced. These definitions are usually incredibly complicated, but very precise.

Second The duration of 9,192,631,770 periods of the radiation corresponding to the transition between the two hyperfine levels of the ground state of the cesium-133 atom.

Meter 1/299,792,458th of the distance light travels in an absolute vacuum in one second. This is based on fundamental unchanging values – how fast light travels in a vacuum, and the oscillations of an existing atom.

Newton The force required on Earth to move a kilogram at one meter per second squared.

Joule The energy expended by one Newton moving one meter on Earth.

Light Year Used in astrophysics – the distance light travels in a single year of 365.25 Earth days: 5,878,625,373,183.61 miles (9,460,730,472,580.8 km).

Mole Used in chemistry, a mole is a number that is roughly 600,000,000,000,000,000,000,000. This is useful in relating numbers of atoms to realistic masses – for example one mole of hydrogen atoms weighs a gram, while one mole of gold atoms weighs nearly 200 grams.

Kilogram The only standard unit of measurement that is defined in relation to an artifact rather than being defined by fundamentals like the speed of light. The IPK (International Prototype Kilogram) is a solid cylinder of a platinum-iridium alloy which is stored in a vault in the International Bureau of Weights and Measures in Sèvres, France. A kilogram is defined as the mass of the IPK. This means that the value of a kilogram actually changes from year to year as gas settles on the IPK and molecules of the IPK leave its surface.

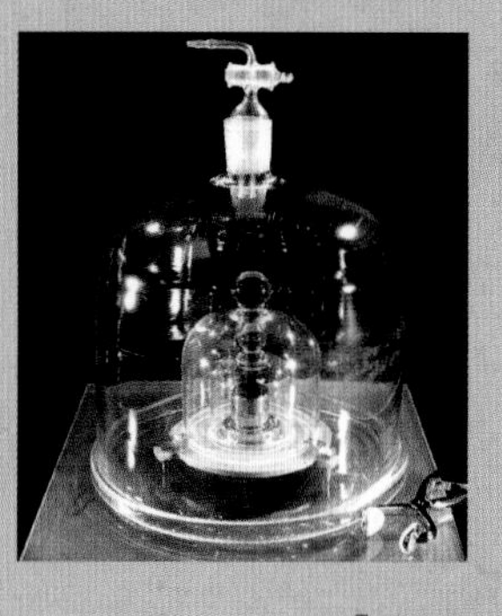

The IPK is kept under bell jars to minimize constant variations in its weight caused by atmospheric conditions.

MATHEMATICS: THE INDESCRIBABLE

Mathematics remains the most purely abstract of sciences. It is used to describe the processes behind physical phenomena that would otherwise remain indescribable. While mathematicians have developed their own 'coded' language of signs, symbols, and numbers to express their ideas (*see also pages 154, 156*), their concern is to decrypt the coded structure of the universe about us, and how it functions, by using pure calculation. In turn, cryptography and mathematics are intricately linked: from the inner workings of the Enigma machine to the latest online banking systems, mathematics is fundamental to describing how coding systems work, and to developing new coding systems.

> "The most incomprehensible thing about the world is that it is comprehensible."
>
> **ALBERT EINSTEIN, 1936.**

Calculus

One of the most important mathematical discoveries of all time is calculus. Formulated by Newton and Leibniz independently of each other, calculus can be thought of as the study of change and the infinitesimal. Below is a very famous philosophical problem that has had thinkers baffled for thousands of years – and its solution, which comes from calculus. This is one of the ancient Greek philosopher Zeno's paradoxes.

Newton's theories were written out (in Latin) without the aid of syncopation.

Achilles and the tortoise

Achilles is going to have a race with his friend the tortoise. To make things a little bit easier for the tortoise, Achilles gives him a lead of 100 meters. How does Achilles ever overtake the tortoise? By the time Achilles has run 100 meters, the tortoise has moved on a bit, say ten meters. By the time Achilles has run the extra ten meters, the tortoise has moved on another meter. By the time Achilles has run that meter, the tortoise has run another ten centimeters – Achilles always has just a tiny bit left to catch up – so how is it that he overtakes the tortoise?

The solution

What Zeno did not realize was that one can actually find a mathematical solution to this problem. Let's simplify by saying that the tortoise has a one-meter head start and runs at half the speed of Achilles – the total distance Achilles must run is:

$$1 + \frac{1}{2} + \frac{1}{4} + \frac{1}{8} + \frac{1}{16} + \cdots$$

Ad infinitum. So an infinite number of little pieces is infinity ... right? Wrong. Adding together the whole of this infinite series gives us exactly two meters. Calculus is the method we use to work out the sums of such infinite sequences.

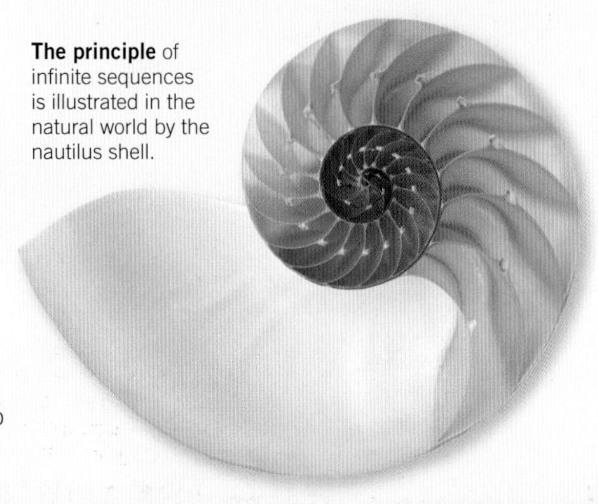

The principle of infinite sequences is illustrated in the natural world by the nautilus shell.

Algebra

Where and when did algebra begin? It is a mathematical and logical means of discovering unknown quantities by examining the function of known quantities, normally expressed by equations. The word algebra comes from the Arabic book *The Compendious Book on Calculation by Completion and Balancing* (AD 820, *above*) written by the Persian mathematician al-Khwarizm – from whose name we also derive the word 'algorithm.' However, the principles of algebra date back to the first counting systems 4,000 years ago (*see page 26*).

For us today, algebra looks like a coded language, in fact the product of 'syncopation.' For hundreds of years all algebra was expressed in sentences. For example, a problem might be written as: 'The number twenty-five is the sum of the number three and some unknown value. Twenty-five minus three produces twenty-two. Therefore the unknown value is twenty-two.'
Using syncopation, the problem can be expressed thus:

$25 = 3 + x$
$25 - 3 = x$
$x = 22$

The 19th-century revolution

From the 1800s there was an explosion of mathematical activity in Europe, as a group of gifted mathematicians revolutionized geometry, number theory, and physics – Joseph Lagrange, Pierre-Simon Laplace, Joseph Fourier, and Bernhard Riemann among them. These mathematicians formulated new types of geometry, moving beyond the precepts of Euclid, such as elliptic geometry (or Riemannian geometry) where parallel lines will always eventually intersect, and derived accurate equations for the movements of the planets and other celestial bodies. Perhaps the most influential was James Maxwell (1831-79) whose equations defining electromagnetism helped Einstein develop his general theory of relativity.

Maxwell's equations

One of the most important discoveries in the history of science, Maxwell's equations completely describe electromagnetism – how electric and magnetic fields interact with each other. These equations not only codify a complicated phenomenon, but employ several mathematical devices; to write the equations out fully would be very time-consuming, so different symbols are introduced to represent more complicated operators.

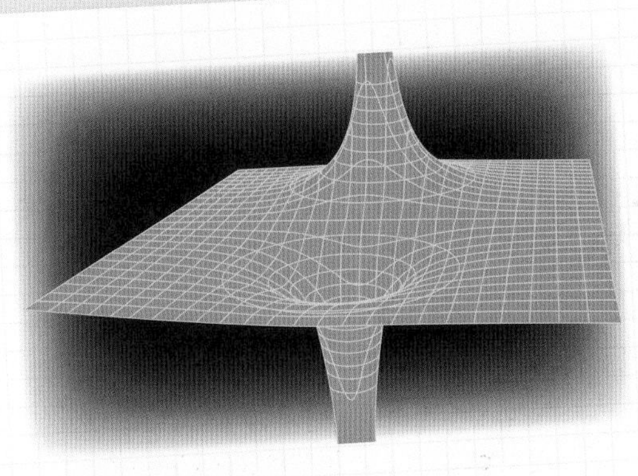

E The electric field E represents the electric field of the system. In the computer-generated picture above we see the electric field created by a positive ion (the large peak) and the field created by a negative ion (the large dip, or sink).

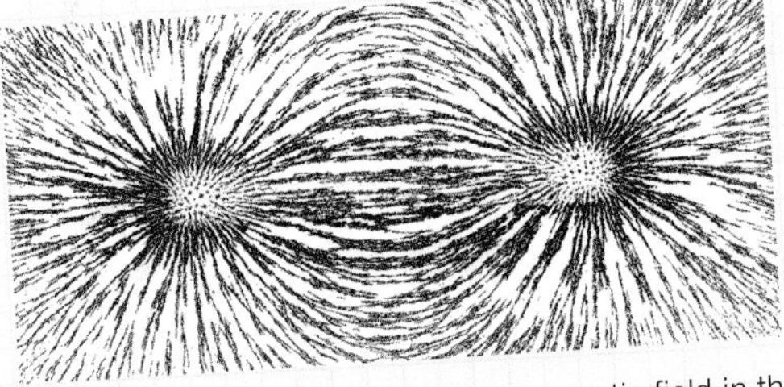

B The magnetic field This is the magnetic field in the system. The image above shows iron filings near a bar magnet – they align with the magnetic field around the magnet allowing us to see what such a field actually looks like.

$$\nabla \cdot B = 0$$

$$\nabla \cdot E = \frac{\rho}{\epsilon_0}$$

$$\nabla \times E = -\frac{\partial B}{\partial t}$$

$$\nabla \times B = \mu_0 J + \mu_0 \epsilon_0 \frac{\partial E}{\partial t}$$

$\nabla \cdot$ **The div operator is** a complicated mathematical operation that essentially tells us whether a field is a source or a sink of some property, so for example in the electric field shown there is a source (the positive charge) and a sink (the negative charge, the dip). Technically the divergence is as shown below:

$$\operatorname{div} F = \nabla \cdot F = \frac{\partial F_1}{\partial x_1} + \frac{\partial F_2}{\partial x_2} + \cdots + \frac{\partial F_n}{\partial x_n}.$$

$\nabla \times$ **The curl operator** This tells us how much the electric field is rotating or circulating at a given point. Again the operator itself looks simple, but is actually far more complicated in its full form.

$$(\vec{\nabla} \times \vec{F}) \cdot \hat{n} \overset{\text{def}}{=} \lim_{A \to 0} \frac{\oint_C \vec{F} \cdot d\vec{s}}{A}$$

$\frac{\partial}{\partial t}$ **The partial derivative with respect to time** Applying this operator to the magnetic field B means instead of looking at B we are looking at how B changes with time.

The permittivity and the permeability of free space ϵ_0 μ_0 These are fundamental physical constants relating to electromagnetism. They also combine to give us the speed of light.

Speaking as a boffin

A lexicon of symbols has been developed since Newton's day to describe a wide variety of functions.

i **The imaginary number** This is defined as the square root of minus one. In some equations we find that the solution is a complex number – a normal number with an imaginary part, such as 12 + 3i.

$\sum$ **The sum of** This indicates that one takes the sum of the expression following it.

$\begin{pmatrix} a & b \\ c & d \end{pmatrix}$ **Matrix** A matrix can be used to transform numbers and vectors by setting them in grids. For example, one could create a matrix that rotates a line in space by 90 degrees around the x axis.

∞ **Infinity** This is the mathematical symbol for infinity. In more complicated areas of mathematics we find that there are actually infinities of different sizes – some infinities are larger than others!

$\propto$ **Proportionality** This symbol indicates that two things are proportional. For example writing: 'Car Speed α Engine Size' means that as one increases the size of the engine, the speed of the car increases also.

$\therefore$ $\because$ ■ **Therefore, because, and Q.E.D.** In mathematical proofs we often have to write therefore and because, and these are shorthand symbols for them. The square is put at the end of the symbol and means Q.E.D., which is short for *quod erat demonstrandum* ('that which was to be demonstrated'), meaning that the proof is complete.

$\mathbb{N}\,\mathbb{Z}\,\mathbb{R}$ **The set of natural numbers, integers, and real numbers** In set theory, which relates to the properties of groups of numbers, we see a few sets regularly. The natural numbers are the counting numbers, 1, 2, 3, 4, and so on. The integers are all of the whole numbers, such as -1, 0, 1, 2, etc. The real numbers are any numbers that have no complex part (i.e. don't contain the imaginary number such as 12.3, 15, -19.2, and so on).

THE PERIODIC TABLE

The *Encyclopédie*

The French philosopher Denis Diderot (1713-84) was the driving force behind the *Encyclopédie*, an enormously ambitious publication which set out to encompass 'each and every branch of human knowledge.' The first volume appeared in 1751, and in all 27 folios were published over the next 20 years. The venture was fraught with difficulties, Diderot's radical views expressed in the political and social entries often threatening the cancellation of the project. The entries on science and the arts were nevertheless ground-breaking.

The Alchemical chart

Diderot's *Encyclopédie* included an 'Alchemical Chart of Affinities,' perhaps one of the earliest attempts to classify chemical compounds according to how they react *(below)*. When this chart was compiled, only some 30 chemical elements were known (many others that were included were in fact compounds – mixtures of elements). Diderot's younger contemporary, the French chemist Antoine Lavoisier (1743-94) also pioneered a classification system based on properties and reactions of known elements. Diderot's chart draws in part on alchemical symbols for the various elements and is laid out in tabular form.

The roots of modern chemistry lie with the medieval alchemists (*see page 52*), whose search for the 'philosopher's stone' and the ability to turn base substances into gold led them to investigate the properties of the chemicals they discovered and used. The development of the more formal science of chemistry in the 18th century saw a more rigorous approach to classifying the elements which make up the universe. The enormity of the problem required a new coded language, one in which names and properties were embedded, and flexible enough to express functions. A variety of tables were drawn up, but the modern periodic table was first developed by Russian chemist Dmitri Mendeleev in 1869; it represented a huge leap forward in understanding chemistry. The table is 'periodic' because elements in the same period (or column) contain similar properties; the modern periodic table sums up an enormous amount of information about the chemistry of the elements. A cursory glance at the table reveals, in coded form, the basic properties of the elements known to science.

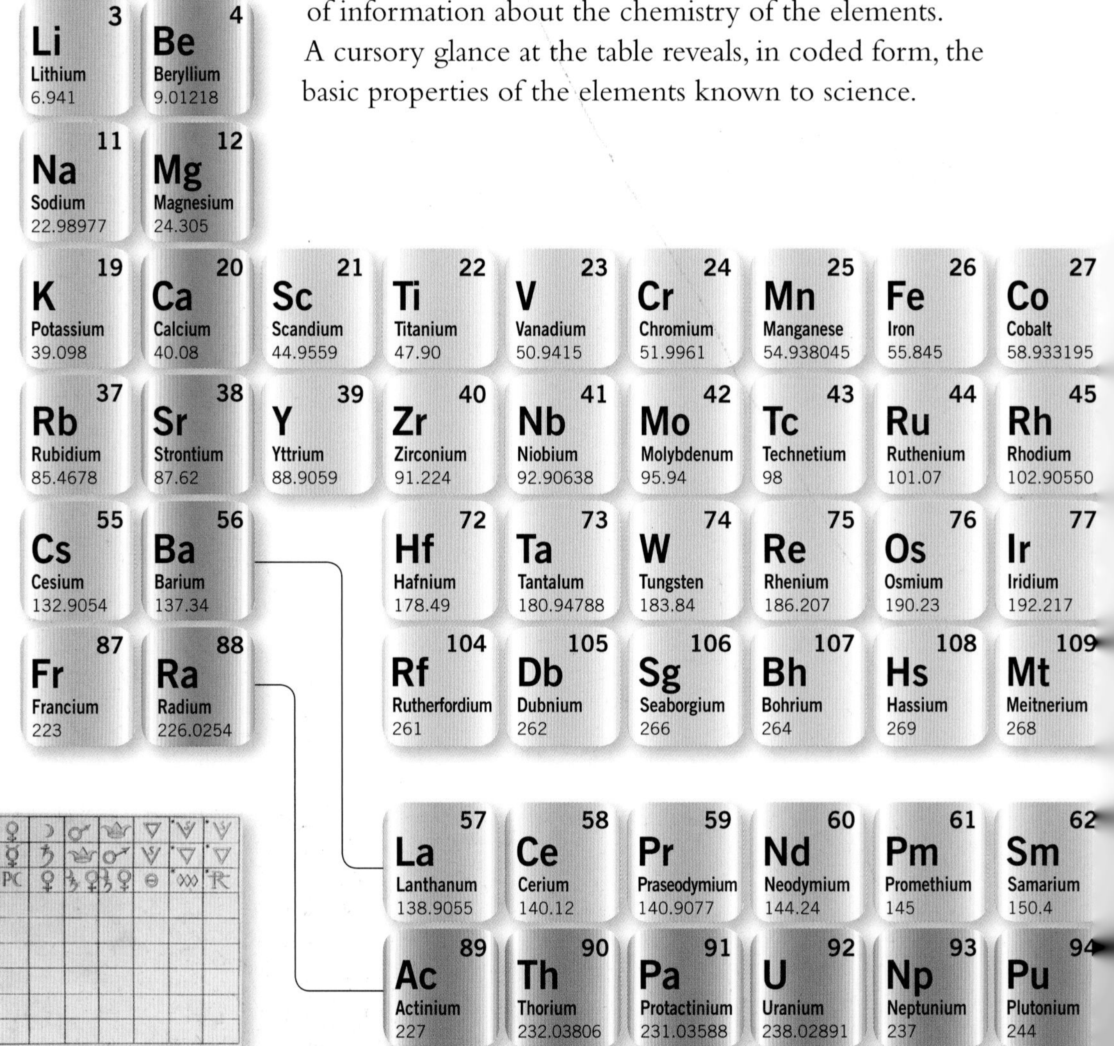

The periodic table

The elements are arranged in 18 columns ('groups') and seven rows ('periods'). Why does a certain element go in a certain group? That is to do with its valence electrons – the electrons that react with other particles. If two elements have similar valence electrons they will react in a similar way. When Mendeleev first created the modern periodic table he included gaps for elements he assumed had yet to be discovered.

The noble gases are the last group in the periodic table and have a completely full outer shell of electrons. This is an extremely stable configuration and as such these gases are extremely unreactive in most conditions. They are used extensively in lighting – argon fills many light bulbs so that the filaments don't burn, and is used to create 'inert' atmospheres when preparing sensitive chemicals. Neon glows when in a bulb and is used in neon lights. Helium is very light and unreactive so is perfect for filling balloons and blimps.

Key

Alkali metals · Alkaline earth metals · Transition metals · Poor metals · Other nonmetals · Noble gases · Lathanoids · Actinoids

Atomic symbol: H · Atomic number: 1 · Atomic name: Hydrogen · Atomic mass: 1.0079

								2 He Helium 4.00260
			5 B Boron 10.81	6 C Carbon 12.011	7 N Nitrogen 14.00674	8 O Oxygen 15.9994	9 F Fluorine 18.999840	10 Ne Neon 20.179
			13 Al Aluminum 26.98154	14 Si Silicon 28.086	15 P Phosphorus 30.97376	16 S Sulfur 32.06	17 Cl Chlorine 35.453	18 Ar Argon 39.948
28 Ni Nickel 58.6934	29 Cu Copper 63.546	30 Zn Zinc 65.409	31 Ga Gallium 69.723	32 Ge Germanium 72.64	33 As Arsenic 74.92160	34 Se Selenium 78.96	35 Br Bromine 79.904	36 Kr Krypton 83.80
46 Pd Palladium 106.42	47 Ag Silver 107.8682	48 Cd Cadmium 112.411	49 In Indium 114.818	50 Sn Tin 118.710	51 Sb Antimony 121.760	52 Te Tellurium 127.60	53 I Iodine 126.90447	54 Xe Xenon 131.30
78 Pt Platinum 195.084	79 Au Gold 196.966569	80 Hg Mercury 200.59	81 Tl Thallium 204.3833	82 Pb Lead 207.2	83 Bi Bismuth 208.98040	84 Po Polonium 209	85 At Astatine 210	86 Rn Radon 222
110 Ds armstadtium 71	111 Rg Roentgenium 272	112 Uub Ununbium 285	113 Uut Ununtrium 284	114 Uuq Ununquadium 289	115 Uup Ununpentium 288	116 Uuh Ununhexium 292	117 Uus Ununseptium Unknown	118 Uuo Ununoctium 118

63 Eu uropium 51.96	64 Gd Gadolinium 157.25	65 Tb Terbium 158.92535	66 Dy Dysprosium 162.500	67 Ho Holmium 164.93032	68 Er Erbium 167.259	69 Tm Thulium 168.93421	70 Yb Ytterbium 173.04	71 Lu Lutetium 174.967
95 Am mericium 43	96 Cm Curium 247	97 Bk Berkelium 247	98 Cf Californium 251	99 Es Einsteinium 252	100 Fm Fermium 257	101 Md Mendelevium 258	102 No Nobelium 259	103 Lr Lawrencium 262

Other codes in chemistry

Most people are familiar with the formulas used to express certain compound substances; these are a shorthand way of describing the molecular structure of the substance, and use the symbols of the elements which make up the compound, and numbers to indicate the proportion of each element. Among the most common are:

Ammonia NH_3 One nitrogen atom and three hydrogen atoms.
Carbon dioxide CO_2 One carbon atom and two oxygen atoms.
Chalk, limestone, marble $CaCO_3$ One calcium atom, one carbon atom, and three oxygen atoms.
Caustic soda NaOH One sodium atom, one oxygen atom, and one hydrogen atom.
Hydrogen chloride HCl One hydrogen atom, one chlorine atom.
Salt NaCl One sodium atom and one chlorine atom.
Sulfuric acid H_2SO_4 Two hydrogen atoms, one sulfur atom, and four oxygen atoms.
Washing soda Na_2CO_3 Two sodium atoms, one carbon atom, and three oxygen atoms.
Water H_2O Two hydrogen atoms and one oxygen atom.

Molecular structure

Chemical structures may be drawn in different ways.

Benzene: C_6H_6
The compound benzene is a ring of carbon atoms bonded to hydrogen atoms.

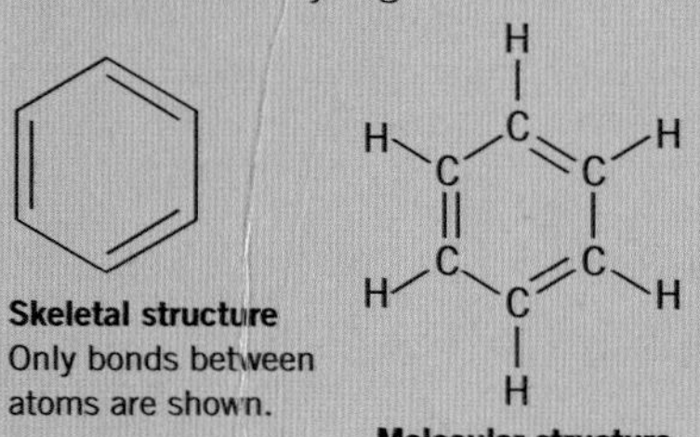

Skeletal structure
Only bonds between atoms are shown.

Molecular structure
All atoms are shown.

Caffeine: $C_8H_{10}N_4O_2$

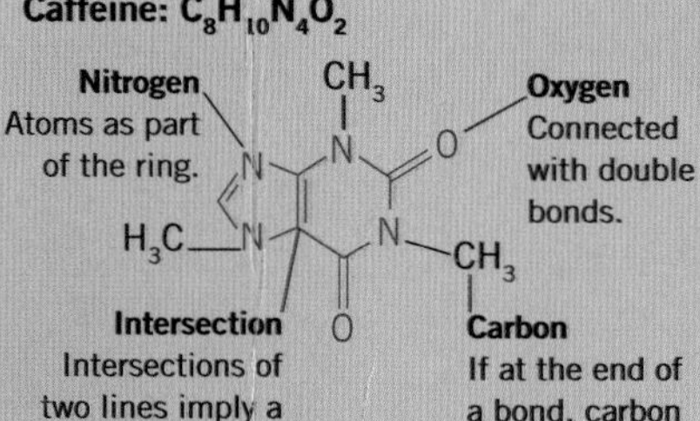

Nitrogen
Atoms as part of the ring.

Oxygen
Connected with double bonds.

Intersection
Intersections of two lines imply a carbon atom.

Carbon
If at the end of a bond, carbon is shown.

DEFINING THE WORLD

The urge to describe in diagrammatic form the landscape beyond the horizon is found in many cultures, from early rock art to classical China, Japan, and Rome. In each instance surprisingly similar solutions to the problem of graphically encoding features like rivers, coasts, mountains, the sea, and settlements were developed.

Scale and orientation

What made each mapping system different was a function of scale and orientation. Today, we are familiar with maps oriented to the north with a specific purpose and scale. In the past, scale was often measured in days of travel, purpose was a reflection of the patron's needs, and orientation often linked to this. The orientation of most Muslim maps of the time was towards the south, but what is interesting is the formulation of a symbolic language for specific features which is readily recognizable today.

The world upside down

Arab geographers are ranked among the most ambitious mapmakers of the pre-modern era. Al-Idrisi produced this map of the known world for Roger of Sicily c.1154. The Muslim realms stretched from the borders of China to the shores of the Atlantic; meanwhile, Muslim travelers such as Ibn Batuta and ambitious Arab traders were exploring opportunities well beyond the limits of the Muslim world in Africa and East Asia.

The Peutinger Table

This 3rd-century AD late Roman map demonstrates, with little contextual geographical information, the adage that 'All roads lead to Rome.' Such 'strip' maps showed the principal road system and how to get from one place to another, with little additional detail (not unlike GPS navigation maps today). In the Middle Ages similar maps were produced to guide Christian pilgrims to Jerusalem.

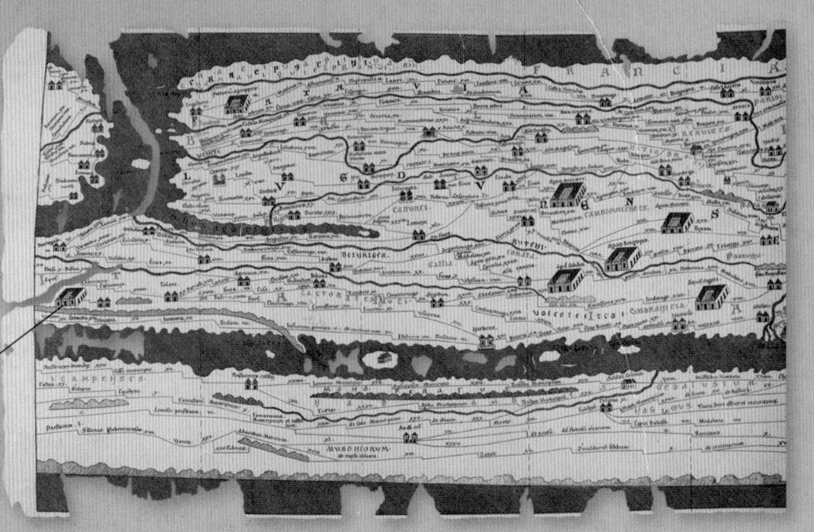

In addition to Roman roads, the map picks out major cities along the traveler's route, using an icon which is fairly regular in form, showing a walled settlement, and which varies in size to reflect the relative importance of each city.

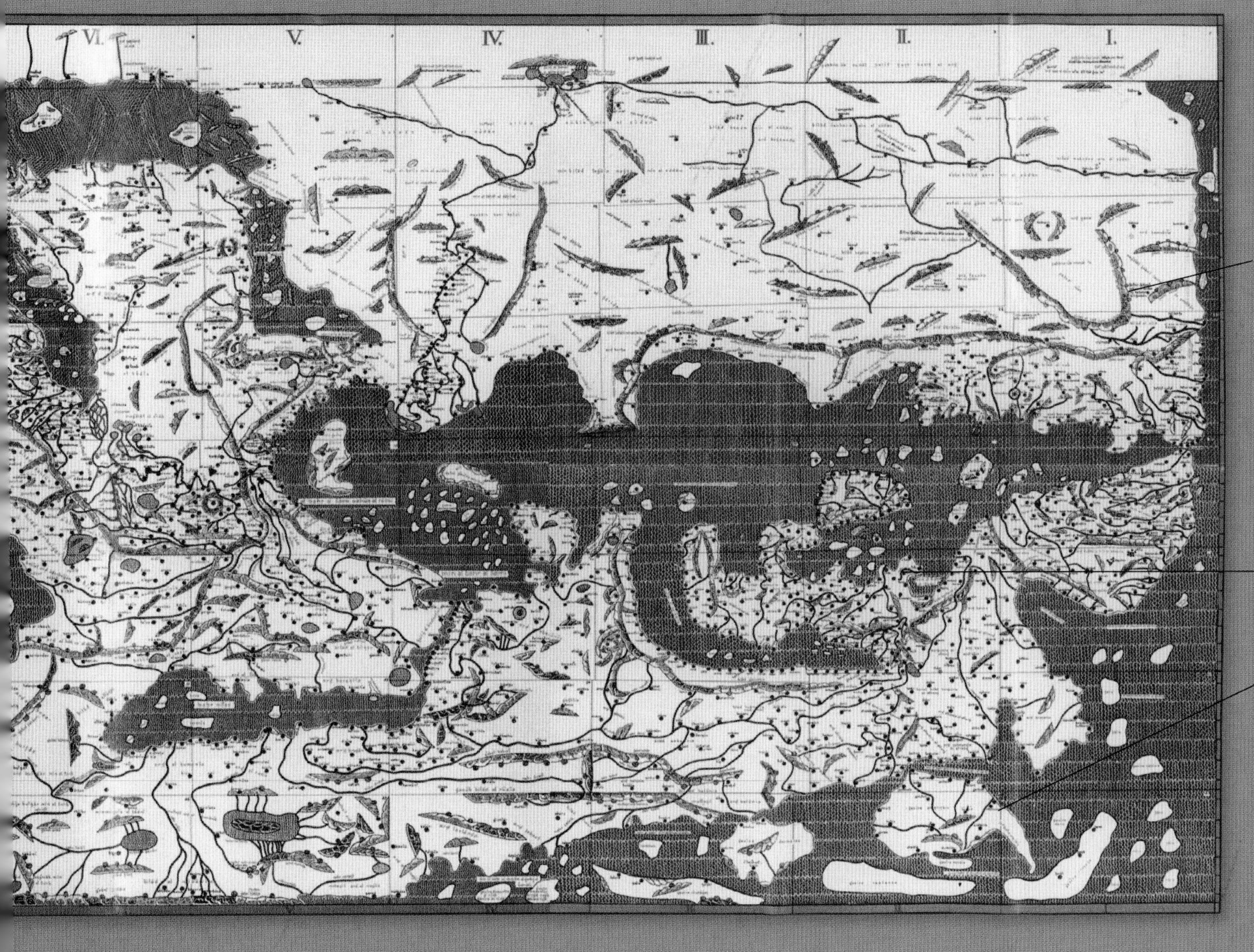

Mountains
Major ranges are picked out as chains, but with little attempt to convey their height or area. They are often accompanied by small tree symbols indicating dense forest.

Dense detail
The shores of the Mediterranean and southwest Asia are filled with detail, and the familiar 'boot' shape of Italy can be clearly made out.

Less information
It is clear the cartographer only has a partial understanding of northwest Europe. The British Isles are very distorted, and several non-existent Atlantic islands have been added.

Portolan charts

As Europeans began to search for new markets and colonizing opportunities, new maps were produced to aid the pioneering navigators. This map is from the Portolan Atlas of 1540, and indicates compass bearings. Often unfamiliar areas were generalized or guessed. Particular attention was paid to coastal topographies, features, and settlements, while unfamiliar tracts and interior regions were often left blank, and sometimes populated by fanciful illustrations.

The dense equatorial rain forest of the Brazilian interior had hardly been explored when this map was produced, but a small vignette indicates that it exists.

Encoding the Landscape

Mapmaking remains one of the most sophisticated systems of encoding information ever invented. As a way of describing the complexities of the world about us – notwithstanding modern advances in satellite GPS and the wonders of Google Earth – cartography remains a coding system unparalleled in flexibility and richness of information. Maps may be said to have one unifying basic function: to show where somewhere is in relation to other places. Nevertheless, maps can convey a massive quantity of information through the use of an array of graphic devices: grids, colors, shading, lines, and symbols, while varieties of typography are used to identify many different features.

Surveying
The first accurate and detailed surveys were conducted to create naval charts. The accurate plotting of coastlines, submerged hazards, and tides and currents was increasingly important for shipping in the early modern era. Land surveys in such close detail began with the establishment of the Ordnance Survey in Britain in 1747 when accurate maps for artillery bombardment whilst quelling the Highland rebellion were needed. A detailed survey of Britain followed, and continues today. The same techniques were used during the Great Trigonometric Survey of India (*below*), when native teams (*above*) were trained to map the subcontinent.

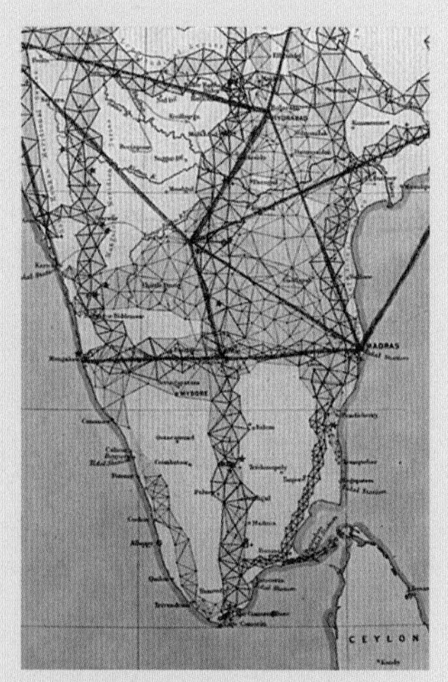

The Survey of India used triangulation on a range of scales to plot both the 'flat land' configuration of the landscape, and the elevations of hills and mountain ranges.

Landscape description

A variety of conventions for graphically describing the many features of the world around us have developed over the past 500 years. Depending on the scale of the map, the cartographer's task is always selective, but modern satellite technology has made it possible to show the physical relief of the Earth's surface in extreme detail.

Early modern maps The first large-scale maps of localities described the landscape, foliage, and distribution of man-made features using a variety of generalized pictograms, the forerunners of modern map icons.

Hill shading and hachuring

The need to render the landscape more accurately on a flat surface led to attempts to create a three-dimensional effect through shading or fine lines (hachuring). Very specific topographic features were carefully named, and spot heights given. A more rigorous method of showing landscape developed with contouring, where lines join points of a common height forming closed loops, the gaps between them often being colored (*see opposite*).

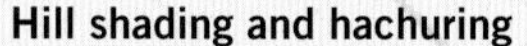

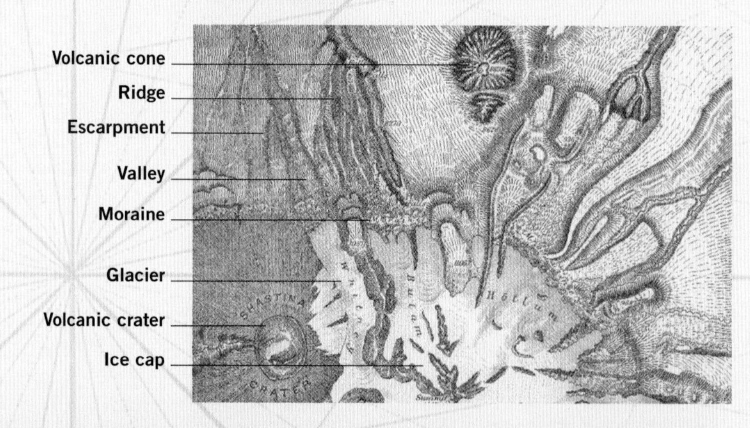

Remote sensing Bouncing infrared beams off the Earth's surface from orbital satellites has provided dense data sets describing the Earth's topography. The signals themselves are in coded form, and need to be interpreted using a variety of algorithms to produce digital terrain models (DTMs), which now form the basis of the most accurate modern mapping. Here an area of Tibet is displayed in false color.

The lines on the map

To locate features, a reference coding system, which relies on artificial lines, is usually drawn over the map. This is commonly based on 'graticules,' selected lines of longitude (meridians which join the North and South poles) and latitude (or 'parallels') projected over the globe. Locations are defined using coordinates expressed in degrees and minutes (or more recently decimal degrees). Maps may also have a simple, geometric grid system.

In this example, the location of Seville may be defined in longitude/latitude coordinates as 37.24°N 5.59°W, or on the publisher's grid (which is, in this case, based on graticules) as 7C.

Latitude
Orientation north or south of the equator (parallels).

Longitude
Orientation east or west of the Prime (or Greenwich) Meridian.

Greenwich Meridian 0°
Longitude is measured west or east of this line.

Publisher's grid
Shown in red numerals and letters, this is an easily understood coded reference to the positions of features on the map.

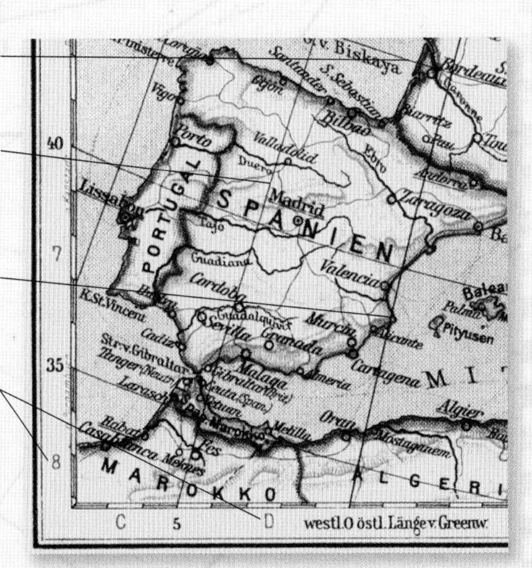

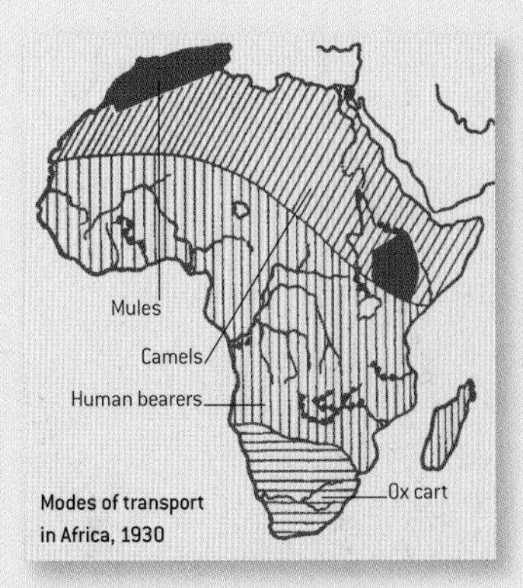

Modes of transport in Africa, 1930

Schematic maps

Maps can be used to show a variety of distributions (*above*), such as population density or land use, known as 'thematic mapping.' Schematic maps (*below*) are increasingly used to display information with underlying geographic features, frequently in a highly stylized, simplified form. These maps are often used to show traffic movement or communication networks, such as rail or subway systems. Both thematic and schematic maps create their own coded language, usually explained in a key or legend.

Modern mapping

Most general maps use a wide range of colors, lines, and symbols to convey information. Conventions may change according to the style and purpose of the map – highway maps for example emphasizing the network and classification of roads.

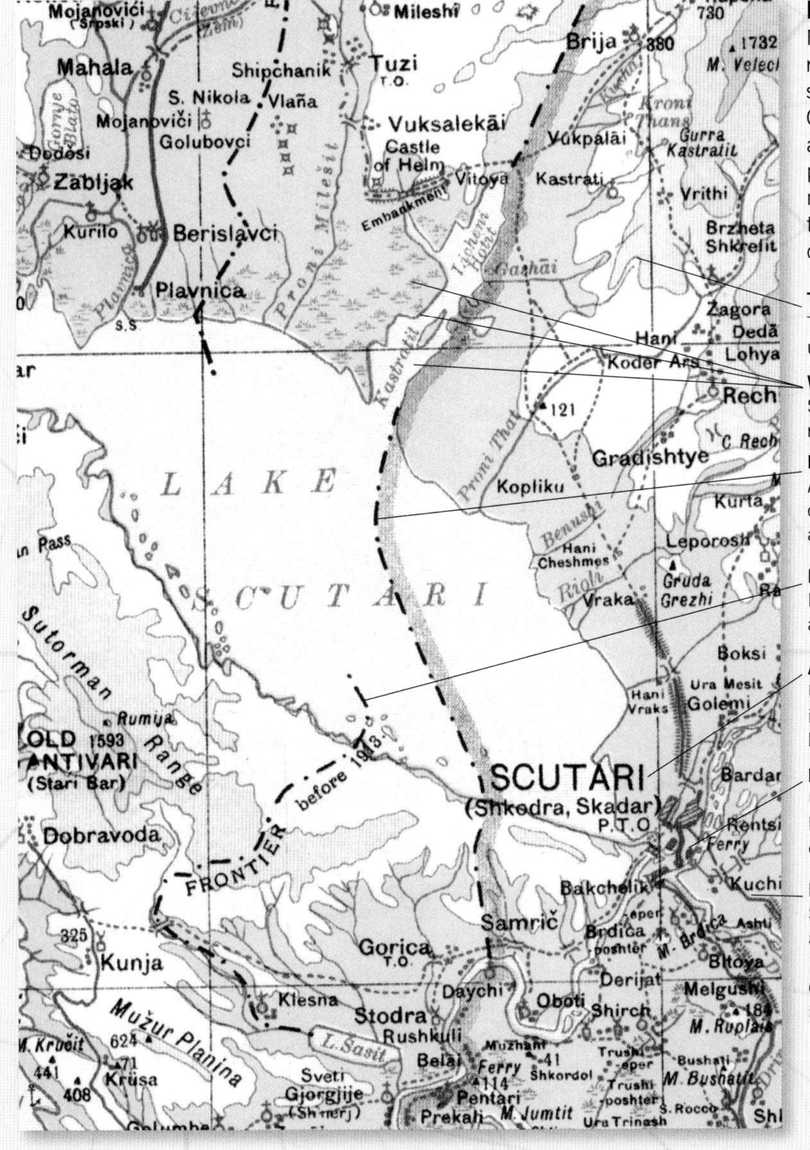

Topography
The shape of the landscape is shown using contours and coloring.

Water features
Seas, lakes, coastline, marshes, and rivers often shown in blue.

Demarcation lines
Although for the most part invisible on the ground, national and other administrative borders may be shown.

Historical features
Here a historical border has been added.

Administrative centers
Usually picked out in larger typeface, if extensive the shape of the city may be described.

Populated places
Usually shown using a town stamp; here a series of red dots is used to demonstrate relative sizes.

Roads
Shown here as dotted or solid lines according to size.

Other features shown on this map include:
♁ Church
Mosque
⌑ Fortification
)(Bridge

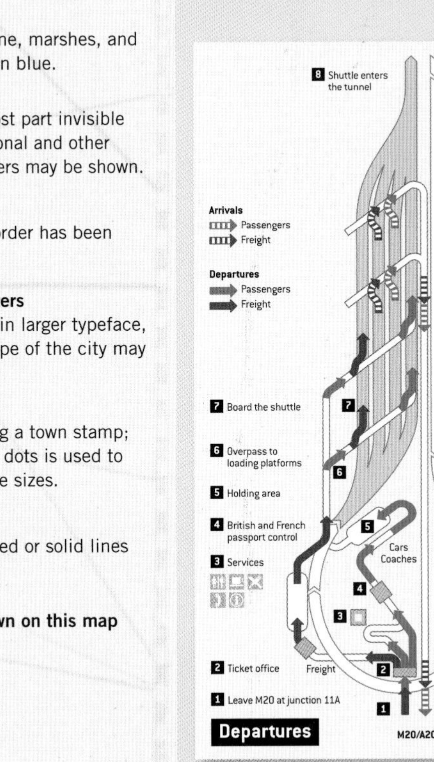

NAVIGATION

Methods of finding your way at sea, avoiding reefs and other hazards, have one of the most extraordinary histories. No one knows how the first navigators found their way across the Pacific to colonize the islands of Australasia and Oceania thousands of years ago, while in the last three decades, the advent of satellite navigation has all but negated the navigational devices and codes developed over the last two millennia. Nevertheless, the techniques that were developed to aid sailors remain a monument to the human ability to find a solution to apparently intractable problems.

Lighthouses and lightships

Beacons as navigational aids date back over 2,000 years, but modern occulting lights were the 19th-century invention of Charles Babbage. Lights are divided into Major (high intensity, marking an important feature) or Minor (secondary intensity, at harbor or estuary entrances). The default color of the light is white, although red, green, or yellow may be used.

Fixed (F) Constant, often colored.

Flashing (FL) Up to 30 a minute.

Quick Flashing (Q) 60 a minute.

Interrupted Quick Flashing (IQ)

Isophase (ISO) Equal long phases.

Group Flashing (GP FL)

Occulting (OCC)

Alternating (AL) Changing colors.

Long Flash (LFL) At least two seconds.

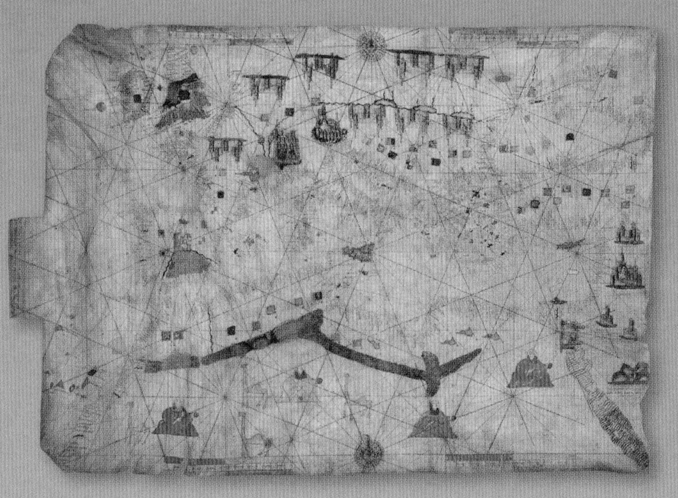

The Arabs and some Mediterranean nations produced rudimentary charts to aid navigators from the 13th century onwards. Although small in scale, these provided some information about the profile of coastlines, landmarks at sea, and compass bearings. Portolan charts (*left*) indicated the location of coastal towns and distances between them using a scale.

Charts and pilots

The outstanding difference between maps of the landscape (*see page 164*) and charts of the sea is that the former represent what can be readily seen by the user; maritime charts, and associated lighthouses, lightships, and buoy markers, need to inform the sailor about the coastline and what is invisible under the ocean, information of critical importance to the navigator. The British Admiralty was the first to produce detailed navigation charts from the 17th century (although localized Portolan charts, largely for Mediterranean and inshore European and African navigation had been produced in Iberia from the 13th century). The Admiralty also produced detailed 'pilots' describing the coastline, identifying features like mountains, ports, and other landmarks.

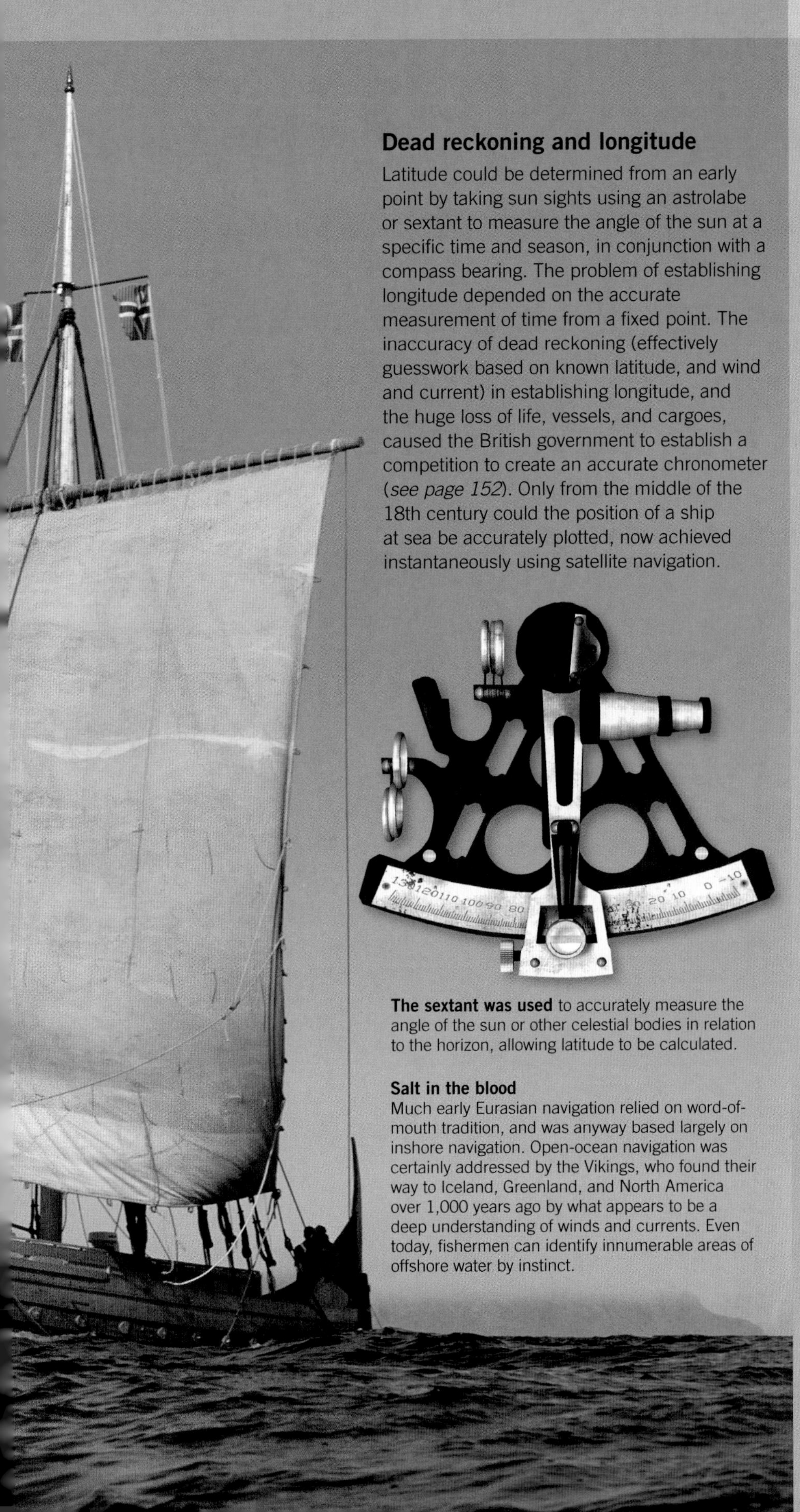

Dead reckoning and longitude

Latitude could be determined from an early point by taking sun sights using an astrolabe or sextant to measure the angle of the sun at a specific time and season, in conjunction with a compass bearing. The problem of establishing longitude depended on the accurate measurement of time from a fixed point. The inaccuracy of dead reckoning (effectively guesswork based on known latitude, and wind and current) in establishing longitude, and the huge loss of life, vessels, and cargoes, caused the British government to establish a competition to create an accurate chronometer (*see page 152*). Only from the middle of the 18th century could the position of a ship at sea be accurately plotted, now achieved instantaneously using satellite navigation.

The sextant was used to accurately measure the angle of the sun or other celestial bodies in relation to the horizon, allowing latitude to be calculated.

Salt in the blood

Much early Eurasian navigation relied on word-of-mouth tradition, and was anyway based largely on inshore navigation. Open-ocean navigation was certainly addressed by the Vikings, who found their way to Iceland, Greenland, and North America over 1,000 years ago by what appears to be a deep understanding of winds and currents. Even today, fishermen can identify innumerable areas of offshore water by instinct.

Buoys

Since 1977, there have been two internationally recognized systems for buoy markings, IALA A (Europe and most of the world) and IALA B (the Americas, Japan, Korea, and the Philippines), although the variations today are minimal. Buoys are anchored, floating markers, often with phased flashing lights, providing specific information for the navigator concerning hazards and channel passages. Location, shape, color scheme, and other markings provide a coded message.

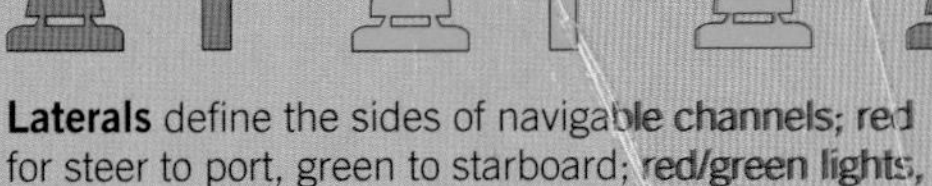

Laterals define the sides of navigable channels; red for steer to port, green to starboard; red/green lights, no specific phase.

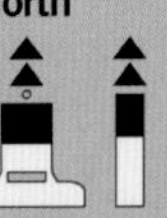

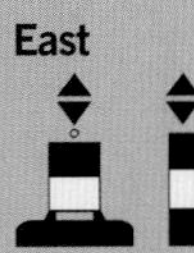

Cardinals show the direction from the mark in which the most navigable water lies, or a required change of course; lights, always white, phases vary.

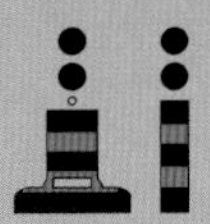

Isolated danger shows a small hazard with navigable water all around; lights white, group flashing.

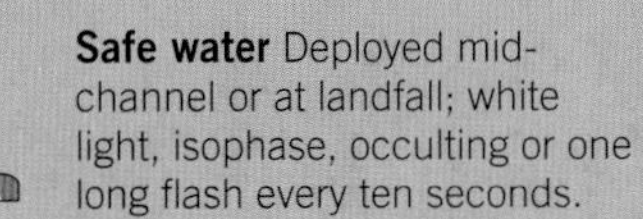

Safe water Deployed mid-channel or at landfall; white light, isophase, occulting or one long flash every ten seconds.

Special No specific purpose or shape, just a general beware of inshore hazard; colored light, no specific phase.

Navigation conventions It has long been established that approaching vessels should pass starboard to starboard. At night, vessels mount a green light on the starboard bow, a red on the port, and a third, white light on the mast to identify their direction.

TAXONOMY

Archaeopteryx lithographica (extinct)

A fundamental 'code' in terms of our understanding of ourselves and our fellow species is the taxonomic 'code,' or biological organization, of all forms of life on Earth. Since classical times, humans have attempted to create order out of the countless organisms on Earth in order to understand how various plants, animals, fungi, and bacteria are potentially related to each other. It wasn't until the early 18th century, when Swedish naturalist Carl Linnaeus devised a system that classified organisms based on physical similarities, that a system was established which is still used today. Indeed, Linnaeus' classification system, known as 'binomial nomenclature,' helped pave the way for Darwin's ideas in the 19th century. Analysis of the DNA code (*see pages 170-175*) has allowed for even greater understanding of how all forms of life on Earth are related.

Classical roots

Attempts to understand how distinct organisms are related to each other are quite ancient. Probably the best known example is the Greek philosopher Aristotle, who was also most likely the first to classify each known organism, or 'being.' It is Aristotle's writings on how to divide and classify organisms that brought us such words as 'substance,' 'genus,' and 'species.' Aristotle used several features to classify organisms, such as whether they had 'red blood,' how they reproduced, and what kind of habitat they lived in. Aristotle's work also paved the way for naturalists like Linnaeus, who – as a student of the Enlightenment – would have read his classical works.

Medieval observations

Medieval scholars still utilized portions of Aristotle's ideas in their search for the relationships among all living things. Aristotle's 'being' therefore permeated most medieval works on the subject. In fact, the scientist and philosopher St. Thomas Aquinas developed the idea of the 'analogy of being,' which later became the field of ontology (the metaphysical study of how separate entities, or organisms, are connected). The 13th-century Franciscan friar Roger Bacon has often been identified as the author of the so-called Voynich manuscript (*left*). This mysteriously enciphered document has never been decrypted, but it appears to contain systematic analyses of plants, among other natural phenomena; several of the species included, however, are thought to have been unknown before European contact with the Americas, which may rule out Bacon as the author.

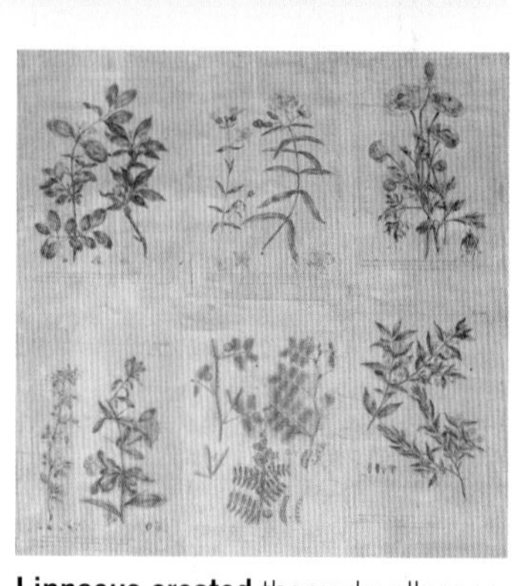

Linnaeus created themed wallpaper for his house from his plant drawings.

Linnaeus and taxonomy

Carl Linnaeus (1707-78) is often known as the 'father of modern taxonomy' – and for good reason. An eclectic polymath, he drew heavily on the research of his peers to create a robust system of codified classification for all living things. His *Systema Naturae*, first published in 1735 and continually updated, utilized a new system of classifying organisms based on a hierarchy that we still use today. The categorization by kingdom, phylum, class, order, family, genus, and species was pioneered by Linnaeus, and gave rise to the system of binomial nomenclature to differentiate every organism, such as humans (*Homo sapiens*) from chimpanzees (*Pan troglodytes*). The organization of how he grouped various organisms has since been heavily modified due to advances in developmental biology.

Linnaeus' family was named after a giant linden tree.

Species: lythographica
Genus: Archaeopteryx
Family: Archaeopterygidae
Order: Archaeopterygiformes
Class
Sub-phylum
Phylum
Kingdom

Geospiza fortis
Medium Ground-finch (extant)

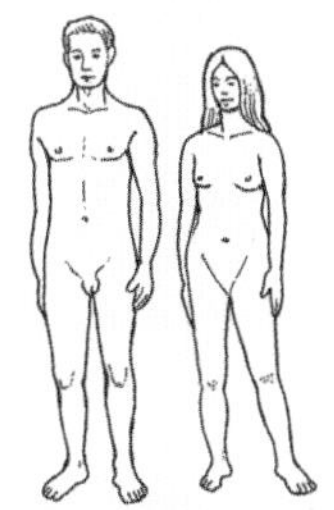

Homo sapiens
Human being (extant)

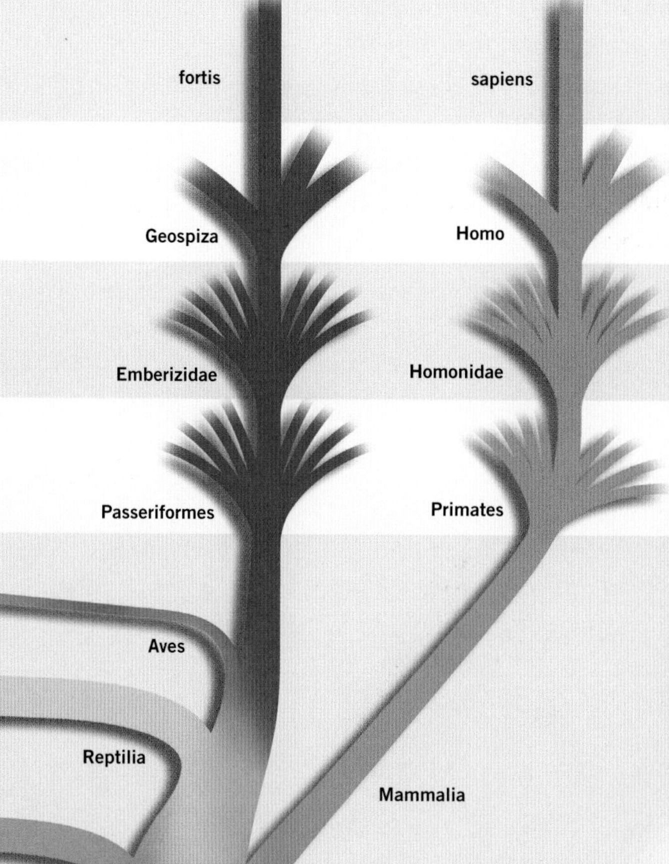

This artwork demonstrates the classification of three animals into the major taxa following Linnaeus' structure. The 'class' level has been expanded to show all seven recognized classes arranged as a 'phylogenetic tree' hinting at the evolutionary relationships between the groups.

Archaeopteryx lithographica (*top left*) is as yet the sole occupant of its own order. Although classified as a bird (Aves) it shares many characteristics of small dinosaurs (Reptilia), and is probably closely related to the original bird ancestor.

How do scientists classify organisms?

In simple terms, scientists divide and classify organisms based on the similarities or differences between them. However, the actual process is much more complicated. This is because two animals can look alike, but have vastly different evolutionary histories. For example, both birds and bats have wings, but are not closely related at all, whereas dinosaurs are now thought to be the ancient ancestors of birds. So, these scientists – called 'systematists' – must distinguish between similarities that are meaningful, and similarities that simply arose by chance. In order to do this, systematists examine very closely such features as anatomy, development, and methods of sexual reproduction, as well as the fossil record. Recently, advances in genetics (*see pages 170-175*) have allowed taxonomic classification at the genetic level.

Darwin and taxonomy

As Charles Darwin (1809-82) developed his theory of evolution by natural selection, the ability to utilize the classifications of Linnaeus and his successors to develop a 'tree of life' *(below)* connecting various organisms and lineages was immensely useful. Indeed, in Darwin's *On the Origin of Species*, published in 1859, he showed that not only can organisms be classified taxonomically (as Linnaeus had done), but that they all share a common genealogy. After Darwin, systematists began to understand that one could use evolutionary relationships to deduce how closely various species were related to each other – that the taxonomic categories of class, order, and genus would reflect a similar pattern in evolutionary relationships – leading to an even more detailed picture of how the various forms of life on earth are related.

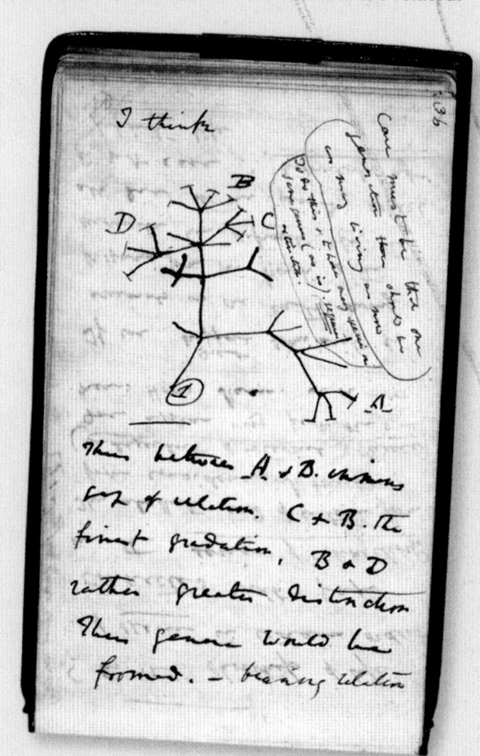

The future of taxonomy

The taxonomic 'code' has evolved much since Aristotle first attempted to classify organisms over 2,000 years ago. The exploration of the New World and Linnaeus' hierarchical system of classification, as well as the biological advances of the last 200 years, have helped us to understand how organisms evolved and how they relate to each other with an increased degree of accuracy. However, science itself is always evolving, and recent research in evolutionary genetics means that we are beginning to understand the relationships between organisms at a genetic level. For example, we now know that humans and chimpanzees share 98.5% of the same DNA, confirming the theory that we share some sort of common ancestry. But then again, humans and bananas share a high level of genetic similarity too, which is perplexing.

THE GENETIC CODE

Of all the codes in existence, perhaps the most fundamental is the genetic code. This code, imprinted in the DNA of every organism alive today, contains a list of instructions for how we function and reproduce, not to mention deciding the color of our hair, and if we like brussels sprouts. The organization of each organism's DNA determines whether we are a human, a chimpanzee, or a banana, as well as whether we are at greater risk for heart disease, diabetes, and breast cancer. 'Cracking' this genetic code has been the task of scientists for the past 50 years, so that we may gain insight on similarities we have to other animals, as well as our similarities to each other.

Watson and Crick

Perhaps the two names most associated with DNA are James Watson (*left*, b.1928) and Francis Crick (*right*, 1916-2004). In 1952 both Watson and Crick were researchers at the Cavendish Laboratory at the University of Cambridge with the goal of determining the structure of DNA. In 1952 there was no definitive understanding as to how DNA was structured, organized, or how vital it was in determining our genetic code. Watson and Crick attempted to discover the structure of DNA by playing with scale-model atoms. Soon they discovered how the four bases of adenine, thymine, cytosine, and guanine fit together. They noticed that the molecular structure of each of these bases was such that adenine only fits together with thymine, while cytosine only fits together with guanine. Using this information, they decided to stack these bases on top of each other to see the entire structure. The result was the now-famous 'double helix,' most often compared to a winding staircase. Watson and Crick won the Nobel Prize for their work, together with colleague Maurice Wilkins, in 1962. Though their discovery has been colored by controversy, such as the role of fellow researcher Rosalind Franklin's previous findings, and Watson's statements on race and gender, they are still lauded for bringing to public light the structure and function of DNA.

How the genetic code works

The instructions, or 'blueprints,' that determine how our bodies are constructed and function are housed within each of the trillion cells of our bodies. Each cell's nucleus (excluding the germ cells) contains an identical set of structures called chromosomes. The chromosomes in turn consist of a compound called deoxyribose nucleic acid (DNA). It is the number of chromosomes and variety of genes within each chromosome that makes a human a human, a gorilla a gorilla, and a banana a banana. For example, humans have 46 chromosomes, gorillas have 48 chromosomes, and bananas have 33 chromosomes. Furthermore, although all members of the same species have the same number of genes housed in the same number of chromosomes, many genes have a number of variations (e.g. genes for eye color, hair color, etc.), and it is the specific combination taken from the overall 'gene pool,' that makes each of us a unique being.

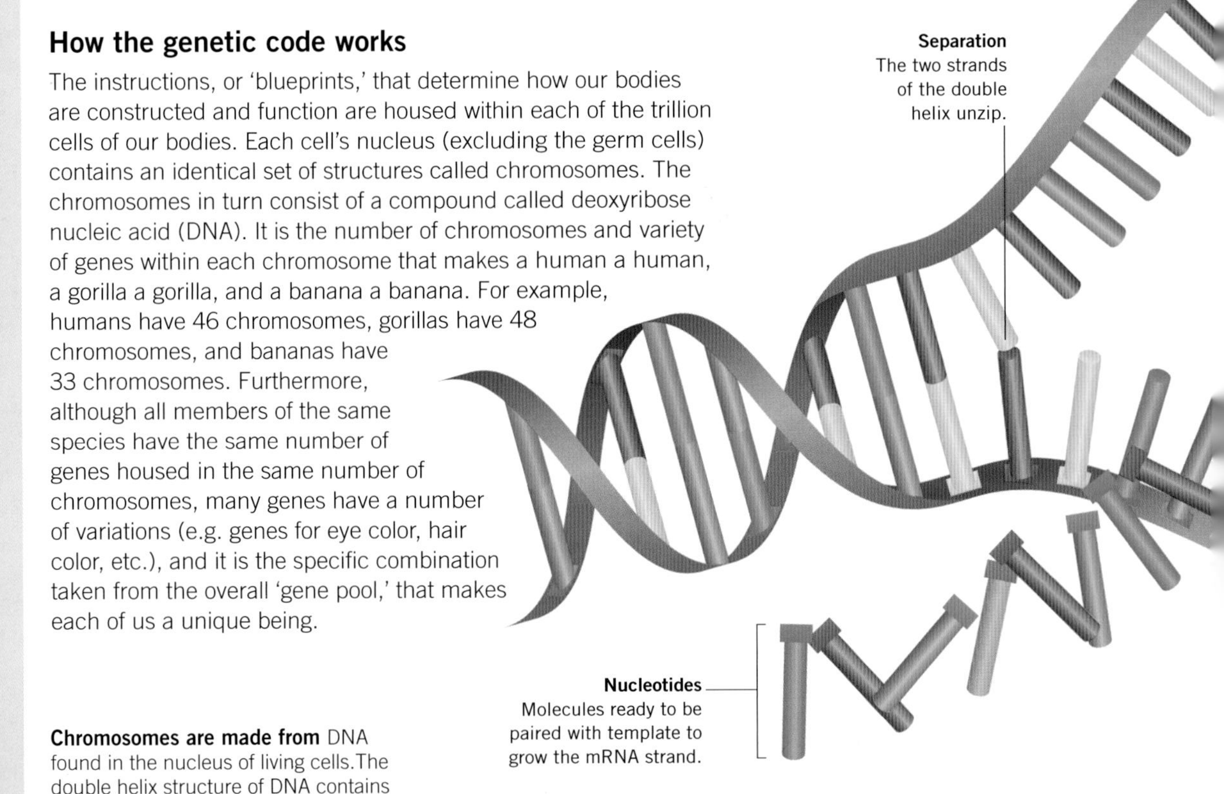

Chromosomes are made from DNA found in the nucleus of living cells.The double helix structure of DNA contains the blueprint for life.

Cell containing nucleus.

Nucleus containing chromosomes.

Chromosome made from DNA.

DNA double helix.

DNA is made from four molecules, adenine (A), thymine (T), cytosine (C), and guanine (G) that make up the letters of the DNA code. These four 'bases' are connected to a support structure to form a 'nucleotide' and then strung together to form pairs – adenine with thymine, cytosine with guanine, like the rungs of a ladder. Within the mRNA strand, thymine is replaced by uracil.

Thymine
Cytosine
Guanine
Adenine
Uracil

Genes and proteins

Living things break down nutrients into their constituent parts and synthesize what they need according to the template supplied by their genes. A gene is a length of the DNA strand containing anything from 500 to 10,000 base pairs that provide the code for an individual protein. The order of the base pairs within the genes forms a 'template' or 'code' that determines how the proteins of our body are manufactured, the primary task of genes. Proteins are constantly being produced in order to regulate our bodies' functions and build or repair tissue/muscle.

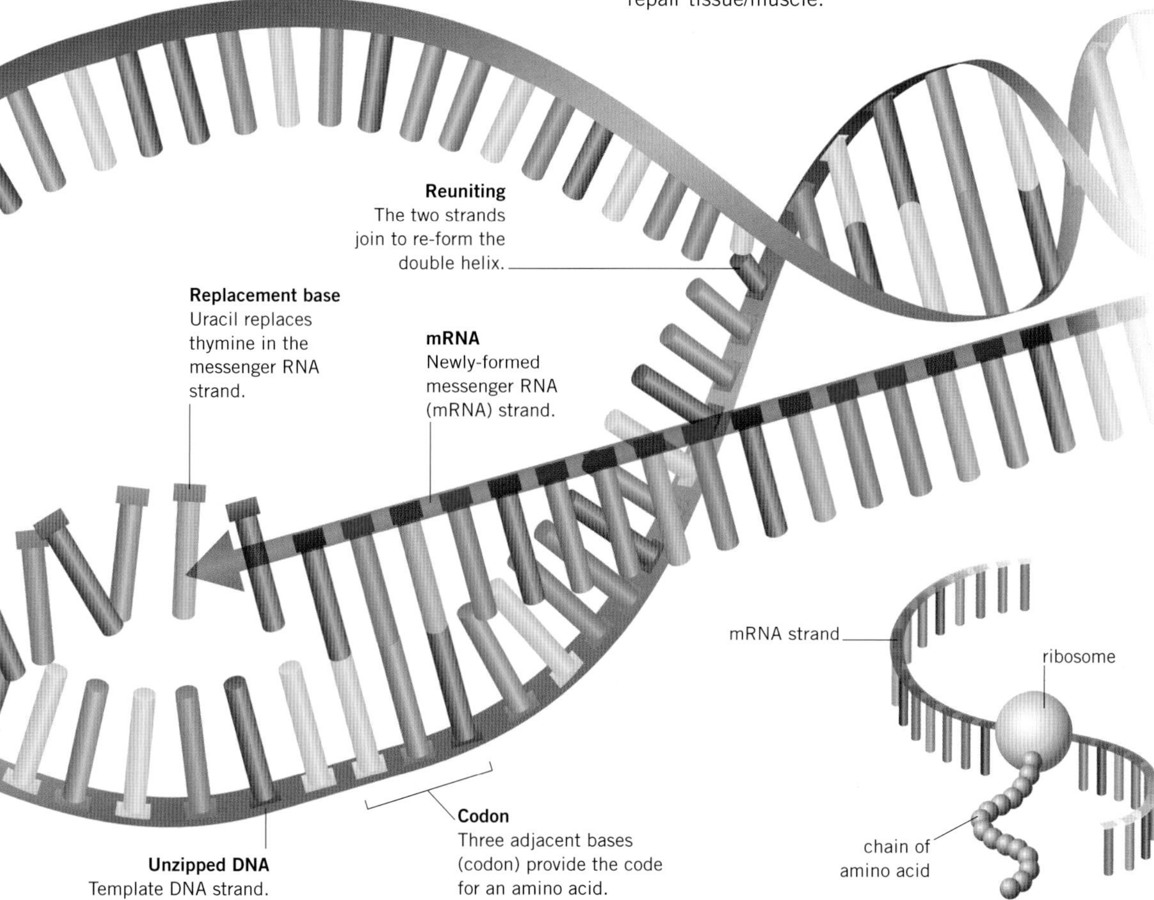

Transcribing the code When genes are being read, the two sides of a section of DNA 'unzip.' One of the DNA strands acts as a template. Nucleotides align themselves sequentially by base-pairing along the 'template' strand, forming a 'messenger' RNA (mRNA) strand. The sequence of bases on the new mRNA strand thus matches the sequence on the DNA strand that was previously paired with the template (with the exception that RNA uses a base called 'uracil' instead of thymine to pair with adenine). This process is called 'transcription.' The newly-formed strand of mRNA detaches and migrates out of the nucleus to a cellular structure called the 'endoplasmic reticulum,' which is the site of protein synthesis.

Translating the code Proteins are essentially long chains of molecules called amino acids. Only 20 types of amino acid exist. Each amino acid is specified for by three adjacent bases (called a 'codon') on the mRNA strand. There are four different bases so 64 different codons are possible. During protein synthesis a cellular organelle, called a 'ribosome,' works its way along the strand reading or 'translating' the codons. Another type of RNA molecule, transfer RNA (tRNA), attaches to the required amino acid and delivers them to the ribosome where the protein is built up, amino acid by amino acid, according to the code originally inscribed in the DNA.

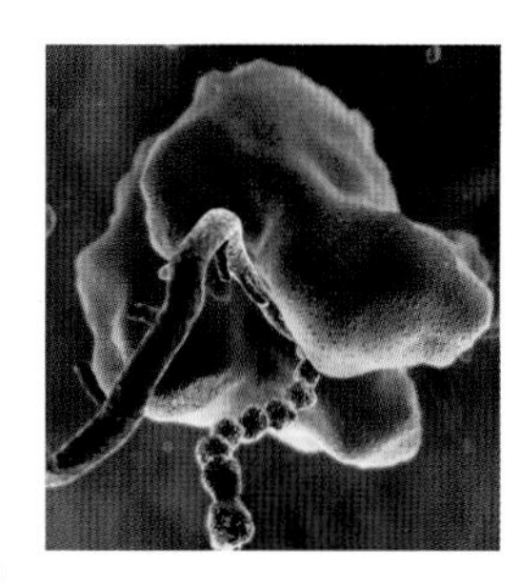

A ribosome reading along an mRNA strand, attaching amino acids to each other to build a protein.

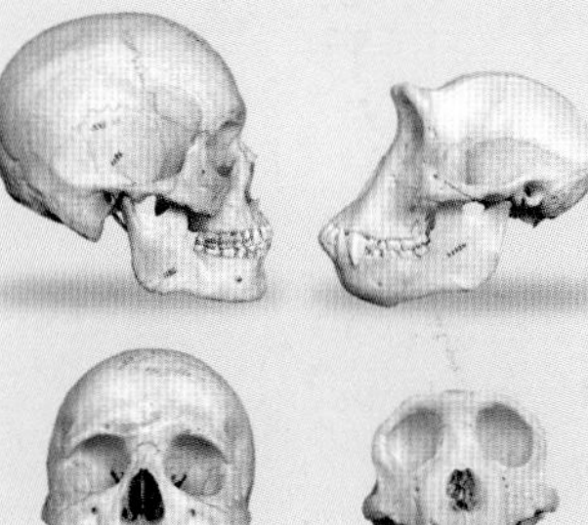

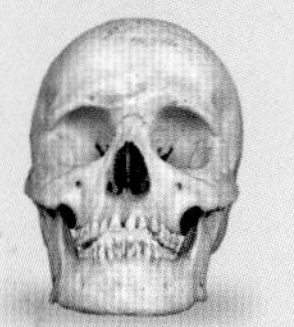

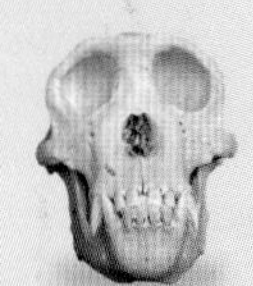

Humans and our closest relatives

One of the most important pieces of information we can learn from studying the genetic code is how humans are related to our closest animal relatives. A comparison of human *(above left)* and chimpanzee skulls *(above right)* suggests marked similarities, but many differences. In fact, comparisons of our genetic code to that of chimpanzees, for example, has shown that we share approximately 98.5% of our genes, and that it is only the 1.5% difference between us and chimps that makes us human. In addition to discovering how closely related humans are to our closest relatives, genetics studies can also determine the approximate date of the most recent common ancestor between humans and any number of animal relatives. This information can be extremely useful as a complement to fossil or archaeological information. Research in this area has revealed that the most recent common ancestor between chimps and humans was approximately 5-7 million years ago, a date that is also supported by fossil data on the earliest human ancestors.

GENETIC ANCESTRY

Long before Watson and Crick discovered the structure of DNA, and set the pattern for the Human Genome Project (*see pages 170, 174*), scientists – and before them agriculturalists – had understood something of the pattern of the genetic code and how it functioned. The development, through selective breeding, of hardier and more productive strains of food crops dates back to the earliest times, as does the selective breeding of animals, but both accelerated in the 18th century as European agriculture sought new efficiencies to feed a burgeoning population. During the 19th century, in the twin shadows of Baron Frankenstein and Charles Darwin, scientists began to speculate about how such techniques of genetic engineering might be applied to human beings.

Selective breeding

The idea of the genetic code (if not a detailed understanding of how it functioned) was central to the early modern agricultural revolution.

The development of crop rotation and an understanding of the need to provide both plants and soil with the appropriate cropping cycles to maximize fertility and productivity developed in the 17th century, mainly in the Netherlands and then England. The product of patient empirical research and cross-breeding and cross-pollination, these experiments were to produce an extraordinary range of flowers and vegetables (with new varieties imported from the New World and Asia), and new breeds of prize meat-bearing livestock *(above)*.

The Dutch were also keen to develop ways to maximize their investment in reclaiming land from coastal inundation, and experimented with crop rotation to provide nutrients for the soil in order to avoid the tradition of leaving agricultural land fallow for one year in three.

By the middle of the 18th century new integrated systems were in place in northwest Europe, combining crop cycles and the products of animal husbandry, which resulted in a massive growth in food production, which in turn fueled the so-called Industrial Revolution.

The Hapsburg jaw
Among the most selectively bred human pedigrees were the European royal families. Due to centuries of intermarriage, traits such as mandibular prognathism (a pronounced lower jaw) became common among royal families. This was particularly true of the Hapsburgs. Charles II of Spain (1661-1700, *left*) is said to have suffered the most pronounced case of the Hapsburg jaw – so deformed that he was unable to eat properly.

Hemophilia: the royal disease

Understanding how genes are passed down through generations has helped us understand how diseases are inherited. One special case involved Queen Victoria and the genetic disease hemophilia. Carried on the X chromosome it results in a reduced ability for blood clotting. Females carry two X chromosomes; if either carries the hemophilia gene the 'healthy' X chromosome ensures that the female has no symptoms. As males have one X and one Y chromosome, if they inherit an affected X chromosome they will suffer from the disease. Victoria unknowingly carried one affected X chromosome and passed it on to several of her children and via them to her grandchildren, many of whom became members of royal families across Europe. After four generations, Victoria's descendants did not pass the hemophiliac gene any further. Today's royal families do not appear to carry the trait.

Alfred (1844-1900), married Princess Marie of Russia.

Victoria Princess Royal (1840-1901), married Frederick of Prussia.

Alice (1843-1878), married Prince Louis of Hesse-Darmstadt. Hemophilia transmitter. Mother of Tsarina Alexandra of Russia.

Louise (1848-1939), married the Marquis of Lorne.

Queen Victoria's family of nine all married into the nobility or royal houses of Europe. Three carried their mother's hemophilia gene, and two transmitted it to a further nine family members across the courts of Europe.

DNA and the Romanovs

The best-known hemophiliac descendant of Queen Victoria was Alexei, son of Tsar Nicholas II of Russia and his wife Alexandra (granddaughter of Victoria). They were murdered after the Bolshevik revolution in 1918. The remains of Tsar Nicholas, Alexandra, and three of their children were found in a shallow grave in 1991. In 2007 further bone shards were DNA tested and found to be the remains of Alexei and Marie.

The missing remains of the Romanovs *(left)* provoked numerous theories about survivors who might claim their fortune. Recently DNA tests have proved that none survived.

A forensic scientist examines a skull believed to be that of Anastasia Romanov, later confirmed by DNA profiling.

Prince Albert (1819-61)

Beatrice (1857-1944), married Prince Henry of Battenberg. Hemophilia transmitter.

Queen Victoria (1819-1901). Hemophilia transmitter.

Albert Edward (Bertie) Prince of Wales (1841-1910), married Princess Alexandra of Denmark.

Helena (1846-1923), married Prince Christian of Schleswig-Holstein.

Arthur (1850-1942), married Princess Louise of Prussia.

Leopold (1853-84), married Princess Helena of Waldeck-Pyrmont. Hemophiliac.

Vivisection and eugenics

When the writer H.G. Wells published his novel *The Island of Dr. Moreau* in 1896 about surgical intervention in animal and human development, he was merely reviving a nightmare ignited by Mary Shelley's *Frankenstein:* that scientists could potentially exert God-like control over the destiny of the human race. It was at a point when the validity of such surgical experiments – vivisection – were being seriously questioned. Nevertheless, enough had been learned of how genetics worked for a much more dangerous pseudo-science – the genetic manipulation of the human race known as eugenics – to gain credibility. Wells himself was in favor, saying "it is in the sterilization of failure, and not in the selection of successes for breeding" that he was interested.

The Nazis were advocates of social engineering via euthanasia. Long before the extermination camps were built, they promoted 'mercy killing' for the 'subnormal' or 'degenerates,' here claiming that keeping this individual alive would cost the state 60,000 Reichmarks: "Fellow Germans, that is your money too."

USING THE GENETIC CODE

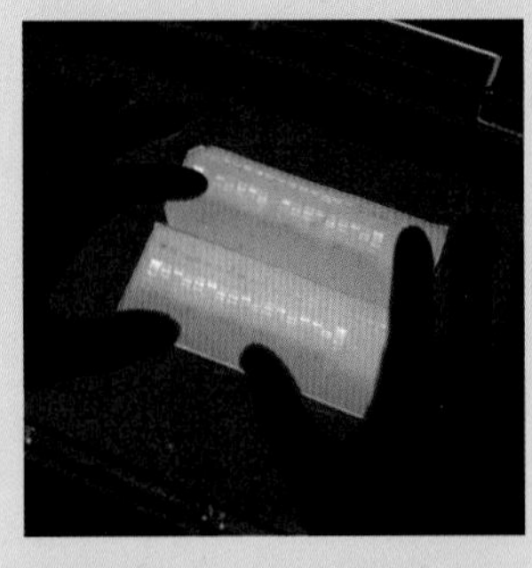

DNA and crime-fighting

The use of DNA profiling to identify a potential perpetrator's presence at a crime scene has increased dramatically since its first use by police and forensic scientists, in Britain, in the 1980s. The DNA code provides a unique 'fingerprint' for everybody. Saliva, hair strands, perspiration, and other secretions inadvertently deposited at a crime scene can be detected, analyzed, and compared with swabs taken from suspects (or compared to records on increasingly rich databases) to provide investigators with highly accurate DNA profiles.

The first criminal investigation to use DNA sampling was in 1987 in Leicestershire, UK. Two similar rape and murder cases, in 1983 and 1986, led the police to accuse a local 17-year-old, Richard Buckland who, under pressure, admitted to one of the killings. DNA researchers from Leicester University offered to compare residues from both cases to establish his double guilt. They proved the perpetrator was indeed the same person, but not Buckland. Samples were then taken from 5,000 local males, but no match was found. However, one sample donor admitted that he had given his sample in the name of Colin Pitchfork, a local baker. Pitchfork was arrested, tested, and a DNA match was found. He admitted his guilt, and was sentenced to life.

The discoveries of Watson and Crick and others at the Cavendish Laboratory in Cambridge, UK (*see page 170*) proved much more significant than merely establishing how genetics worked in relation to DNA. The Human Genome Project was established in the 1980s in Cambridge, but also in the USA, China, France, Germany, and Japan, when computing power had reached a point where the many calculations required to provide a complete 'map' of the human genome using sequence analysis became possible. A preliminary announcement of the completion of the 'map' was made in 2000, and the last chromosome was finally identified and mapped in 2006. The impact was enormous: while the uniqueness of each person's DNA profile had been used in criminal detection since the 1980s (*left*), the completion of the map raised a variety of ethical issues concerning health, insurance, genetic engineering, and biometric passports and security databases. Suddenly, humans were trapped by their own, unavoidable code.

Preparing DNA for analysis

DNA taken from chromosomes is chopped into pieces. The pieces are separated and cloned to make numerous solutions, each containing identical strands of DNA up to 4,000 base pairs long. The solutions are heated to split the double strand and one of the single strands is used as a template. This is copied by adding enzymes, primers, the four bases A,G,C,T, (*see page 170*), and 'special' versions of each base. These special bases stop the copying process when incorporated into a growing strand and are marked to show up under specific conditions. Copying starts at the primer (which always attaches at exactly the same spot on the template strand), and continues by base pairing until a 'special' base is incorporated, stopping the process. Ultimately a solution is obtained that contains a mixture of billions of varying length copies of the template that all start at the same point and end with a marked 'special' base.

Comparing two strands of DNA using an autoradiogram developed from gel electrophoresis. In this method separate solutions are prepared for each base (C, A, T, G), which are run in parallel lanes of gel. The same section of DNA from two samples are run side by side to identify differences in the positions of their bases, which could rule out the possibility that both samples came from the same individual.

Reading the code

The mixture is now sorted into fragments of equal length by a process called gel electrophoresis. A length of gel is subjected to an electric field. The solution of DNA fragments is added at the negative end and, as DNA has a negative charge, it moves through the gel toward the positive end. The rate at which each fragment moves is related to its size, smaller fragments moving faster; fragments of equal length travel along together. These clumps of identical fragments show up due to the markers attached to their special bases (when struck by a laser, A may fluoresce green, G red). In automated systems, detectors read colors as the clumps pass by and a computer records the sequence. Alternatively, the electric field can be switched off, stranding the various clumps at their positions along the gel.

The Cohen mystery

Scientists can trace genetic lineages back tens of thousands of years, producing often controversial results. It has long been hypothesized that the popular Jewish surname 'Cohen' is associated with a Jewish priestly caste, the 'Kohanim' – descendants of Aaron, brother of Moses. In 1998 scientists analyzed the DNA of hundreds of male Cohens to assess the genetic link between the surname and paternal inheritance. Results indicated that there did seem to be such a link, but analysis showed that this particular lineage dated back 3,000 years ago to the Arabian peninsula – far from the Levant – raising the question of how these findings fitted with Biblical tradition.

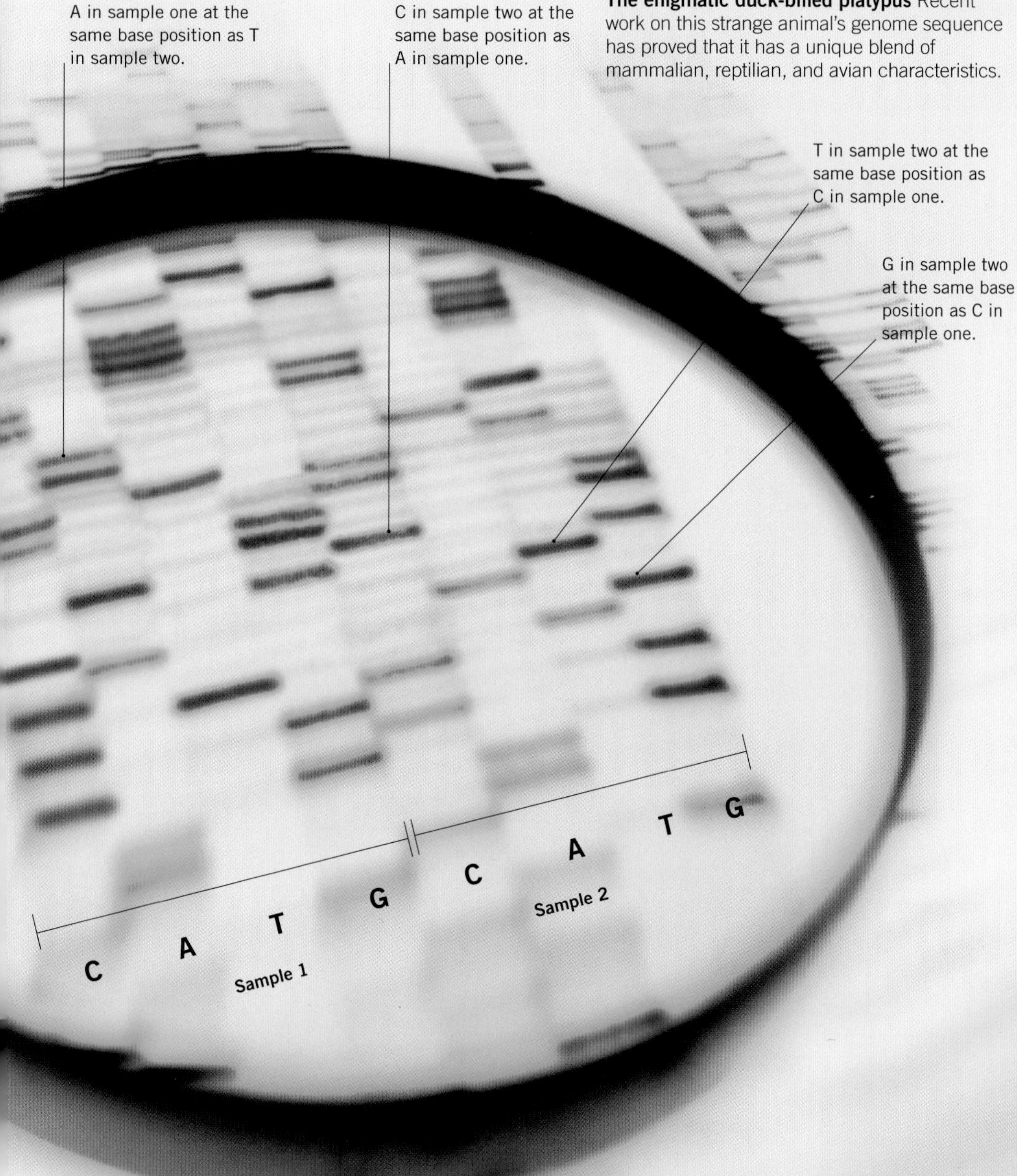

The enigmatic duck-billed platypus Recent work on this strange animal's genome sequence has proved that it has a unique blend of mammalian, reptilian, and avian characteristics.

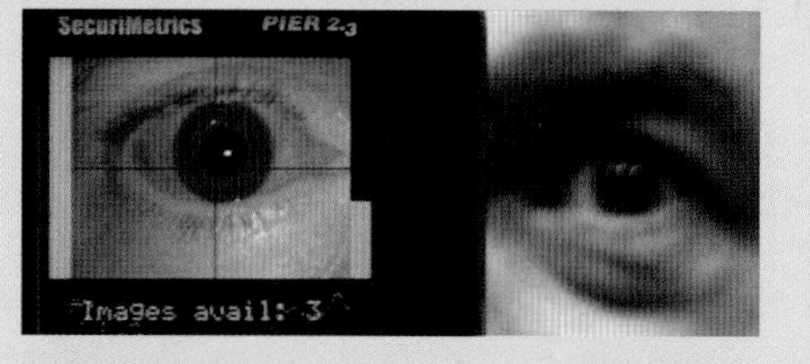

Big Brother's DNA

Fingerprint analysis for criminal detection was among the first instances of unique genetic profiling (*see page 138*), but DNA profiling has revolutionized security in other ways. Currently, biometric passports, issued by some 40 countries, contain tiny chips holding an image of the owner, and increasingly their fingerprint and iris scan *(above)*. There is increasing pressure to include DNA information as well, making passport counterfeiting all but impossible. Equally, police and security forces argue that national DNA databases should be established, possibly linked to chips on ID cards as well as passports, while civil libertarians see this as a huge encroachment on individual privacy.

Insurance and health

The decoding of the human genome has led to fears that insurance companies will deny health coverage to individuals who exhibit genes associated with conditions such as diabetes, heart disease, Alzheimer's, or cancer. As a result, many people are wary of having their DNA tested for such genetic markers, or do so in secret. This issue has many legal and ethical implications. There are various laws in existence, as well as several soon to be enacted, that address the issue of 'genetic confidentiality.'

Genetic engineering

The use of genetic technology to manipulate living organisms has caused great controversy in recent years. The potential benefits of genetically engineered crops holds promise for feeding the world's growing population, although many feel uncomfortable about unknown long-term effects. The genetic modification of animals and humans, heralded by the creation in 1996 of the first cloned animal, the short-lived sheep, Dolly (*right*), has since provoked many ethical, moral, and legal arguments.

08

Cultural groupings frequently develop verbal, literary, and graphic shorthands – ways of communicating ideas and sending messages often so deeply embedded in a shared history that their origins may have been forgotten, although their meanings live on.

codes of civilization

Such devices may be used to convey discrete but complex ideas, often expressed in the form of allusions or iconographic references. Many of these remain recognized within communities where continuity has been provided by religious faith or a shared set of values but, equally, many have been obscured by the passage of time.

CODES OF CONSTRUCTION

Until the rise of Modernism in the early 20th century, Western architecture was dominated by two contrasting architectural traditions, each with its own formal language and structural style – traditions so powerful that they remain vibrant even today. On the one hand there was the classical column-and-lintel tradition established and elaborated in Greece and Rome, preserved in a simplified form in Romanesque architecture of the first millennium AD, but a language revived and reinvented by Renaissance antiquarians and humanists; on the other was the Gothic, a graceful, organic, skeletal style which developed during the European Middle Ages.

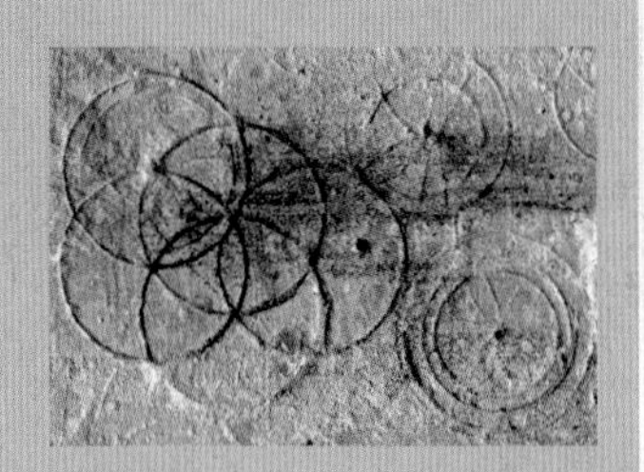

Masons' marks

Marks, either a monogram or, more often, a symbol, are found cut into stone blocks from the Roman period onward and are very common throughout the Middle Ages. Their purpose and meaning have never been clearly established. There seem to be two types of marks: laying-out marks, indicating where the stones were to be placed, and signature marks identifying who had worked the stone, possibly to tally up how much work had been done by a particular mason – in effect a double code.

Orientation

Most Christian basilicas – whether classical or Gothic in style – conform to a cruciform ground plan, and have a processional format which is oriented toward the East. Mosques are less frequently oriented, although an elaborate niche, the *mihrab*, is always provided, indicating for worshippers the direction toward Mecca.

Gothic

The restoration of the abbey church at St-Denis near Paris, under the auspices of the Benedictine Abbé Suger (c.1081-1151) inspired an architectural style that spread rapidly across Europe. Originally conceived as a way of introducing light into a church, by reducing the structural elements to a minimum – the pointed arch, the fan vault – the effect produced a sense of celestial wonder and awe – a building that resembled heaven. Although at first stressing simplicity of form and decoration, and used only for ecclesiastical architecture as at Salisbury cathedral (*above*), the style became increasingly elaborate, and by the 15th century was used widely for buildings of all types. Gothic architecture developed its own vocabulary (*right*).

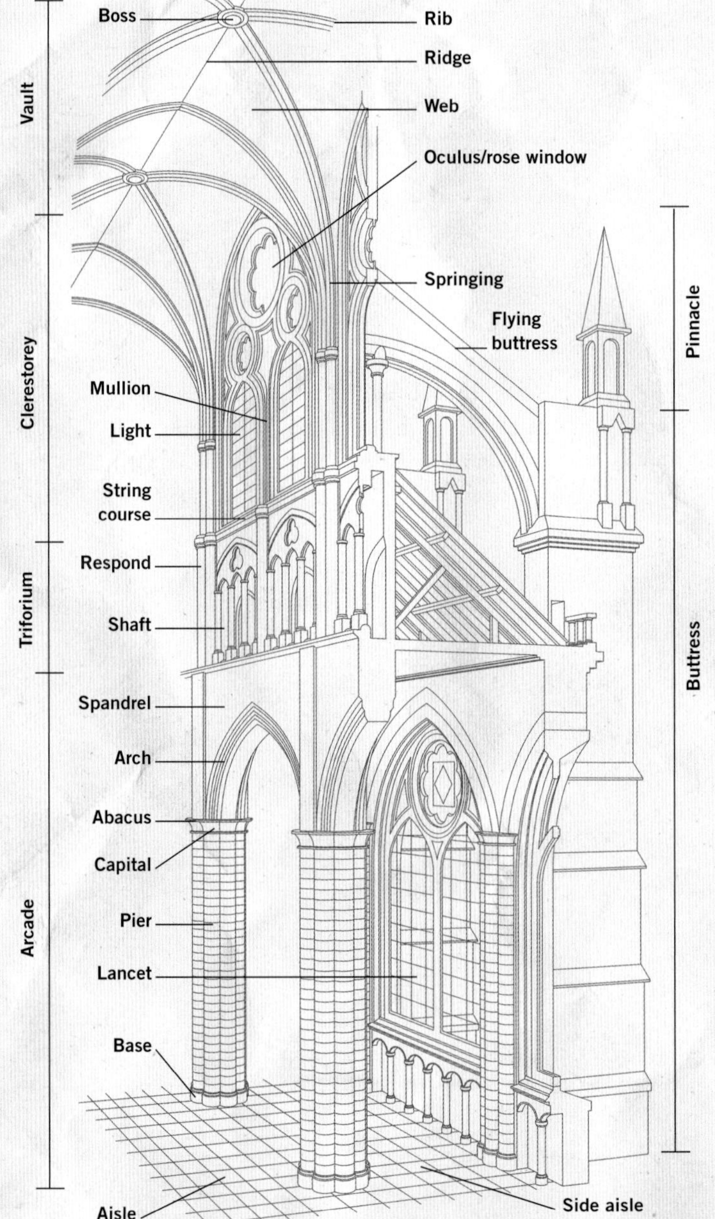

Classicism

The canon of Greek and Roman architecture was passed down to the Renaissance via two channels: the many largely ruined remains of classical buildings, and the writings of the Roman architect, Vitruvius (c.80-15 BC). In his *Ten Books on Architecture*, Vitruvius covered subjects from temple building to civil engineering and landscape gardening. His writings were edited and popularized by Leon Battista Alberti (1404-72), and were promoted further by later Italian architects such as Serlio, Vignola, and Andrea Palladio. By 1500, classical architecture had become the accepted style for most public buildings, palaces, and even private houses throughout Italy, and would spread north to France, Britain, and America. Vitruvius described the proportions, characteristic, and appropriate use of classical architectural elements, the most important of which were the three Greek columns or 'orders,' and their correct use to organize or 'articulate' a building.

One of the best preserved classical buildings in Rome, the Pantheon with its beautifully proportioned exterior, elegant circular concrete vault, and bold inscriptions was a source of inspiration to innumerable Renaissance and post-Renaissance architects.

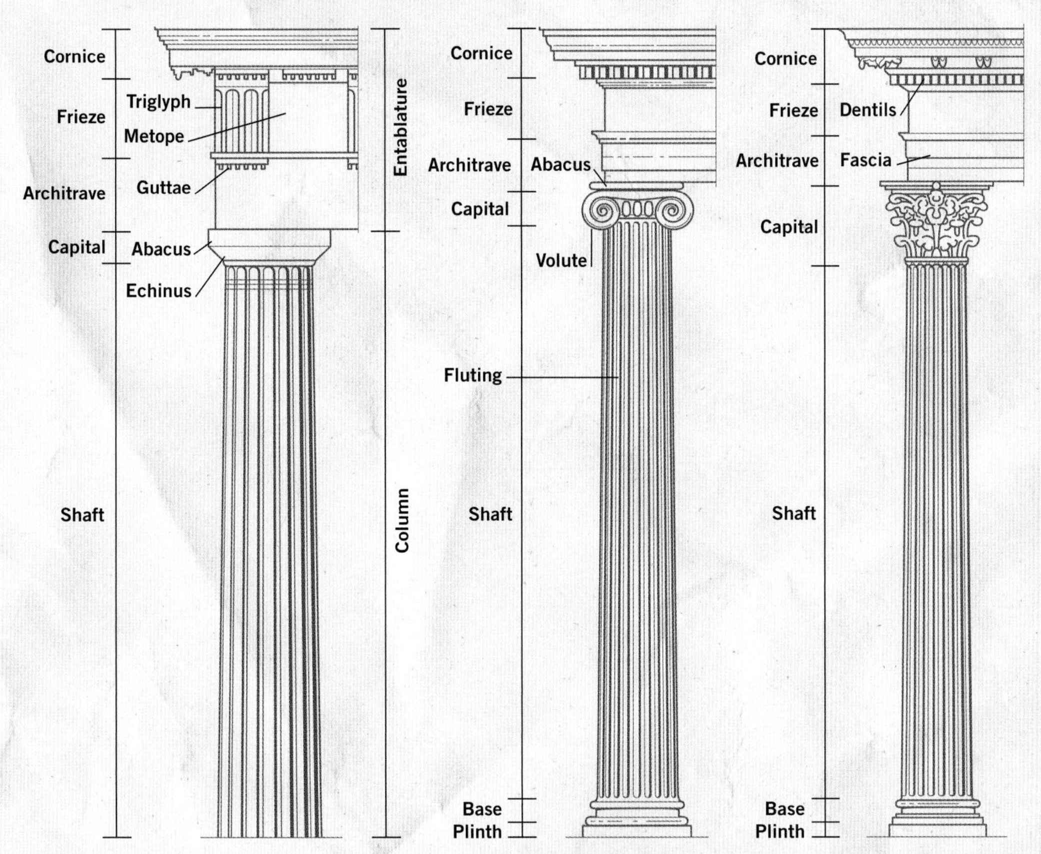

Doric The most ancient, the Greek version having no plinth or base; proportions 6/7:1; masculine; primitive simplicity, seriousness of purpose, noble sobriety; public buildings, public utilities.

Ionic Proportions 8:1; feminine but matronly; grace and authority; libraries, courts of justice, universities, and colleges.

Corinthian Proportions 8/9:1 (the nearest to the ideal proportions of the human body); gracefully feminine; decorative and decorous; government buildings, places of entertainment.

The *Hypnerotomachia*

One of the most influential architectural texts of the Renaissance, this mysterious illustrated book, possibly by a Franciscan friar, Francesco Colonna, was printed with extraordinary elegance by Aldus Manutius in Venice in 1499. It tells of the erotic dream of Poliphilus in various enchanted pagan settings adorned with rich architectural fantasies drawn not just from the classical world, but from further afield such as Egypt and the Middle East. Many of the buildings are ruined, and most have enigmatic relief carvings.

Seville cathedral, the largest in Europe, comprises a unique blend of Moorish, Gothic, and post-Renaissance classical architectural styles, one overlaying the other across the centuries.

Taoist Mysticism

Chinese philosopher and mystic Lao-tzu (6th century BC) is regarded as the founder of Taoism, a largely animistic religion, which seeks to find unity in opposites: between heaven and earth, order and chaos, and men and women. Taoism is all-embracing, and its roots appeal in China has meant that, despite occasional suppression, it has survived in many forms, influencing neo-Confucianism and the propaganda imagery of the Communist era. Almost all features of Taoist art – color, shape, materials, calligraphy – have symbolic meanings for the initiated. Some, such as the use of jade (believed to derive from divine dragons), are widely recognized, others less so.

Landscape art

Taoism celebrates the majesty of nature and nowhere more than in the representation of the world. Carefully composed landscape paintings, in delicate washes highlighted with ink strokes, were not conceived as representations of observed topography, but as a means of reflecting the all-embracing power of nature. Compositions usually comprise sinuous 'power' lines representing elemental forces of energy, often in the form of rocky outcrops, rivers, and waterfalls. Humans and their buildings are shown as being overwhelmed by the natural world around them. The paintings are frequently accompanied by poetic texts.

The yin and yang symbol of unity at the center of this imperial relief illustrates the inseparable interaction of heavenly (divine) and earthly (temporal) powers.

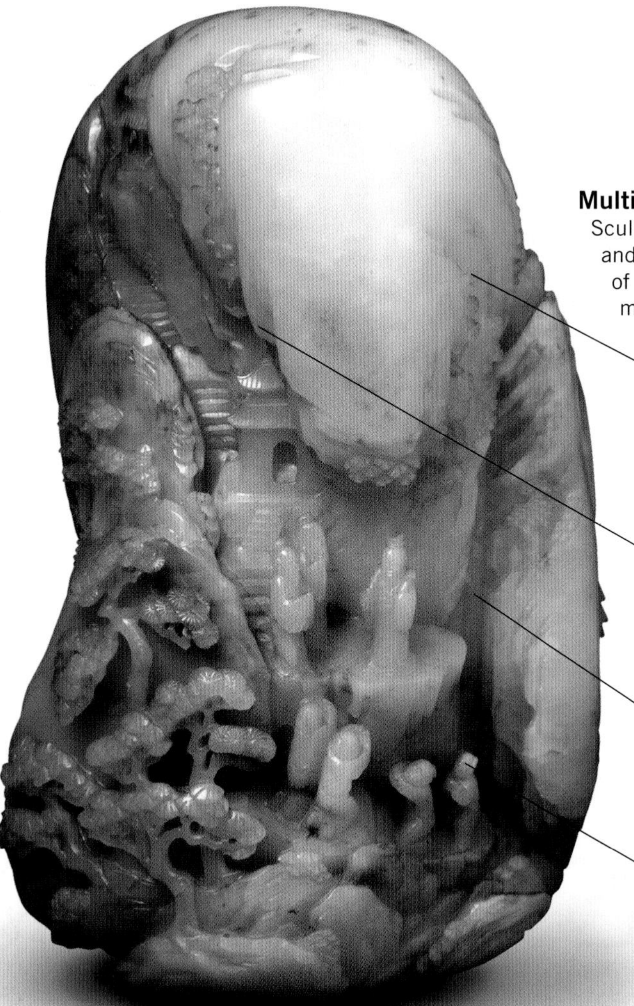

Multilayered meanings

Sculpture remains probably the most versatile and perfect artistic medium for the expression of core Taoist ideas, by using a multilayered matrix of symbolism.

Material
The choice of jade is both critical and symbolic. Regarded as the finest material for medium- to small-scale sculpture, the stone is seen as a direct link with the world of gods, creation, and dragons.

Form
The kidney shape is intentionally organic and ambiguous, both suggestively phallic in profile and, with the deep incision, evocative of the female vulva.

Natural markings
The veins in the jade have been retained and where possible emphasized by the sculptor to suggest lines of natural energy and dynamism, both elemental and ethereal.

Figurative representation
The diminutive, and again suggestively phallic, pilgrims seem overawed by the feminine actuality of the chasm, and the majestic landscape around them.

Heaven The I Ching trigram representing heaven or creation appears on each shoulder panel.

Cranes Messengers of the gods.

Dragons Within Taoism, these fantastic beasts were regarded as intermediaries between heaven and earth.

The moon A powerful I Ching trigram, also symbolic of water.

The hem Decorated with symbolic flowers, including the lotus, a homonym for harmony, peonies for wealth and honor, and orchids representing wisdom and virtue.

The Sun Balancing the moon, this I Ching trigram also symbolizes fire.

Formal robes

Taoist symbolism was frequently incorporated into the decoration of sumptuous courtly costumes, such as this 14th-century robe, rich in coded messages, reflecting the status of the wearer.

Secret writing

The communication of mystic Taoist formulas, charms, and spells to the initiated, whilst retaining an air of mystery, saw the development of a range of calligraphic styles. Their extreme cursive form was designed to make the scripts only recognizable to an 'informed' elite.

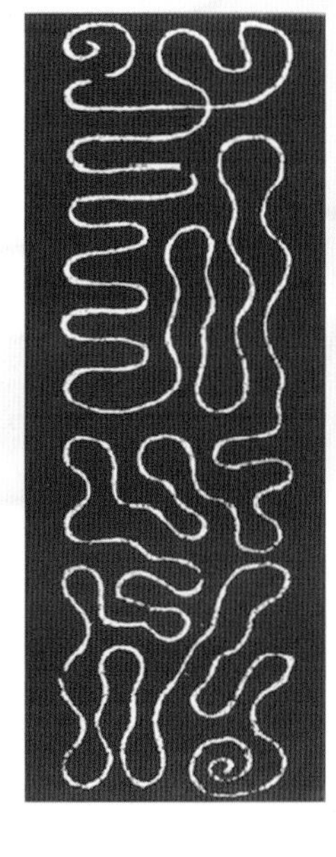

Magic diagram A single wandering line (*right*) describes a magic formula. The looped shapes reflect the intuitive yang element in Taoist thought.

Grass script This highly stylized method of representing Chinese pictograms (*left*) had a magical significance. Here the character '*shou*' (long life) has been represented in a disguised form, and then printed. The date of the work, 1863, and the artist, Yen Chih, have been printed on the image.

The I Ching

The dominant mystical text at the heart of Taoism is the I Ching. It is effectively a series of oracular formulas which act as a complex decoding device used for prophecy, divination, and ordering the natural sciences. The principal method of consulting the I Ching oracles is by casting yarrow sticks or three coins, to build up trigrams (*below*), symbols which can have multiple meanings.

- heaven, creation; energy, conflict, strength; jade, ice; father; head; the horse
- wind, wood; gentleness; thighs; the cockerel
- the abyss, the Moon; water; toil; ears; the pig
- mountain, beginning/end, birth/death; seed; remaining still; hand; black-beaked birds, dog
- Earth, receptive, nourishing, yielding; mother; belly; cow
- thunder, arousing; Earth forces; green bamboo; foot; dragon
- Sun, fire, light, clinging, consciousness; eye; cock pheasant
- lake; joyful, bursting; concubine; mouth; sheep

Divination diagrams
Trigrams are then combined to form schematic hexagrams, which diviners use to provide an 'answer' or prophecy when read against the I Ching.

South Asian Sacred Imagery

The great religions of the Indian subcontinent, Hinduism, Buddhism, Jainism, and Sikhism, all share iconographic elements, some going back to the religion of the Vedic period, around 1000 bc. They are of great complexity and open to a wide range of interpretations. For the Hindus, religious symbols are often considered to actually embody the divine, and hence seen to be sacred in themselves.

Divine symbols

Many symbols appear in various South Asian religions, although the Hindu usage tends to be predominant. Buddhism, which evolved much later, adapted some Hindu ideas (the footprint and the lotus, for example) to meet its own aniconic requirements (*see page 184*). Some of the most important recurring symbols in South Asian religion include the following, which may be found in a variety of widely stylized and locally adapted forms.

Lamp (*dipa*)
A symbol shared with Islam, representing enlightenment.

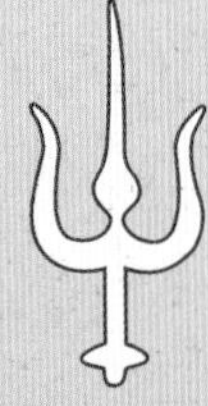

Trident
The three-pronged weapon represents the power of Shiva.

Coconut
The coconut (*kalasha*), surrounded by mango leaves and arranged in a pot is a favorite ritual offering; embodies fecundity.

Conch shell
This symbol is closely associated with Vishnu. It was perceived as a call to battle, spiritual as well as temporal.

Lingam
The carvings vary widely in size and shape, and tend towards abstract stylization.

Canopy
This giant *lingam* is protected from the elements by a seven-headed cobra, or *naga*.

Yoni
The base of the *lingam* often resembles a serpent or coils of cloth.

Sacred practices

Religious observance is part of the fabric of everyday life for many throughout the Indian subcontinent, ranging from body decoration to daily ritual. Both body decoration and the creation of *kolams* are seen as sacred acts, as well as expressions of religious observance and allegiance.

Lingams and *nagas*

The *lingam* or *linga* is a Hindu phallic symbol associated with the god Shiva, and procreation. It normally rests on a pedestal representing the female sexual organ or *yoni*. Although there are precise rules about form and proportions, they are often so stylized as to be hard for an outsider to recognize. This example (*left*) is at Lepakshi, where it is protected by a seven-headed cobra, or *naga*. The *nagas* associated with Shiva symbolize death. But *nagas* can also represent cosmic power; and as a manifestation of the Vedic god Agni (or fire) they are also guardians. The *naga* is also associated with Vishnu and then represents knowledge, wisdom, and eternity. *Nagas* are divine in their own right. They are depicted as either fully human, fully snake, or humans with cobra heads and hoods.

Shiva
Shiva is shown bearing his trident, with his consort Parvati and the elephant-headed god Ganesha.

Ritual decoration
Shiva and Parvati display *tilak* and *bindi* forehead decoration respectively.

Om (Aum) A most important concept is the representation of a vocal expression associated with the sound of the creation of the world; it opens many prayers, mantras, and rituals, and is frequently represented above shrines (*left*). Although often associated with transcendental meditation, its roots lie in Hindu ritual, and it is common to most of the major religions of South Asia.

Tilaks and *bindis*

Marking the forehead is common in South Asia. For men, the *tilak* is used when participating in rituals or by devotees of a particular deity. The *tilak* may be ash from a sacrificial fire, cow dung, turmeric, or charcoal. Followers of Shiva, the destroyer, use ash, sometimes from a funeral pyre, to mark three lines across the forehead, or the trident, or a crescent moon. Followers of Vishnu use sandalwood to make a symbol like a U, representing his footprint. For women, most common is the *bindi* – a dot worn between the eyebrows, the site of the 'third eye,' often shown in representations of the deities. Traditionally worn by married women (and abandoned if widowed), the *bindi* is now often treated as a decorative accessory, worn also by unmarried girls. Another symbol of rejoicing worn on the wedding day and festival occasions is kumkum – vermilion representing fertility and strength – in the parting of the hair.

Kolams (*rangoli*)

Traditionally, each morning south Indian women would draw elaborate geometric patterns in front of their houses in rice flour. The *kolams* were made up of auspicious signs and symbols of the gods. This work of art would be progressively destroyed throughout the day to be renewed the following dawn. Ideally, the *kolam* would be drawn without interruption in a single gesture, and learning to make the innumerable variants was thought to be training in concentration, dexterity, and skill. For festivals and religious celebrations vast colored *kolams* are drawn at temples, often the work of a number of women. Now, frequently, a single simplified design is painted in front of the door of the home or temple.

The daily creation of *kolams* involves intense concentration, and is regarded as a religious rite.

The Language of Buddhism

The birthplace, or 'hearth,' of Buddhism lies in the foothills of the Himalayas in northern India, focused around the places associated with the life of its founder, Siddhartha Gautama (c.566–c.483 BC). As a consequence, Buddhism shares some of its symbolic language with the other established religions of South Asia, principally Hinduism. The missionary activity of Buddhist monks from the first century AD spread the faith north to the Himalayas, Tibet, China, and eventually to Japan, and south to Ceylon, Southeast Asia, and the Malay archipelago. As Buddhism took root in these different cultures, so localized styles and symbols developed alongside regional schools of thought, but the core iconography of the religion can be found throughout the Buddhist world.

Buddhist mudras

Mudras are hand gestures found in Hindu and Buddhist iconography. They symbolize particular aspects of the Buddha's teachings and help define a particular image. One 7th-century sutra enumerates 130 separate mudras and, while there are local variants, a Buddhist, seeing any of the mudras, can interpret the spiritual lesson that each implies.

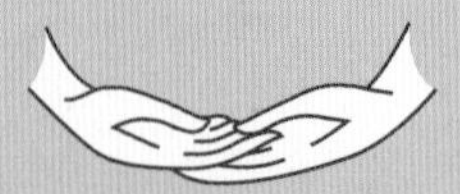

Dhyani Mudra Meditation gesture.

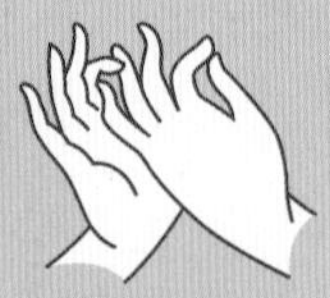

Dharmachakra Mudra Turning the wheel of the law.

Vitarka Mudra Teaching gesture.

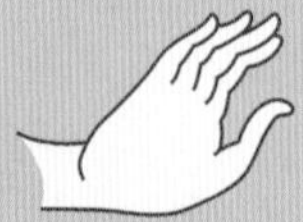

Abhaya Mudra Fearlessness and granting protection.

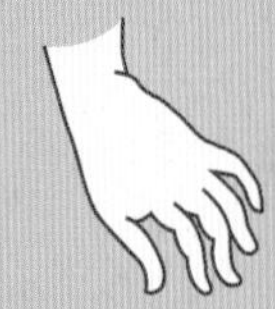

Varada Mudra Compassion and the granting of wishes.

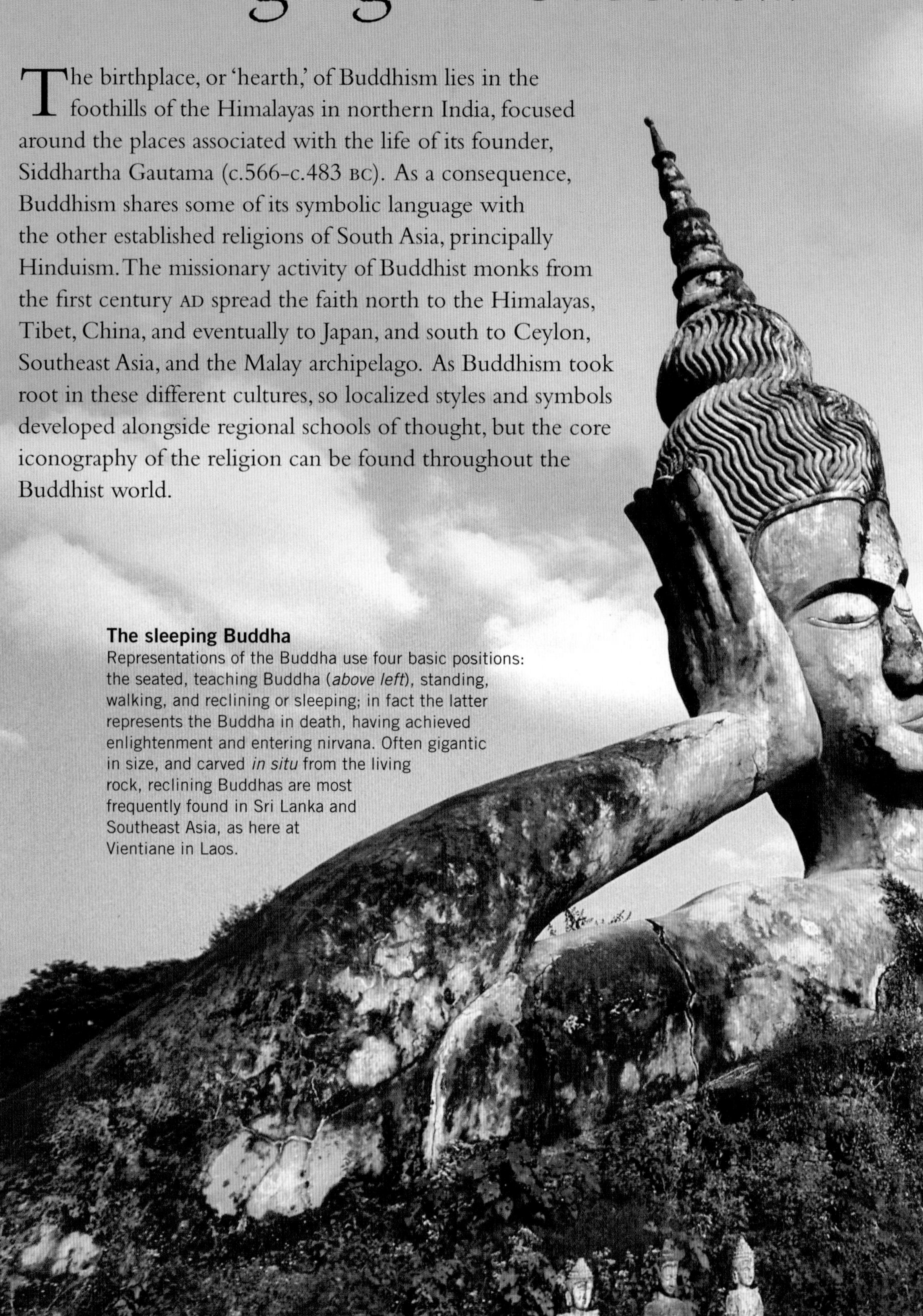

The sleeping Buddha
Representations of the Buddha use four basic positions: the seated, teaching Buddha (*above left*), standing, walking, and reclining or sleeping; in fact the latter represents the Buddha in death, having achieved enlightenment and entering nirvana. Often gigantic in size, and carved *in situ* from the living rock, reclining Buddhas are most frequently found in Sri Lanka and Southeast Asia, as here at Vientiane in Laos.

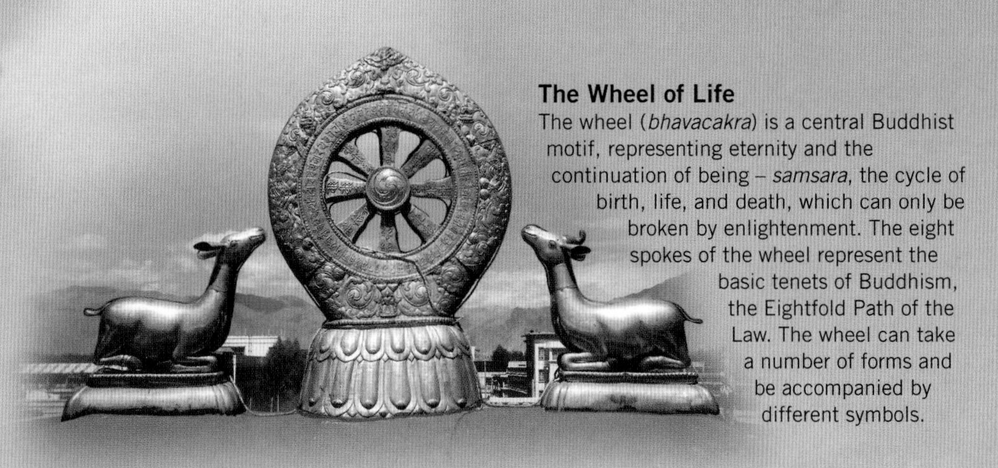

The Wheel of Life

The wheel (*bhavacakra*) is a central Buddhist motif, representing eternity and the continuation of being – *samsara*, the cycle of birth, life, and death, which can only be broken by enlightenment. The eight spokes of the wheel represent the basic tenets of Buddhism, the Eightfold Path of the Law. The wheel can take a number of forms and be accompanied by different symbols.

Mandalas

The word mandala comes from the Sanskrit for 'circle,' but also for 'connection' or 'completion,' linking these religious drawings to the Buddhist concept of the wheel (*above*). Their elaborate symbolism aids meditation and creates a 'sacred space.' The center generally shows the Buddha or a symbol for him, such as a lotus, surrounded by layers of squares within circles representing the world and the paths towards enlightenment. Drawing them forms part of monastic training and tantric initiation rites, and takes hundreds of hours. Often created in colored sand or the dust of precious stones, their evanescence provides a lesson on the fragility of the world. But even in ink, the colors used are symbolic (*right*).

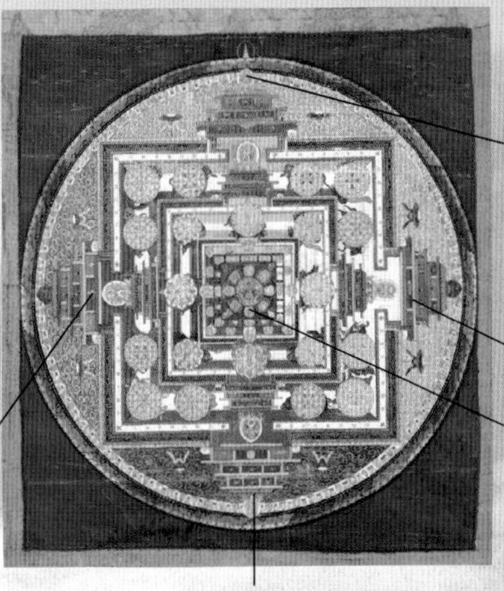

Green
North, air, Amoghasiddhi, jealousy becomes creativeness.

Blue
East, water, Akshobya, anger becomes wisdom and peace.

White
Center, ether, Vairocana, ignorance becomes wisdom.

Red
West, fire, Amithaba, attachment becomes discernment.

Yellow
South, earth, Ratnasambhava, pride becomes abnegation.

Buddhist symbolism

At first, representations of the Buddha were discouraged, leading to the use of a number of symbols to mark the passage of his life and his teaching. Many of these had their origins in Hindu iconography (*see page 182*).

Swastika The word *swastika* comes from the Sanskrit for 'all is well.' It is shared with Buddhism and Jainism, and appears in many other cultures. The right-handed swastika is a good luck charm, the symbol of Vishnu, the preserver of the universe, and sun and light. The left-handed swastika is the emblem of Kali, the terrifying goddess of death and destruction, and the powers of darkness.

Footprints At its inception, Buddhism was aniconic, for Gautama did not wish to be portrayed, so at first he was represented only by his footprints.

Lotus (*padma*) For Buddhists, the lotus represents the Buddha himself, or the four elements: roots – earth, stem – water, leaves – air, flower – fire, and is often used to represent the progress of the soul from mud – materialism, through water – experience, to light – enlightenment. The number of petals are significant, from the eight petals of the Eightfold Path to the 1,000 or 10,000 of the Supreme Being, as are the colors:

White Bodhi – purity, the Eightfold Path of the Law.
Red Avalokitesvara – compassion.
Blue Manjusri – wisdom, the victory of the spirit over the senses.
Pink The lotus of the historic Buddha.

The Buddha is often represented sitting on a lotus plant, with a single elaborate leaf behind him.

The Patterns of Islam

Geometric patterns go back to the beginning of humanity's desire to decorate and adorn, and basketwork and weaving in themselves form geometric patterns. Complex variants soon evolved, such as the carved stone threshold imitating a rug from Ashurbanipal's palace at Nineveh, c.645 BC, with its intersecting circles forming a daisy pattern, a clear forerunner of Islamic geometric decoration. There has always been a connection between different media and designs, and motifs move from one to the other.

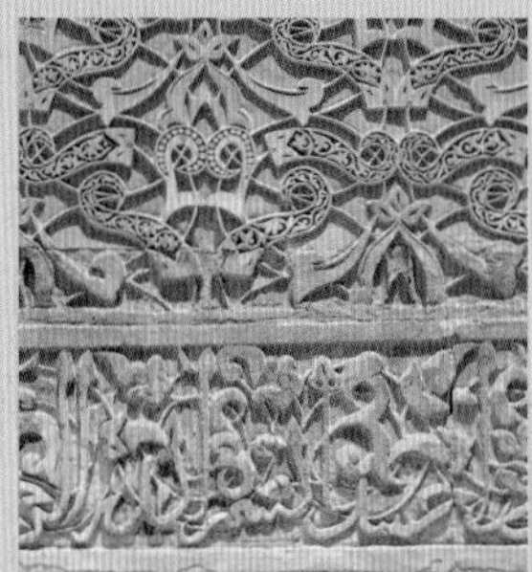

Infinite patterns

Dense patterns in tiles, stone, stucco, and brickwork clothe the surfaces of many mosques. It has never been satisfactorily explained how the very complex space-filling patterns, often made of polygons or cells with infinite variants, were worked out. Whether they were originally designed by mathematicians for craftsmen, subsequently passing into the aesthetic repertory of the Islamic world, or whether they were elaborated by the craftsmen themselves, is unclear. Although some of these patterns existed earlier, the insistence with which they occur and reoccur from one end to the other of the Islamic world has led them to be instantly identified as part of Islamic culture. Together they create an extraordinary textbook of geometrical possibilities.

Islamic iconoclasm

With the coming of Islam in the 7th century AD, geometric patterns gained immensely in importance, passing from the essentially decorative to the profoundly symbolic. This was in part because of the Islamic prohibition on representing living creatures, but it has also been suggested that the endlessly repeating and impersonal geometric patterns are an effort to represent the immensity and omnipresence of Allah. "You shall define the Infinite by creating a beautifully ordered symmetric structure which repeats itself for ever and is bejeweled with Divine words" is an observation attributed to Jalal ad-Din Muhammad Rumi.

Symmetry
For Islam, symmetry reflects perfection. Quite how Arab craftsmen calculated and achieved these patterns on the curved surfaces of their domes and spandrels still fascinates mathematicians today.

Floral motifs
Although the representation of humans was not permitted, the incorporation of often highly stylized floral designs provided a way of celebrating the glory of Allah's creation.

Symbols in Islam

Islam discourages symbols as it discourages all attempts at representation. Preferably, its presence should only be announced by the name of Allah or a pious phrase – hence the great development of decorative calligraphy on innumerable manuscripts, especially the '*unwans* – or 'portals' – of finely decorated *Qur'ans.* Those symbols most associated with Islam are either late arrivals or part of folk customs, and in both cases are disapproved of by the orthodox.

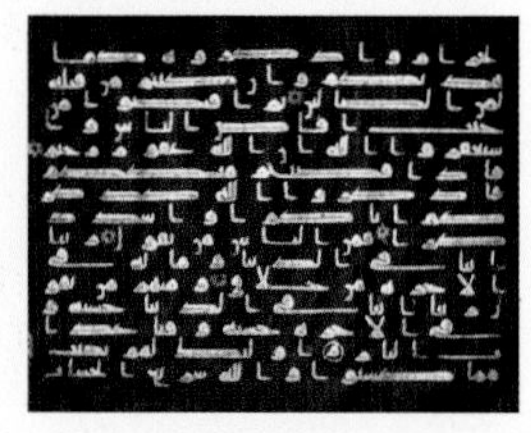

Dhu al-Faqar (Zulfikar)
The double-bladed sword given by the Prophet Muhammad to Ali is of great symbolic importance, particularly to the Shi'a. Swords in general are also often represented, for example on the Saudi flag, as a reminder of all Muslims' duty to *jihad* – war against polytheists.

Khamsa ('The Five') This symbol, also known as the 'Hand of Fatima,' is very popular in North Africa. The hand is one of the most archaic symbols of protection, particularly from the much-feared 'Evil Eye.' It is sometimes associated with the fish and the eye – again pre-Islamic symbols of protection which persist today.

The all-seeing eye Another archaic protective symbol, it is widely used, particularly in Turkey and the Eastern Mediterranean, for example in the form of the omnipresent blue glass beads against the 'Evil Eye.' Turquoise, both stone and color, is likewise believed to be a protection.

Green The color is particularly associated with Islam, and wearing it is the prerogative of descendants of the Prophet Muhammad.

The crescent moon The symbol is most closely identified with Islam. Originally the symbol of the pagan moon goddess, it passed, together with the star, to the iconography of the Virgin Mary and was used by the Byzantines as the symbol of Constantinople. In 1453, the Ottoman Turks took over the symbol together with the city. Perhaps because the Islamic world uses a lunar calendar, the moon is of great importance and has become accepted as the prime symbol on top of mosques and on many flags.

The star Often appearing with the crescent moon, this motif is also of great importance in Islamic geometric design, for its symbolic value, because of its endless possible variants, and because it makes the perfect focal point within a repeat design – for example at the summit of a dome or the center of a wall hanging.

MYSTERIES OF THE NORTH

Kennings

The oral tradition of sagas (largely only transcribed into manuscripts from the 9th century onwards), relating the mythic adventures of Norse forebears and their pantheon, are particularly rich in 'kennings.' Derived from *kenna* – Old Norse for 'to know' – kennings are circumlocutions much used in Norse poetry; the expressions would have been immediately understood by the listeners or readers.

Whale's road Sea.
Forest's grief Axe.
Earth's winter raiment Snow.
Wound-hoe Sword.
Swarm of angry bees A flight of arrows.
Healer of the wolf's grief A great warrior, who left the battlefield covered in corpses, to be eaten by wolves.
He fed the ravens He killed many men; often in runes on warriors' tombstones.
Carving the blood eagle A method of torture and execution.
Valkyrie The name of these women, present at every battle and originally mounted on wolves and attended by eagles and ravens, means 'choosers of the slain.' The 'battle web' they are sometimes described as weaving was made of the entrails of the dead.
Baldur's bane Mistletoe, the only living organism that could kill Odin's son, Baldur.
Ægir's daughters Waves, the nine daughters of the god of the sea.
The tears of the goddess of the chariots Gold or sometimes amber – the goddess is Frejya, goddess of love and also death, a kenning for her is 'possessor of the dead,' whose chariot was drawn by wild cats.
Fire of the serpent's lair Gold, because dragons were believed to horde gold.

The peoples of northern Europe – the Norse, or Vikings, or Varangians – renowned for their trading, raiding, and colonization skills, sustained a highly individual culture that existed from the last centuries BC until their conversion to Christianity around AD 1000. Much of their culture, linked to the so-called Celtic world, is enshrined in elaborate sagas, and in their unique form of writing – runes. As a trading nexus, these roving peoples made contact with many other cultures – not least across Eurasia, but also forming the first European settlements in North America. Much of the meaning of their culture has been lost to us today, while much of its mystery remains. The supposedly magical quality of runes found its way into Anglo-Saxon Christian imagery, and persists in aspects of modern soothsaying and prediction, finding a new life reinvented in the novels of J.R.R. Tolkien (*see page 262*).

Runes persisted until c. 1100. The late 19th-century revival of interest in Germanic nationalism was taken up by the Nazis, together with much Norse mythology; the Sig rune, for example, was adapted as the official form of the SS badge from 1933.

Runes

Runes were a writing system in use across northern Europe from about AD 150, varying in form in different countries at different times; they also appear to have had magic implications. In the first century AD, the Roman historian, Tacitus, described Germanic people using pieces of wood with symbols on them to predict the future, perhaps rather like the *omikuji,* or fortune-telling sticks, found in Japanese temples today. It is not known whether the symbols on the sticks were runes, but later references in the sagas to 'casting the runes' implies that something similar may have been in use. In any case, the signs would have been interpreted by a priest or other person who understood the significance of the code.

The legend of the god Odin hanging for nine days and nights on Ygdrassil, the Tree of the World, in order to learn wisdom and gain mastery of the runes is not a simple reference to attaining literacy; the runes are clearly perceived in terms of spells and also as symbolizing power, or as a source of protection.

Runes were used for writing inscriptions, especially on gravestones and memorials, and for completely mundane purposes; but many objects found in Scandinavia are marked with a few letters rather than a complete inscription, which may have been abbreviations, or code, for some well-known phrase: a wish, curse, or prayer.

The Gundestrup Cauldron

Found in a peat bog in Denmark, and dating from about the 1st century BC, this great silver vessel is covered with enigmatic scenes. If, as is probable, it was intended for ritual use, each element of the elaborate decoration would have held coded references only understood by initiates. Even though their message cannot be read today, the mixture of different elements provides a series of clues to the cultural origins and far-reaching contacts of the makers and owners.

Horned god
Holding a serpent in his right hand and a torque in his left, this is believed to be Cernunnos, the god of fertility.

Initiation
A god or giant immerses a warrior in a cauldron, while others ride off to battle.

The griffin
A common motif across the Persian and Scythian world, seen as an emissary between this world and that of the gods.

A Celtic conundrum

Although much about the cauldron has been debated, it clearly was intended for use in a Celtic, perhaps Druidic, context, but it has been argued that the workmanship is Thracian (around the Black Sea). Various influences from across Eurasia can be traced, and several of the figures have been tentatively identified.

Bearded god
This figure, with small attendants, appears several times and may represent the sea god Manannan.

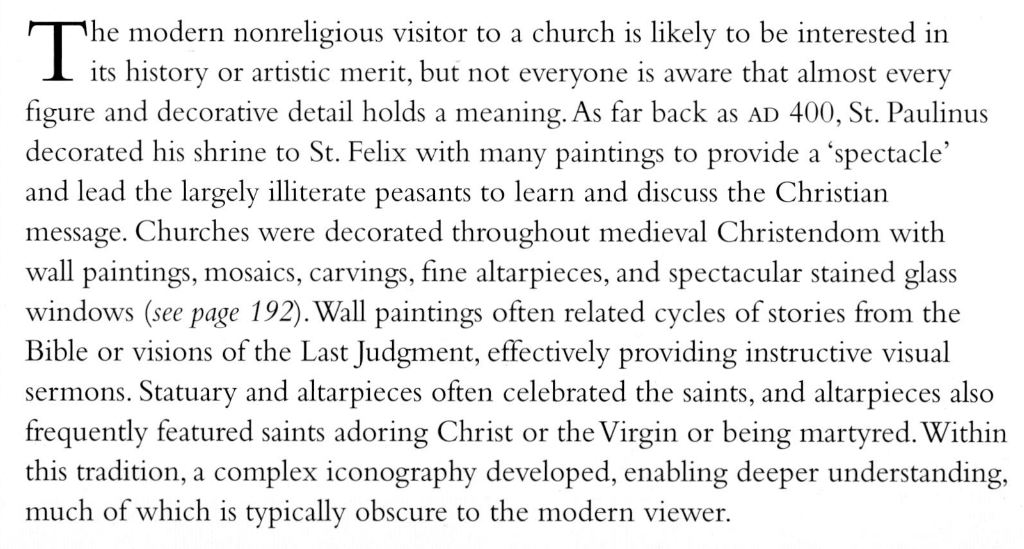

MEDIEVAL VISUAL SERMONS

The modern nonreligious visitor to a church is likely to be interested in its history or artistic merit, but not everyone is aware that almost every figure and decorative detail holds a meaning. As far back as AD 400, St. Paulinus decorated his shrine to St. Felix with many paintings to provide a 'spectacle' and lead the largely illiterate peasants to learn and discuss the Christian message. Churches were decorated throughout medieval Christendom with wall paintings, mosaics, carvings, fine altarpieces, and spectacular stained glass windows (*see page 192*). Wall paintings often related cycles of stories from the Bible or visions of the Last Judgment, effectively providing instructive visual sermons. Statuary and altarpieces often celebrated the saints, and altarpieces also frequently featured saints adoring Christ or the Virgin or being martyred. Within this tradition, a complex iconography developed, enabling deeper understanding, much of which is typically obscure to the modern viewer.

The Evangelists

The emblems of the four Evangelists are found everywhere across the Catholic world, often completely detached from their human counterparts, serving as shorthand for four of the key tenets of the faith, and sometimes associated with four of the tribes of Israel:

Matthew Angel; incarnation; Reuben.
Mark Lion; resurrection; Judah.
Luke Ox; sacrifice; Ephraim.
John Eagle; ascension; Dan.

Saints and their attributes

Once, most people would have been able to recognize the saints from their attributes, and often only the attributes were shown. Some are still remembered today – St. George and the dragon (and St. Michael, trampling a dragon with his cross-like sword), St. Mary Magdalene with her flowing hair and pot of ointment, or the giant St. Christopher, the patron saint of travelers, carrying the child Christ across a river – but many are no longer recognized and their stories have been forgotten. Many images of saints also include the palm of martyrdom.

Saints may often be identified by their dress. This group from the north portal of Notre Dame, Paris, are (*from left*): St. Maurice dressed as a Roman soldier with his lance; St. Stephen, the first martyr, in deacon's robes with a tonsure; St. Clement, the fourth bishop of Rome, shown in his robes with stave and miter; St. Lawrence, like St. Stephen, portrayed as a deacon – they are frequently grouped together.

An anchor St. Clement, who was drowned with one around his neck.

Bees St. Ambrose, his words were as sweet as honey.

A tower, cannon (often seen over the main entrance to arsenals) St. Barbara, the patroness of all those who work with fire.

A wheel St. Catherine of Alexandria, who was broken on a spiked wheel before being beheaded; the patroness of mathematicians, scholars, and lawyers.

An apron of bread and flowers St. Casilda, patroness of all who work with prisoners and for charity.

A cockleshell St. James the Great, associated with the pilgrimage to his shrine at Santiago de Compostela.

A cloak cut in half St. Martin of Tours, a young Roman soldier who shared his cloak with a beggar, patron of soldiers and those who help others.

A lamb with a flag St. John the Baptist.

A gridiron St. Lawrence, who was martyred on one.

Arrows St. Sebastian, a Roman soldier tortured by being shot with arrows.

Keys St. Peter the Apostle, the gatekeeper of Heaven.

Eyes St. Lucy, whose eyes were torn out before her martyrdom; patroness of the blind.

Disembodied breasts St. Agatha, whose breasts were cut off before she was burned.

A dog, plague sores St. Roch, believed to have the power to dispel plague.

Money bags, three gold balls St. Nicholas, who provided dowries to save girls from prostitution, also patron saint of pawnbrokers.

A vision of hell

The detailed paintings of Dutch artists, such as Hieronymus Bosch and Pieter Brueghel the Elder, are rich in symbolic references to mystical and theological concepts combined with local folklore and proverbs – a particularly effective way of communicating ideas to the illiterate onlookers. In the right-hand panel of Bosch's *The Garden of Earthly Delights* (c.1500), he unleashed his imagination in showing the damned reaping their just deserts for their vices in life.

Folk instruments
Bagpipes were associated with discordance, debauchery, and dunces.

Sacrilege
The knight devoured by pack animals is guilty of sacrilege; he holds a chalice in his fist.

Avarice
The miser is hoisted on the keys to his strongbox.

Skater
A lifelong chancer skates on thin ice.

Musical instruments
Symbols of carnal love and lust.

The bird
A cooking pot on its head, eating and defecating his victims, represents greed.

Vanity
The proud lady is condemned to contemplate her features on a devil's bottom.

Sloth
The lazy man is tormented in his bed.

A glutton
Forced to vomit into a pit.

Revenge
Hunters become the hunted as a giant rabbit drags his victim and hounds devour his companion.

Gamblers
Beaten with backgammon boards.

The Church
The avarice of the clergy is represented by a pig in nun's clothing, amorously luring a man to sign away his possessions.

Stained Glass Windows

With the development of the Gothic style of ecclesiastical architecture in 12th-century Europe *(see page 178)*, stained glass windows replaced the wall paintings of Romanesque buildings and the mosaics of Byzantine in providing a spectacular means of telling the story of the Gospels and the saints to an often illiterate congregation. They also transformed the light entering the building, providing an ethereal quality that complemented the transcendental mystery of Gothic structural techniques. Further, 'reading' a stained glass window was an exercise in understanding the mysteries of the Christian religion. Like wall paintings, some windows illustrate familiar episodes (the Nativity, the Last Judgment), others much more complicated theological ideas. Among the finest and earliest examples of the art form are found in the rose windows of Chartres cathedral, near Paris.

Virgin Mary
An image of the Virgin Mary with the Christ child on her knee resides in the center of the window, where a larger proportion of clear glass provides a visual focus. Fleurs-de-lis (lilies) – a symbol of the Trinity also associated with the Virgin – rim the roundel. The fleur-de-lis was also the heraldic device of Queen Blanche of Castile, who donated this window.

Rosettes
Used in antiquity and many other cultures as a motif to represent perfection and unity, the rose (like the lily) is frequently associated with the Virgin Mary. Here, rosettes contain yet more fleurs-de-lis.

The north transept: The Virgin in Glory

The Virgin sits at the center of this rose window at Chartres, framed by a variety of panels each comprising a series of twelve. Twelve was an important symbolic number in medieval theology, linked to the Twelve Disciples and the Twelve Tribes of Israel, and divisible by two (Duality), three (the Trinity), four (the Evangelists), and six. The panels are all designed to draw the eye in towards the Virgin at the center.

Four white doves
Immediately above the Virgin, the doves represent the Holy Spirit delivering the four Gospels.

The kings of Israel
The square panels show and name the twelve kings of Israel, cited by St. Matthew as being the ancestors of Joseph.

The prophets
Twelve Old Testament prophets perch on the outer edge of the window, again named, and fittingly inaugurate the movement of the eye from the perimeter of the rose toward the center. They in turn are fringed by fleurs-de-lis.

Angels
The remaining eight panels immediately surrounding and protecting the Virgin show a selection of archangels and angels with various attributes to identify them.

RENAISSANCE ICONOGRAPHY

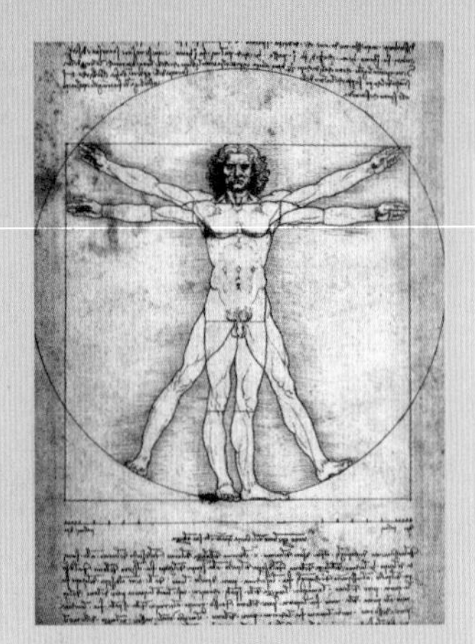

Vitruvian man

A singular aspect of the Renaissance was the melding of science and the arts. The ultimate Renaissance man, humanist, painter, sculptor, scientist, and inventor, Leonardo da Vinci apparently saw little distinction between the two. In order to accurately represent the human form in painting and sculpture, he needed to perform anatomical studies to understand how the body worked, and to enable him to design war machines he needed to understand mechanics (*see page 76*). At the center of much Renaissance thought was the quest to define the ideal. This linked to the rediscovery of the writings of Plato and Aristotle, and in one way found its expression in alchemy (*see page 52*), and the search for the Philosopher's Stone. Da Vinci, however, produced an image in which the human form could be demonstrated as a proportional archetype, a coded expression of ideal proportion, inscribable in both a true square and a perfect circle. Leonardo based these principles of proportion on those described by the Roman architect Vitruvius (*see page 179*).

The rediscovery of the classical world, which provided much of the impetus to the 'rebirth' or Renaissance in Europe, provided a new secular (and sometimes profane) dictionary of symbolism and allusion for artists, architects, and poets. One of the significant impacts of the Renaissance was the revival of humanism and a renewal of the empirical sciences. These often sat rather uneasily with the sacred requirements of the Church, which nevertheless remained one of the principal patrons of the arts. But alongside the Church there emerged a new class of wealthy and powerful princes, keen to display their wealth, importance, and erudition by commissioning ambitious artistic projects.

The hidden message of the Three Kings

Until the Counter-Reformation, which strongly discouraged the use of religious painting for personal or political promotion, a commissioned work of art could carry a very strong contemporary message, although many details are lost to us today. Benozzo Gozzoli's *Procession of the Magi* (1459-60, in the chapel of the Medici Palace in Florence) is one that has been in part deciphered, with many key characters from the Bible story represented by contemporary figures. At the forefront is Lorenzo de' Medici, as Caspar. Painted across three walls, the side walls include portraits of the Byzantine emperor as Balthasar (*above right*), the previous Holy Roman Emperor Sigismund as Melchior, and shows the Sienese Pope Pius II as merely a member of Lorenzo's retinue. The landscape is not an imagined Holy Land, but an exotic view of a Tuscany that the Medicis ruled.

Balthasar is a likeness of the Byzantine emperor, John VIII Palaeologus, who visited Italy in 1438 to attempt a reconciliation between eastern and western Christendom.

Rival powers
North Italian princes from the powerful Malateste and Sforza families are at the forefront of Lorenzo's train.

The painter
Gozzoli did not fail to include his self portrait in the throng.

Piero the Gouty
Lorenzo's father is shown following Lorenzo along with other members of the Medici court.

Caspar
An idealized portrait of the young Lorenzo de' Medici – a daring statement of the pecking order of the age, aligning himself with the emperors of the East and West, and the Vicar of Rome (who was in debt to the Medicis).

The Ambassadors

Hans Holbein the Younger's double portrait of two French courtiers, Jean de Dinteville (left, who commissioned the painting) and Georges de Selve (Bishop Elect of Lavaur), ambassadors to the court of Henry VIII of England, was painted in 1533, at the crisis point of the Reformation as Henry threatened to break away from Catholic Rome. It is rich in coded messages, and is not just as it seems about two noblemen displaying their erudition, but also about the political crisis at hand.

The globe is positioned to show French territorial and diplomatic interests. It also names Dinteville's château at Polisy.

The Crucifix
Partly concealed behind the hanging, this reminds us of Christ's significance both in the lives of the sitters and the momentous events of the day.

Celestial globe
The brass frame has been set to the latitude of Rome, rather than London, betraying the Catholic faith of the two subjects.

Sundial
Shows the date to be April 11, 1533, the day when England effectively broke with Rome.

Book
The fore-edge of the book under de Selve's elbow is inscribed with his age.

Lute
Often a symbol of harmony, its broken string suggests the Protestant discord with the Catholic Church. The adjacent set of flutes is also missing one instrument, meaning that a harmonious effect cannot be achieved when played.

Hymnbook
Open on a page of hymns used by both Protestant and Catholic churches (Come, Holy Ghost and The Ten Commandments) translated from Latin into German by Martin Luther.

Set square
This, holding open a German treatise on applied mathematics for merchants – on a page dealing with 'Division' – along with other scientific instruments on the table top, indicates the sitters' mastery of modern ideas.

Dagger
This is inscribed with Dinteville's date of birth in a Latin abbreviation.

The skull
Painted in extreme perspective, this acts as a disguised memento mori. The skull was also part of Dinteville's personal insignia.

Cosmati mosaic
An accurate representation of the floor of the Sanctuary at Westminster Abbey. Cosmati mosaics were common in Roman churches.

THE AGE OF REASON

Between 1600 and 1900 a series of revolutions in all aspects of European life – religion, science, and politics – brought about the birth of the modern world. A new rationalism emerged, which inevitably affected the arts in a number of ways. New codes were promulgated to control legal and political institutions, while a host of new codes were developed to describe the world, from the natural sciences to the applied sciences (*see pages 154-161*). The period began with the Enlightenment, and drew to a close with the Age of Reason.

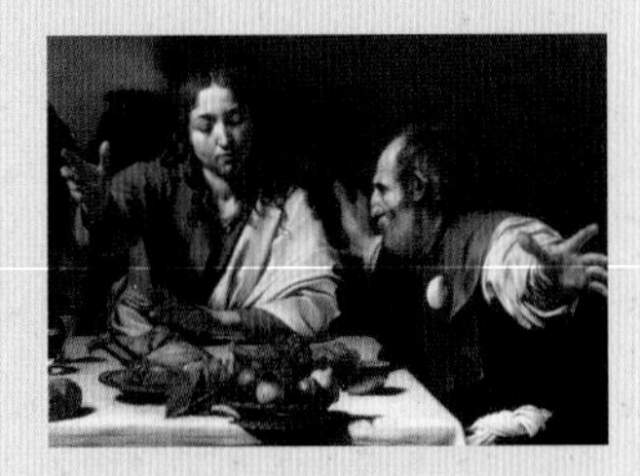

A new simplicity

In 16th-century Europe, both the Protestant Reformation and the Catholic Counter-Reformation rejected elaborate iconography in art. For most Protestants this meant a ban on religious imagery of all kinds, and for Rome a concentration on a directness and simplicity in devotional art. The Italian painter Caravaggio imbued his paintings with a simplicity and naturalism that did not preclude the use of theatrical lighting, contemporary costume, and the use of working-class models to represent sacred figures. In *The Supper at Emmaus* (1601, *above*) the scene could be a Neapolitan inn.

A visitor to Rome might also wonder about various incongruous symbols incorporated into paintings, sculpture, and architecture. These were frequently the family emblems of the incumbent Pope, the most famous being the bees associated with the Barberini Pope Urban VIII.

Hogarth: the moral maze

During the 17th century, a new style of secular art developed, dealing with the modern world with an apparent naturalism not seen since the paintings of Brueghel. In England, a major proponent of the new secularism was the British painter and printmaker, William Hogarth (1697-1764). Conventionally trained, but a devoted political and social critic, he peppered his intricately designed works with both contemporary and historical allusions. His various morality series, including *Marriage A-la-Mode* (c.1743, *below*), anticipate the narrative techniques of comics and the cinema, and innumerable knowing references enrich the onrush of his narratives.

Debts
The threadbare butler is dismissed with a handful of unpaid bills.

Debauchery
Musical instruments and an upturned chair indicate her previous night's revels.

Broken nose
A classical bust with a broken nose signifies impotence.

The sword and the dog
Symbols of both devotion and lust, the sword (broken) shows the husband has been unfaithful, as does the dog sniffing another's handkerchief.

Neo-Classicism

Archaeological excavations in the mid-17th century created a new enthusiasm for classical ideals. Stoicism was especially admired, and a style of Neo-Classicism developed across the arts in Europe, adopted most significantly in Revolutionary France and newly-independent America. Implicit in this was the admonishment of examples of political and moral rectitude. The paintings of Jacques Louis David (1748-1825, *above*) exemplified this, selecting key moments from the classical canon, painted realistically, as if a classical frieze had been brought to life.

The Sleep of Reason Produces Monsters (*left*) One response to the new rationalism was Romanticism. Artists such as William Blake and Henri Fuseli investigated dream scenarios. The Spanish artist Francisco Goya (1746-1828) witnessed the reality of the 'new thinking' when Napoleonic troops invaded his country, perpetrating appalling atrocities. His ironic image illustrated the dichotomy between the normal and the abhorrent.

The *Code Napoléon* After the dissolution of the Directoire in 1799, Napoleon introduced institutions in France and across his conquered territories to create a society based on wealth and merit rather than on tradition and inherited privileges. Secular in inspiration, the *Code Napoléon* influenced the development of political thought in Western society fundamentally, and was also adopted in Egypt, Japan, and the Ottoman empire, and many of the emerging independent states of Latin America later in the 19th century.

CODE CIVIL
DES
FRANÇAIS.

ÉDITION ORIGINALE ET SEULE OFFICIELLE.

À PARIS,
DE L'IMPRIMERIE DE LA RÉPUBLIQUE.
AN XII. 1804.

The decimal revolution

Apart from the extreme rationalization of French society performed by the post-revolutionary Directoire, which saw many of the aristocracy under the blade of the guillotine, there were many less extreme measures to codify the world anew. Decimalization was seen as the logical answer, and a search for the most scientifically appropriate order for many aspects of life was promoted by the government. One enduring legacy was the establishment of the metric system for weights and measurements.

The Revolutionary calendar

After much deliberation, a new French Republican calendar was instituted to date from January 1, 1792. It comprised an unavoidable 12 months due to the lunar and solar cycle (each month was renamed, and the calendar started at the Fall equinox), but each month comprised three ten-day weeks, each day divided into ten hours, made up of 100 minutes of 100 seconds each. Decimal clocks were manufactured, and contemporary dating organized to reflect the 'Year Zero' era of the institution of the Republic.

Fall	
Vendémiaire (grape harvest)	from Sept. 22/23/24
Brumaire (fog)	from Oct. 22/23/24
Frimaire (frost)	from Nov. 21/22/23
Winter	
Nivôse (snow)	from Dec. 21/22/23
Pluviôse (rain)	from Jan. 20/21/22
Ventôse (wind)	from Feb. 19/20/21
Spring	
Germinal (germination)	from March 20/21
Floréal (flowering)	from April 20/21
Prairial (pasture)	from May 20/21
Summer	
Messidor (harvest)	from June 19/20
Thermidor (or Fervidor, heat)	from July 19/20
Fructidor (fruit)	from Aug. 18/19

Victoriana

Perhaps because so many things could not be said openly, and also because it was the beginning of great social mobility, the Victorians had a passion for signs and symbols, and for coded messages that only the 'right' people would understand. Many of these symbols reappear all across the Victorian arts, from fine art to embroidery samplers, and from tombstones and garden design to jewelry.

Symbols for the dead

Following the death in 1861 of Queen Victoria's adored husband Prince Albert, the British developed a cult of the dead. The tombs in any older graveyards today bear a wealth of coded messages.

Anchor A disguised cross in the time of the Roman persecutions of Christians, it survived as a symbol of hope and, of course, on the tombs of seamen. With a chain it implies a faith in salvation.
Broken column Grief and loss.
Cherubs Graves of children.
Draped urn Generally used for an older person.
Hands Clasped hands symbolize love and friendship. Whichever one is holding the other (women have frilly cuffs), represents the spouse who died first, leading their partner to Heaven. A hand with a heart represents charity; a hand pointing down could be a Freemason; hands with the thumbs touching indicated a Jewish family.
Hourglass Transience of life.
Lamp Knowledge, hope, guidance, immortality.
Peacock Immortality, a pre-Christian symbol.
Scallop Pilgrim, especially one who had made the pilgrimage to Santiago de Compostela, but used also among the Puritans in North America.

Jewelry

Victorian jewelry, like so much else from the period, carried a coded message in both form and choice of gems. The anchor stood for hope and steadfastness, ivy for evergreen memories, and clasped hands for friendship. Some very ancient images were revived: the butterfly representing the soul, the snake symbolizing eternity (Queen Victoria's engagement ring from Albert was in the form of a snake), the crowned heart meaning love triumphant, and the fly humility.

Gemstones were worn or given for their coded meanings, sometimes using their initial letters to spell out a word:
Diamond
Emerald
Amethyst
Ruby

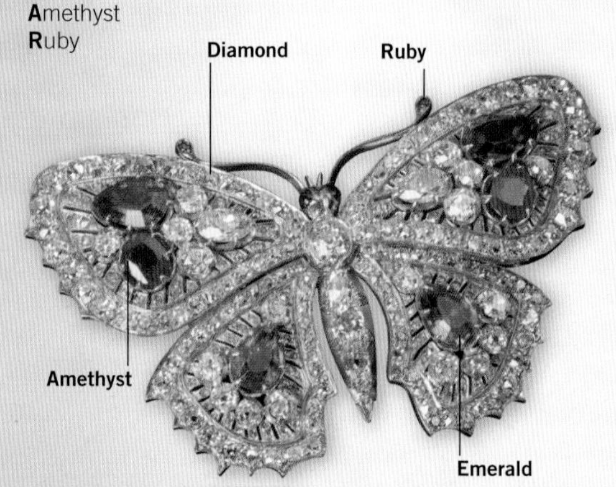

Agate Health.
Amethyst Devotion; soothes violent passions.
Carnelian Prevents misfortune.
Chalcedony Banishes sadness.
Diamond Purity, constancy.
Emerald Hope; ensures true love.
Garnet Constancy, fidelity.
Jasper Courage, wisdom.
Moonstone Good luck.
Onyx A happy marriage.
Opal Inconstancy.
Pearl Purity, innocence, tears.
Ruby Passion.
Sapphire Repentance, loyalty.
Sardonyx Married bliss.
Topaz Friendship.
Turquoise Prosperity, selflessness.

The flower cult

The Victorian taste for symbols led to the creation of an elaborate language of flowers, and making decorative bouquets with a hidden meaning was considered a nice accomplishment for young ladies. Tussie-mussies, as the posies were called, are known from the 15th century, although the word came to have an altogether cruder meaning, doubtless ignored by the young ladies. The significance of some of the flowers goes back to the classical world, others are 19th-century inventions.

Color also altered the meaning of a given flower – yellow implying jealousy, white purity, red passion (or sometimes anger), purple capriciousness, and blue faithfulness.

Acacia Secret love.
Anemone Forsaken.
Bay I change but in death.
Begonia Beware.
Bluebell Humility, constancy.
Camellia Perfection, admiration.
Cowslip Pensiveness.
Daffodil Respect.
Flowering reed Trust in heaven.
Four-leafed clover Be mine.
Geranium You are childish.
Grass Homosexual love.
Honeysuckle Devoted affection.
Love-in-the-mist I don't understand.
Marigold Grief.
Petunia Resentment.
Primrose I can't live without you.
Sweetpea Farewell.
Wallflower Faithful in adversity.

Victorian art

The Victorians particularly liked narrative or 'problem' paintings in which a story, usually with a moral message, was embedded in the scene, a story that the viewer had to unravel by identifying various coded clues. The Pre-Raphaelite Brotherhood were particularly adept at such detailed pictures, such as Holman Hunt's *The Awakening Conscience* (1854). Often, given the shameful nature of the implied story line (errant husbands and fallen women), the paintings also needed to appear, superficially, entirely respectable.

The painting
Over the mantelpiece is the 'Woman Taken in Adultery' of the Gospels or the 'Repentant Magdalene,' either highly unlikely in a kept woman's boudoir.

Clock
Reminds us that her youth is passing and she will soon be discarded but it is decorated with 'Chastity binding Cupid' – her fate is not inevitable.

Flowers on the piano
Possibly anemones, symbols of abandonment, but possibly columbine, emblematic of fickleness and male adultery.

The wallpaper
Grapes and corn – symbols of the Communion – left unguarded for the wild birds to eat.

The song
On the piano is Moore's 'Oft in the Stilly Night' – in which a woman meditates on her childhood innocence.

Rings
She wears rings, but no wedding ring.

The dress
A petticoat – very shocking to Victorian sensibilities and immediately proclaiming that she is not respectable.

The hat on the table
It is not his house, he is only visiting.

Tennyson's 'Tears Idle Tears'
A poem on past innocence and present grief.

The cat playing with a bird
A classic image of a woman at the mercy of a predatory male.

A glove
Lying on the floor, an image of a woman used and discarded.

Tangled skeins of wool
Domestic virtue traduced by lies and reduced to chaos.

Hunt rented a room in a *maison de convenance* in St. John's Wood. It symbolizes city life, worldliness, and the power of money; there are class undertones in the dress of the young man and the rather flashy furnishings. The garden, in contrast to the room, seen in a mirror (also symbolic) and filled with sunlight and white roses, represents purity, innocence, a lost Eden or, ultimately, paradise.

TEXTILES, CARPETS, AND EMBROIDERY

Recurring motifs

A number of ancient motifs are shared across Eurasia, although they vary in stylization and precise interpretation. In carpet and rug design the imprint of Islam is particularly apparent, with variations on the *mihrab* (niche), mosque lamps, religious inscriptions, and other Islamic motifs occurring alongside older, traditional symbols.

Animals Stylized domestic animals speak of traditional tribal livelihood; with walking humans may be a reflection of a migrant lifestyle or historic migrations; wild animals celebrate the hunt.

Boteh Teardrop pattern found on paisleys and many carpets, meaning protection and joy; possibly a version of the Tree of Life, or a symbol to ward off the evil eye.

Comb Represents the dowry of which the carpet might form part, and reminds Muslims Allah requires cleanliness.

Ewer Reminds Muslims about washing before prayer.

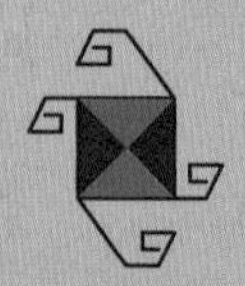

Swastika A good luck symbol across much of Asia, occurring in various stylized forms.

The modern world is inclined to think of fabric patterns in terms of fashion, decorativeness, or originality, but in the past and still in traditional areas of the world, every element, every motif has significance. Not only do particular colors or designs immediately proclaim village, tribe, caste, status, religious affiliation, and so on, but each element in the pattern has a meaning, even if its origins are long forgotten.

Carpets

Apart from the native American tradition, which remains unique, within the 'carpet belt,' which stretches from the eastern Balkans to Western China, carpets have a comparatively standard repertory of designs, usually incorporating a main field, surrounded by borders. Among the most common design is the cross-shaped 'garden' pattern symbolizing paradise and its four rivers. Village and tribal carpets and kelims were, until recently, made for family use, with references which the people who saw it would automatically be able to 'read,' although an outsider would simply see geometric motifs.

Color Rich primary colors are frequently juxtaposed.

Abstract form 'Real' images – animals, humans, and landforms – have often been reduced to geometric shapes.

North America Using a wide range of bright colors, this dynamic indigenous tradition often combines highly geometric abstraction with stylized animal forms (*above*).

The Caucasus At the heart of the 'carpet belt,' Caucasian carpets display strong tribal variations, with highly geometric abstraction but recurring decorative motifs (*below*).

Göl Abstract polygonal motifs, probably once tribal emblems.

Herati A diamond motif with branches, often with four serrated leaves, called the 'fish' in Persia.

Meander Border motif symbolizes both eternity and unity, with flowers or fruit, often a vine, signifying abundance.

Chinese embroidery
Densely patterned but, as with Chinese carpets, the dominant images tend to be plant and animal forms, lucky symbols such as swastikas, and medallions.

Dragons
Some animals are linked to the Chinese zodiac, the most significant being the dragon.

Clouds
A frequent motif, suggesting the ethereal.

Chinese imagery

In China the symbolic code is complex, since it works both on symbols and puns. It builds up to a language that can be read across all the Chinese arts, but especially in embroidery and ceramics. In many ways the canon was absolutely fixed; the five-toed dragon was the symbol of the Emperor and only the royal family could use it, along with the color yellow. The rank of officials and soldiers was indicated by an embroidered square worn on the front of the robe. Some animal symbols are linked to Chinese zodiac symbols, others are prized for their sound, such as the bat –'*fu*,' a homonym for happiness. 'Red bats' sounds like 'great happiness.' Others link to folk beliefs: the Mandarin duck, believed to pair for life, is the symbol of married happiness. Among the standard elements in the repertory are flowers and plants.

Fungus Long life.
Narcissus New Year.
Orchid Scholar, virtue.
Peach Longevity.
Peony 'Flower of wealth and honor' – a happily married woman.

The 'Three Friends of Winter'
A group appropriate for old friends:

Plum Courage, flowering in the cold.
Bamboo Resilience.
Pine Endurance and fidelity.

Borders and hems
Often include lucky butterflies and flowers.

Spanish shawls

In the early 19th century shawls came into fashion across Europe. Kashmir shawls were beyond the means of most people and silk shawls of the type now called 'Spanish' or 'piano' shawls flooded in from China and the Philippines. They were an immense success, but not all of the Chinese motifs were appreciated by Western women: bats (happiness) were quickly transmuted into another favorite Chinese symbol, the butterfly, meaning joy; the magic toad spitting gold coins was simply ignored, as were cockroaches, symbols of plenty to share; rats, energetic accumulators of wealth, became squirrels; while the cloud ear fungus of long life simply became a cloud. When, after 1911, civil war in China led to the manufacture of these shawls in Spain, Andalusian women reinterpreted the Chinese symbols to fit their own code: the peony became a rose, a symbol of love; the trailing gourd, emblem of long life and many descendants, became the vine; and sheaves of grass or rice became wheat.

'Underground Railroad' quilts

In the 19th century slaves escaping from the American South to the Union states and Canada were helped by a number of people and institutions, many of them Quaker. Later this network, never formally organized, came to be known as the 'Underground Railroad' and a number of related terms were used as code: safe houses were 'stations,' guides were 'conductors,' and the slaves themselves were 'cargo.'

In the 1990s a story began circulating that quilts were used to help the slaves escape. The claims were vague – maps were somehow worked into the quilts, they were displayed at certain points to convey messages, different patterns had specific meanings: 'safe house,' 'head north,' and so on. A crafts revival in the 1980s made quilting a multimillion-dollar business and by the time the 'quilt code' theory appeared, black, slavery, and women's studies were growth industries. The story flourished and became more elaborate. Books were written and 'quilt code' kits became available; antique dealers happily charged inflated prices for what they claimed were 'code quilts' made by slaves. But there is no evidence that such quilts ever existed: no firsthand accounts of former slaves mention such quilts; a number of the patterns seen as significant in fact date from the 1920s; and the very few extant quilts actually made on plantations, probably by and for slaves, are far from the elaborate luxury productions now ascribed to them.

Clearly, no one can prove that a quilt was not occasionally hung out of a window to provide a simple 'yes/no' code: 'safe, come/danger, stay away,' but the 'quilt code' has moved from being a nice story to a full-blown modern commercial myth.

Trade, finance, industry, and many of the artisan crafts upon which these functions rely, have evolved coded systems and languages to ensure the efficiency of their enterprises.

codes of commerce

The commerce and products of the modern world are bound up in codes of every sort from catalogs and listings to brands, trademarks, bar codes, and sell-by dates. These codes have been developed to ensure a flow of information between the supplier, retailer, and consumer and are designed to guarantee quality, consistency, and availability. Unfortunately, like all coding systems, they are not invulnerable.

COMMERCIAL CODES

The first numbering and writing systems were records of stock and trading transactions (*see pages 20, 26*). The invention of weights, measures, coins, and assay systems followed rapidly (*see page 212*). But the idea of monitoring stock, administrative overheads, and detailed accounting came of age with the development of banking systems, stock exchanges, and intercontinental trade from the 17th century onwards. By the beginning of the 21st century, the depth and flexibility of coding applied to everything from a tax return to the purchase of a bar of soap from the local grocery store meant that commerce and administration at every level was bound up in a spider's web of codes.

ZIP codes

The introduction of postal zone codes began in 1943 in the USA, for use in large cities where areas of a city were broken down often just by single or double digits. In 1963 the system was extended across the nation, in the form of ZIP (Zone Improvement Plan) codes and began to be taken up by other countries, notably in Europe. In the US, the first three numbers identified the regional post office the mail was to be sent to, the last two numbers were for sorting thereafter. In 1983 ZIP+4 introduced add-on codes, a separate set of four numbers to identify a more specific address within the local postal zone. This ingenious Geographical Information System (GIS) soon proved an invaluable data resource exploited by many other industries for personal identification, targeted mailing, consumer surveys, censuses, courier deliveries, and assessing household insurance. ZIP codes are increasingly being digitized into bar codes known as Postnet, allowing optical character recognition to sort mailings.

Telegraphic codes

The rapid establishment of the electronic telegraph by the mid-19th century was driven in large part by the demands of long-distance commerce (*see page 94*). But telegrams were charged by the letter. Soon industries realized that abbreviated messages would save money. Several systems were invented, notably the A.B.C. Telegraphic Code and Bentley's Second Phrase Code. In part these allowed companies to preset a formula for words or phrases commonly used in their line of business (textiles, shipping, etc), broken down into parameters comprising fixed code words. They also allowed the user to encrypt messages for security reasons. Thus, Bentley's 5-bit code included code words such as 'ATGAM' meaning 'have they been authorized' or 'OYFIN' meaning 'has not been reinsured.'

Teleprinters

The development of automatic teleprinters allowed typed messages to be sent and received without manual encryption into Morse code. In order to increase bandwidth efficiency the messages were initially compressed into fixed 5-bit strings, known as Baudot code, invented in 1874. Demand for more characters led to the development of the 6-bit TeleTypeSetter (TTS) system and the Western Union's International Telegraph Alphabet No. 2 (ITA2), forerunners of 7-bit ASCII and 16-bit Unicode (*see page 273*).

According to Hollywood, Thomas Alvar Edison (1847-1931) invented just about everything, from the electric light bulb to the telephone. The ticker tape, which printed out commercial code messages on strips of paper, and also provided spectacular confetti for celebration parades, is one of his legacies, which lives on in the running displays of stock prices on TV news channels.

Tiffin

Every day in Mumbai (Bombay), India, millions of office workers are supplied with literally home-cooked lunchtime meals (tiffin) at their desks. One team collects tiffin containers from the home and these are loaded onto trains from the suburbs; a second team unloads the train; a third delivers the tiffin to the desk; and the empty containers are collected and delivered home again. It is a miracle of organization, reputedly 100% efficient, and employs thousands of couriers *(dabbawallas)* and an elaborate but obscure coding system. It works rather like a ZIP code; every tiffin container is marked with a colored circle or flower, and carries an identity number, for example:

K-BO-10-19/A/15
K is the dabbawalla's identity code.
BO denotes the district from which the tiffin is to be collected.
10 indicates the destination district of Mumbai (Bombay).
19/A/15 identifies the specific address, building, and floor for delivery. The code works in reverse for returning the empty tiffin containers.

Unlike airline baggage handling, this system is uniquely efficient, and both former US President Bill Clinton and Bill Gates of Microsoft have requested presentations of how the system works.

Tiffin containers being loaded for delivery in Mumbai (Bombay).

Parcels of laundry were carefully packed and marked with individually coded labels.

Laundry marks

One of the mysteries of modern life remains the laundry marker code, designed to ensure that commercial laundries return the correct items of clothing to their owners. In fact there is no set pattern for these, but the technique was developed in the 19th century when many migrant Chinese opened commercial laundries in the USA, Europe, and colonial Asia. Small glued labels carrying a unique number or combination of characters, sometimes in different colored inks, would be applied to each garment in each delivery. In India today, clients are assigned, quite literally, a pin code, a unique pattern of dots tattooed or punched in an inconspicuous area on each garment.

Bar codes

The idea of bar codes applied to products to accurately record stock control levels was patented in 1952, but was not introduced commercially until the mid-1960s, nor widely used until the 1980s. It was first designed for identifying and tracking railroad cars, and was then applied to automobiles using toll bridges. The first retail item to be sold using a bar code was a multipack of chewing gum, at Marsh Supermarket in Troy, Ohio, in 1974. This machine-readable system of tracking sales has transformed the efficiency of the retail industry whilst also opening other opportunities: when linked to credit card sales, or customer loyalty card schemes, consumer buying patterns can be assessed to produce individually targeted marketing campaigns.

The bar code system:

0 0001101	**3** 0111101	**6** 0101111	**9** 0001011
1 0011001	**4** 0100011	**7** 0111011	
2 0010011	**5** 0110001	**8** 0110111	

The most widely used bar code symbology is the UPC (Universal Product Code) used throughout North America, which is entirely digital, encodes up to 12 digits, and comprises a total of 95 bits; in addition to the start and end bars, there is a middle or guard bar. Each digit is coded in seven bits (a system not dissimilar to Francis Bacon's code, (*see page 82*).

How bar codes work

Reading a bar code To the left of the middle or guard marker, the printed bars represent 1, and the spaces represent 0s; after the guard bar, the system is reversed and appears in negative. Although there are variants in different countries and trading zones, the basic principles remain the same.

Brands and Trademarks

Pears' Soap
One of the first commercial branding ventures was launched by Pears' Soap, combining a distinctive logotype and an image by the leading British painter, John Everett Millais, in 1886. Pears also invented 'brand extension' (positioning an innocuous product alongside another set of values or ideas) by sponsoring an enormously popular annual encyclopedic almanac, thereby associating their soap with the cherished Victorian virtue of self-improvement in the home.

One of the most high-profile codes that surround us today is that of the 'brand.' The roots of the idea lay in the indelible branding on the skin of livestock and slaves to display ownership. The modern concept of a brand, in which a product can be reduced to a simple image or trademark that embodies certain proprietary core values of quality, style, and ethical values, developed in the age of competitive mass-manufacturing and mass-consumption that began in the late 19th century. In Japan, the feudal *mon* has been readily adapted into contemporary commercial branding (*see page 130*), but building a brand is much more complicated than simply coming up with a memorable design or logo: the successful reduction of a matrix of values into a single, often abstract, image remains a testament to the power of positive marketing.

Albrecht Dürer's combination of his initials into a simple design is an early example of a publishing colophon.

Publishing ownership: the colophon
In Europe, the first instances of mass-manufacturing occurred with the advent of printing, and the German artist and printmaker Albrecht Dürer (1471-1528) soon seized the idea of branding his prints with a unique signature. Printers and publishers ever since have sought to establish a branded quality through a simple printed image, or colophon. One of the most successful at this was the UK paperback imprint Penguin. Although the concept of paperback or softcover publishing cannot be claimed by them, the name and the logo became synonymous with 'quality' but affordable literature from its launch in 1935, and soon branched into a variety of linked imprimaturs and color-coded products.

The Penguin imprimatur developed in several directions: the Pelican imprint denoted scientific/nonfiction titles, while Puffin came to represent children's books. Within the Penguin list, colors were used to denote different literary genres, with the original orange for literature, green for crime/mystery, blue for nonfiction, and purple for travel titles.

From idea to image

The ideal aim of modern branding is to codify the 'message' of a product to the barest minimum. This can take many forms, from the distinctive Coca-Cola bottle and its 'Dynamic Contour Curve' logotype to McDonald's 'Golden Arches.' Nike achieved brand perfection in the 1970s by referencing the ancient Greek goddess of victory in the company name, with a simple tick motif implying swiftness, linked in 1988 by the 'inspirational' but nevertheless meaningless phrase 'Just Do It.'

Trademarks and quality

Since 1875 it has been possible to trademark (™) the name of a company or product in much the same way as registering a patent. Occasionally the name of an innovative product can become a generically-recognized term, such as 'Hoover' for vacuum cleaner. The recent enthusiasm for 'designer labels,' of little or no value in themselves, has led to a burgeoning piracy industry for the black market (*right*).

Branded bands

While certain music labels and artists have sought brand recognition through typographic and product design – the 'cool' design of Blue Note albums, Vertigo's psychedelic spiral disc centers, and The Who's distinctive 'mod' typography, for example – some have sought to reduce themselves to a minimal visual code. The Rolling Stones' Mick Jagger commissioned the famous 'lapping tongue' logo in 1970, at the pinnacle of their success, and it remains universally recognizable. Some went much further: Led Zeppelin's fourth album had no title or artist on the sleeve, merely four mysterious symbols on the spine, each apparently representing a band member.

Page Jones Bonham Plant

Guitarist Jimmy Page even insisted that no catalog number appear on the sleeve, but relented when it was pointed out that, without it, retailers would be unable to order the record.

'The artist formerly known as Prince' reduced himself to a cipher just as his chart popularity began to wane, but even his longevity as a symbol was limited.

MAKERS' MARKS

Marks, in the form of codes, have been used for over 1,500 years to guarantee the quality of valuable items. They were essentially the first instance of consumer protection, providing a means by which traders could be certain of the value or standard of goods. They were first used by the Byzantine empire in the 4th century AD to denote the quality of silver items: five small punch marks can be found on many silver items dating from this period, and although archaeologists cannot be certain of their meaning, it seems likely that these were forerunners of modern hallmarks, given the economic importance of silver at the time. Since these first crude codes, systems of valuing precious metals, principally silver but later gold as well, have developed enormously. Over the last few centuries, the practice has also spread to fine porcelain and, more recently, to the use of proof marks on firearms.

Carats

The term 'carat' can be used to describe either the mass of gems and pearls, or the purity of gold. The maximum 24-carat gold must be at least 99.9% pure gold, 22-carat 91.6%, 20-carat 83.3%, and 18-carat gold 75% pure. In order to receive a CCM hallmark, gold items must consist of at least 18-carat gold. When used to describe gems, a carat is simply a measurement of weight equal to 0.007055 ounces (200 mg), and has nothing to do with the quality of the gem. This is a metric carat, and is used universally; it can be divided into one hundred points of two milligrams each. A 24-carat diamond for example, would weigh 0.169 ounces, or 4.8 g.

Hallmarks

Some centuries after the Byzantines, France was the first European country to standardize its marking of silver, in 1275, then gold in 1313, with the *poinçon de maître*, or 'maker's mark.' In 1300, Edward I of England decreed that all silver articles were required to meet the 'Sterling Silver' standard, equal to at least 92.5% pure silver, and introduced the lion passant guardant symbol. Articles that met these criteria were marked at the Assay Office, in the Goldsmiths' Hall of the Worshipful Company of Goldsmiths, with a leopard's head. The word 'hallmark' has its origins here – a mark made in the Goldsmiths' Hall, although other assay offices were established in nine other cities in the British Isles. Additional marks usually include the 'maker's mark,' which simply shows which craftsman created the article, often a set of initials or coat of arms, and a 'date mark,' in the form of a lower case alphabetical letter. Date marks vary from one assay office to another. British silver objects created between 1784 and 1890 also carry a 'duty mark,' the head of the reigning monarch.

The four oldest assay offices are indicated by symbols: the leopard's head (London), the anchor (Birmingham), the Yorkshire rose (Sheffield), and the castle (Edinburgh). Others include Chester, Exeter, York, Newcastle, Glasgow, and Dublin. The lion passant guardant means sterling silver.

British hallmarks typically include (*from left*) a maker's mark, the assay office mark, the CCM (*see below*) or the lion passant guardant, and a date mark, in the form of a lower case letter. Occasionally extra marks may be included, such as the Millennium mark, used in the year 2000.

Standardization

Different nations developed their own systems and symbols for marking products over the centuries, and no real attempt at standardization was made until 1972, when the seven EFTA nations drew up the Vienna Convention on the control of the purity and hallmarking of precious objects. This resulted in the use across signatory nations of the CCM, or Common Control Mark, on gold, silver, and platinum articles. Although this has gone some way to help standardize the marking of precious metals, a complete international hallmarking system has not yet been achieved, due to differences in standards and enforcement both within and between nations.

The modern CCMs for gold, silver, and platinum. The numbers indicate the fineness – 750 parts gold to every 1,000 parts, 925 parts silver to every 1,000 parts, and 950 parts platinum to every 1,000 parts.

Fine porcelain

High-quality ceramics were initially imported to Europe from China from the 16th century, the products of the Ming empire bearing painted marks in Chinese characters. Delftware produced at the Moor's Head factory in Holland began to imitate the colors and patterns of Chinese ware when the trade was interrupted in the 1620s. When a way of reproducing the luster and quality of Chinese porcelain had been discovered by the alchemist Johann Friedrich Böttger, factories opened in Dresden (Meissen, 1710) and the technology rapidly spread to France and Britain, and a highly competitive – and lucrative – business developed. Ceramic trademarks are used slightly differently from precious metal hallmarks – there was no need to show the fineness or purity of porcelain, nor its approval by an assay office. These marks are principally maker's marks – they show the buyer that the piece was made by a long-established firm with a good reputation, such as Meissen, Minton, Royal Crown Derby, or Wedgwood. In addition, they can provide information as to the exact date of manufacture (most factories regularly modified their marks), and sometimes the identity of the individual craftsman responsible for the piece. The marks tend to be considerably more ornate than hallmarks, and are applied in four different ways: either incised, impressed, painted, or printed. Incised marks look more individual and spontaneous than impressed marks, and likewise painted marks less complex but more unique than printed ones. Most 19th-century marks are printed, in most cases in blue under the glaze.

The immense popularity of Chinese porcelain led many European manufacturers to emulate not only its quality, coloring, and glazed luster, but also the Chinese system of adding a maker's mark. The Chantilly factory in France developed marks designed to imitate Chinese characters, although soon local symbols developed, such as the crossed swords of Meissen pieces.

Chinese marks from the Ming period | **Chantilly** | **Meissen**

The Meissen factory at Dresden was the first to fully capture the qualities of fine Chinese porcelain, and produced many pieces with a Chinese theme, such as this guitar player, exploiting the enormous popularity of Chinese ware. By the middle of the 18th century many other factories had learned the technology, and many distinctive and local styles developed. Nevertheless, the term 'china' as a generic reference to porcelain is still in use today.

Porcelain marks

Each European factory developed its own maker's marks. Although some used printed stamps, most marks were added by the craftsmen themselves. These were not standardized in any way, and tended to vary enormously within short periods of time. As dates were rarely included in the marks, catalogs of these variations help modern collectors to date many pieces.

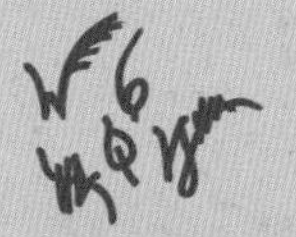

Worcester emulating Chinese

Minton

Derby

Chelsea

The color and style of makers' marks can also aid in dating a piece, as can features like Derby's royal warrant.

Firearm marks

Proof marks are small symbols impressed in the metal of a firearm, often somewhere on the barrel, not visible until the weapon is dismantled. They guarantee that the weapon has passed a proof test, intended to prove the safety of the weapon.

Codes of Work

St. Paul's
The most ambitious single building project undertaken following the Great Fire of London (1666) was St. Paul's cathedral, designed and supervised by Sir Christopher Wren, begun in 1675 and completed in 1710. In addition to elaborate plans and elevations (*right*), a scale model was constructed (*above*), ensuring that the builders on the site had clear guidelines to follow.

For centuries, builders and building technicians have needed to communicate their ideas and designs to others in their own and allied trades. Architects especially had to devise a way to codify their plans for edifices in a way in which master masons and laborers could interpret. As buildings began to involve integrated features such as plumbing and electrical circuits, so new diagrammatic codes needed to be created, not least to inform later generations how buildings and the circuits and systems within them worked. Over the last 150 years, the explosion of new technologies has seen a range of new artisanal skills emerging, which demand a familiarity with a closed world of coded languages.

Architectural plans

While there is little evidence of drawn-up master plans in the creation of Romanesque and Gothic buildings, by the High Renaissance a systematic approach to describing a building, based on a detailed ground plan and elevations, had developed. Reading such plans was often performed in conjunction with a scale model of the building. One of the most astonishing achievements of the early modern era was the rebuilding of London after the Great Fire of 1666. Architects such as Sir Christopher Wren, James Gibbs, and Nicholas Hawksmoor, drew up detailed designs for numerous churches and other buildings to restore the city.

Elevation The plans for the elaborate dome are here shown in cutaway.

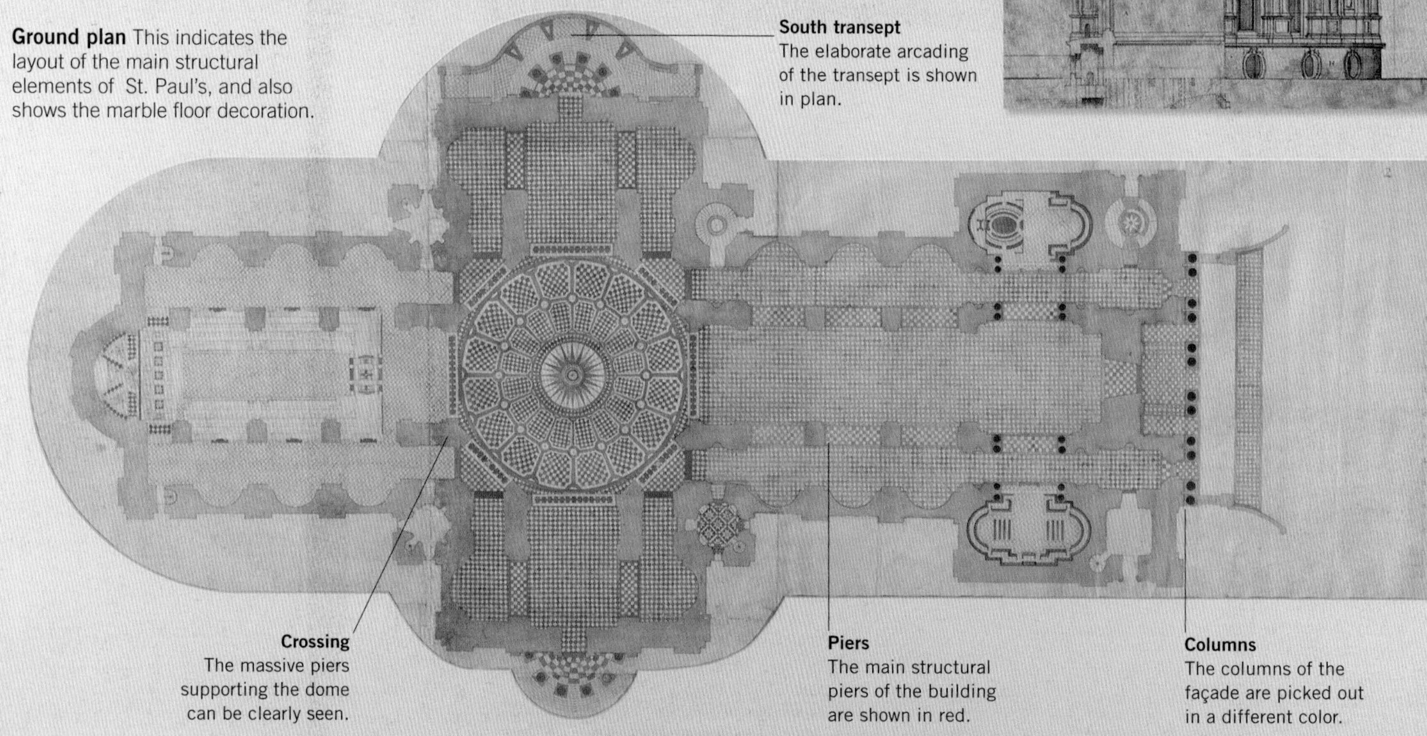

Ground plan This indicates the layout of the main structural elements of St. Paul's, and also shows the marble floor decoration.

South transept
The elaborate arcading of the transept is shown in plan.

Crossing
The massive piers supporting the dome can be clearly seen.

Piers
The main structural piers of the building are shown in red.

Columns
The columns of the façade are picked out in a different color.

Vital circuits

Although both electrical and plumbing circuits vary from country to country, depending on local standards and compliance laws, there is a basic vocabulary of symbols which are widely recognized to describe how such vital closed-system flow circuits are designed and implemented, and how they function.

Electricity The essential aspect of electricity is creating a circuit so that the positive and negative aspects of the circuit run in parallel. There are a number of basic symbols that make up most circuits.

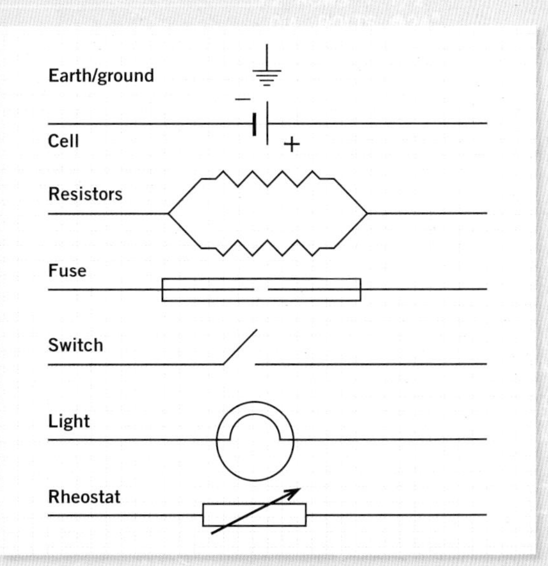

Plumbing Water-flow systems, especially those integrated with central heating and other networks, need to be clearly defined. While most plumbing diagrams display a bewildering array of junctions and joints, each having their own symbol, the principal concern is identifying which plumbing pipe carries what.

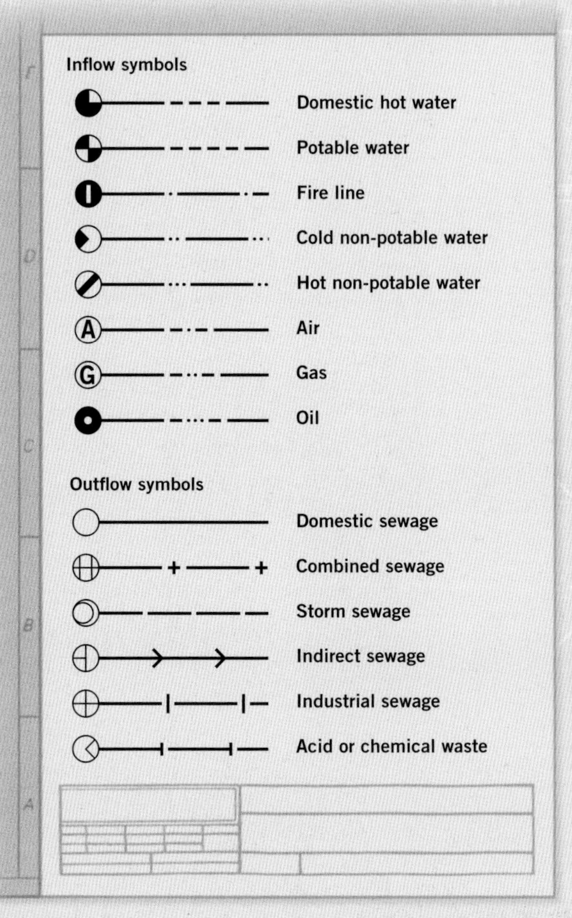

In both industrial and domestic plumbing the need to identify direction of flow, control points such as valves and spigots, and the function of each pipe is essential.

Shorthand

The 'second' industrial revolution, from the middle of the 19th century, was a product of the growth of systems and communications technologies. The invention of typewriters and telegraphy created a massive demand for secretarial skills, people who could process messages quickly and efficiently. The Pitman shorthand system, invented in 1837, provided a way for secretaries to record verbally-dictated information swiftly before it was typed. It was rapidly followed by the French Duployé system, and then the Gregg system in 1888. None of these ideas were that new: Roger Bacon had recommended a variety of shorthand for rapidly transcribing ideas in the 13th century. The Pitman system was entirely phonetic and focused on consonants indicated by strokes, 'vowels' indicated by dots and bars, and four diphthongs. A number of abbreviations of common words was included.

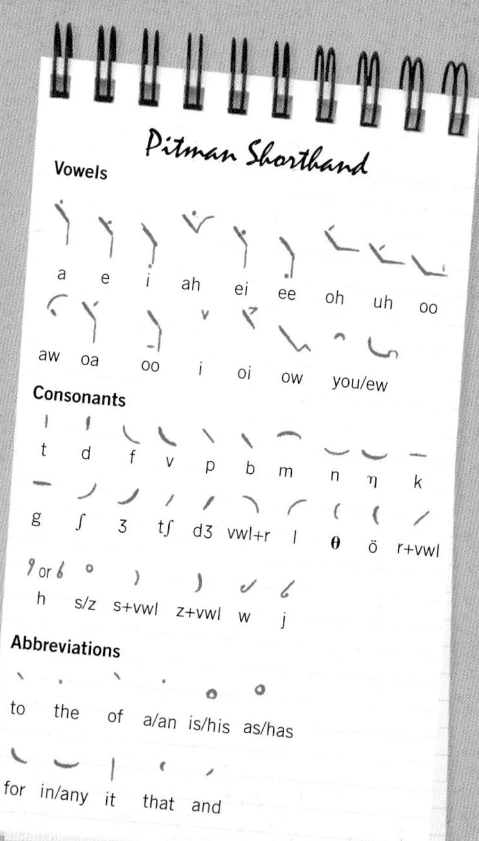

CURRENCY AND COUNTERFEITS

The development of long-distance trade networks in the ancient world led to the need for a form of monetary exchange to replace traditional bartering, the origin of the first coinage. The practical problem of the international control of monetary values really emerged in the 17th century with intercontinental trading houses (such as the English and Dutch East India Companies, founded at the beginning of the 17th century), who wished to avoid transporting bullion to pay for their purchases. Notes of credit, the origin of modern paper currency, became commonplace beyond the *bourses* and trading markets, but with it came the challenge of encoding paper money to guarantee its value and confound counterfeiters.

Printing counterfeit banknotes was long regarded as a capital offense.

The first coins

The Lydian Greeks produced the first known coins in the 7th century BC, but the system soon spread throughout Asia and the Mediterranean. Coins, usually of metal of an agreed weight and value, were impressed with the insignia of a public authority – usually the image of a ruler or the symbol of the city which issued it – accompanied by a suitable motto.

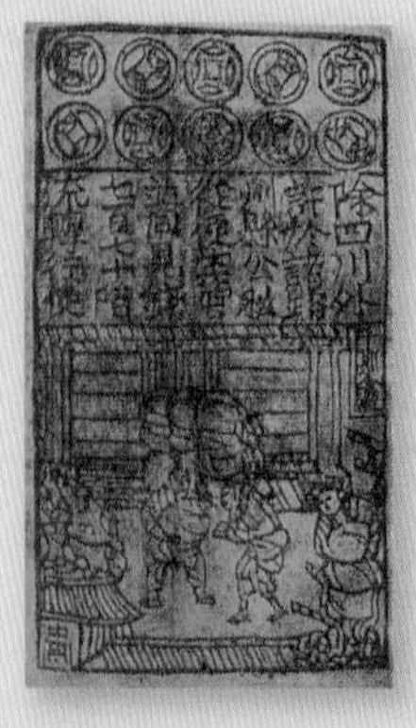

The first notes

The first paper money was introduced in China in the late 9th century under the Song dynasty – a period of great international trade – when the government adopted the practice of issuing paper receipts in exchange for bullion. By the 13th century printed banknotes were widely circulated.

Mints and assay

Counterfeit coinage was a big business in early modern times, when coinage still remained linked to its 'face' value, depending on its metallic composition. 'Clipping,' the art of shaving gold or silver off coins, was widespread, as was melting down gold or copper, then recasting using a cheaper alloy (albeit, as today, both were treasonable offenses). Sir Isaac Newton earned his knighthood not for his scientific achievements, but as controller of the Royal Mint, a role in which his administrative and metallurgical skills ensured that British coinage was accurately assayed.

Printing money

With the advent of paper money bills, counterfeiting took on a new dimension. Initial security printing was based on the quality of the paper used (usually from a single secure source, and elaborately watermarked) and the quality of engraving, ink, and printing. In the late 20th century more sophisticated encryption techniques were introduced. Nevertheless, modern digital scanning technology has meant that paper currency counterfeiting remains a profitable business.

Dollars

Dollar bills only infrequently change in design or format. They have a limited number of security encryptions, although the $5 and $10 bills issued in 2008 have more.

Modern coins In the UK, the cost of manufacturing copper coins was exceeded by the actual value of the metal in 1998, upon which an admixture of iron was introduced to rebalance the assay value. A magnet will rapidly indicate the younger coins in your pocket, and in 2008 the US Mint admitted that a similar problem had occurred with their smaller denomination coins.

Microprinting
Some features have been added that are difficult to replicate due to their size, including a series of yellow '5's.

Metallic security thread
An alternating pattern of 'USA' and '5' can be seen on both sides. The thread shows blue under ultraviolet light.

Watermarks
A new '5' watermark now replaces the former Lincoln portrait watermark, and smaller '5' watermarks are integrated elsewhere in the design.

The euro

The largest issue of both paper and metal currency in recent years was the euro. Used in most of the European Union countries, it was launched in 2002. The larger denomination coins use two metals, while the notes have over 20 security features.

Checksum
Like all banknotes, each euro has a unique serial number. This begins with a letter (identifying the issuing country), and ends with a check digit between one and nine. When the initial letter is converted into its alphabetic position number, then all the numbers are added up: the sum will be a two-digit number. Divide the sum by nine; the remainder from the division should match the sum of the two-digit number when added together.

Registration test
The note value is printed incompletely on each side, but registers perfectly when held to the light.

Raised printing
In certain areas of the notes, ink is applied more thickly, creating a raised texture.

Watermarks In addition to the traditional watermarked paper, a digital watermark is included to make scanning or photocopying often impossible, as well as inked watermarks which only function under infrared or ultraviolet light.

Magnetic strip A metallic security strip, only visible when held to the light, shows the denomination with the word 'euro.'

Smart inks Higher denomination notes use inks that vary in color from different angles, some only visible from certain angles, while areas of the smaller notes use magnetic inks.

Holograms The smaller denomination notes have a holographic strip. 50-euro notes and higher carry a holographic decal.

Few countries publish the numbers or value of counterfeit notes that have been identified, but in its first year of circulation over half a million counterfeit euro notes were removed from circulation, and the annual total continues to rise.

Credit cards

Credit cards have been around in a primitive form ever since the early 1930s, but came into wider use in the 1970s. However, it is only recently that the technology to enable credit cards to work securely has been available.

Holograms These are very difficult to forge and are often used on banknotes as well as credit cards.

Magnetic strips These contain Track 2 data, the personal details of the cardholder which, in conjunction with a Personal Identification Number (PIN), provide access to the cardholder's account.

User history Automated systems can check over the cardholder's history to see if a transaction seems anomalous. Many bank cards will freeze accounts if someone tries to withdraw a very large amount from an ATM that the card owner has never used before.

Smart cards These are the integrated circuits now commonly found on bank and credit cards, and work on the same principle as cell phone subscriber identity module (SIM) cards. They can be used to access, exchange, and store additional data electronically. Some have embedded cryptographic functions, such as 3DES or RSA, carrying a digital signature.

Personal Identification Numbers

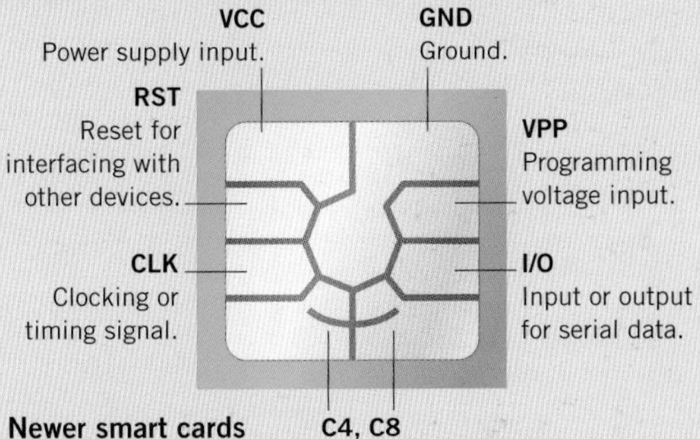

Newer smart cards are contactless, instead communicating with the card reader through RFID.

Four-digit Personal Identification Numbers (PINs) are the most common form of coded authentication for credit and debit card transactions. But how secure are PINs? Any four-digit number has 10,000 possible combinations. Compared to most passwords or pass phrases this is a tiny amount (an eight-character alphanumeric password could have as many as 100 billion possible values). However, normally one can make only three attempts at entering the PIN, which means that 'brute-force attacks' (where every possible number is tried) are not effective: with three attempts there is only a 1/3333 chance that the attack will work.

The Book in Your Hands

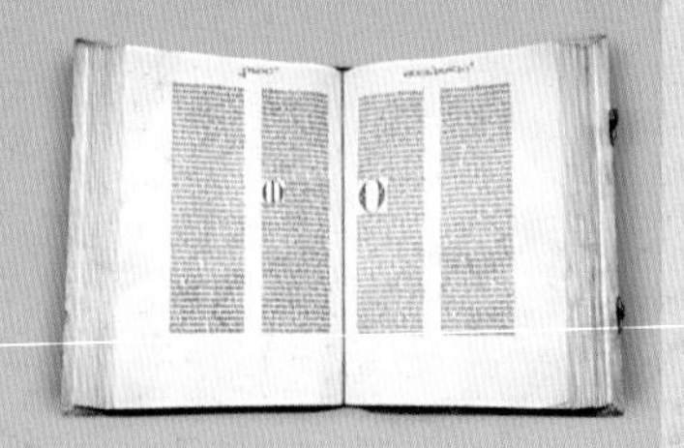

The book

The bound book has remained in its essential construction remarkably unchanged since the inception of movable-type printing in Europe by Johannes Gutenberg (c.1398-1468) in 1450. Most of the organizational techniques Gutenberg developed for producing his famous Bible (*above*) remain largely in use today. This long pedigree means that there are many aspects of this all-too familiar object, whether it is the book in your hands as you read this, a staple-stitched comic, a perfect-bound pulp paperback, a novel, an encyclopedia, dictionary, or atlas, or a numbered limited edition of a fine text, which we accept, read, and handle without thinking. And yet, the book is the product of a dense interaction of trade, craft, cataloging, literary, and other codes.

Most manufactured goods today are identified and characterized by a multiplicity of codes: the brand name, batch identifiers (important for products with variable ingredients like paint and medicines), 'sell/use-by' dates (again important for medicines, and for food), individual serial numbers, usually for electronic and mechanical products for consumer protection and insurance purposes, and bar codes for stock control. However, one of the most significant consumer products to be invested with inbuilt codes is the book. This extraordinary invention remains the most enduring long-distance communication device, crossing barriers of both distance and time, and is entirely a product of coded languages, some dating back many centuries.

The imprimatur page
Commonly known as the 'imprint' page, this displays publishing, copyright, and bibliographic information.

Binding
The two prevalent binding methods are 'gathered and sewn' folios, or 'perfect' binding, often for softcover books, where the folds are cropped and the pages glued directly to the spine.

Folios
Most modern books comprise large sheets of imposed pages, which are then folded and gathered and cropped for binding, often in folios of multiples of 16 pages.

Publisher

Copyright symbol
© indicates the publisher's ownership of the design and contents of the book.

Dust jacket

Endboards

Rear or back cover

ISBN
The Standard Book Number originated in the UK in 1966 as a unique numerical nine-digit code for printed publications, but in 1970 was adopted internationally as a ten-digit code – the ISBN. Since 2007 the code has increased to 13 digits, and a separate code (the ISSN) is used for periodicals. The bar code identifies the group issuing the title, the publisher, the individual title, and a check digit.

Bar code
A bar code (*see page 204*) including the ISBN (*left*) usually appears somewhere on the rear cover.

Spine

Front cover

Fore edge

Library of Congress Cataloging Number
A wide range of data is embedded in this code, including not just the book title and ISBN, but date and birthplace of the author.

Published by Weldon Owen Inc.
415 Jackson Street
San Francisco, CA 94111
www.weldonowen.com

Weldon Owen inc.
Executive Chairman, Weldon Owen Group John Owen
CEO and President Terry Newell
VP, Sales and New Business Development Amy Kaneko
Senior VP, International Sales Stuart Laurence

VP and Publisher Roger Shaw
Assistant Editor Sarah Gurman

VP and Creative Director Gaye Allen
Art Director Tina Vaughan

Production Director Chris Hemesath
Production Manager Michelle Duggan
Color Manager Teri Bell

Conceived and produced for Weldon Owen Inc. by Heritage Editorial
Editorial Direction Andrew Heritage, Ailsa C.Heritage
Senior Designers Philippa Baile at Oil Often, Mark Johnson Davies
Additional Design Bounford.com
Illustrators Andy Crisp, Philippa Baile at Oil Often, David Ashby, Mark Johnson Davies, Peter Bull Art Studio
Picture Research Louise Thomas, cashou.com
DTP Manager Mark Bracey

Consultant editors
Dr. Frank Albo MA, MPhil., Ph.D. candidate History of Art, University of Cambridge
Trevor Bounford
Anne D. Holden Ph.D. (Cantab.), 23andMe Inc., San Francisco, CA
D.W.M. Kerr BSc. (Cantab.)
Richard Mason
Tim Streater BSc.
Elizabeth Wyse BA (Cantab.)

A Weldon Owen production
© 2009 Weldon Owen Inc.

All rights reserved, including the right of reproduction in whole or in part in any form.

Cataloging-in-Publication data for this title is on file with the Library of Congress
ISBN 978-0-520-26013-9 (cloth : alk. paper)

Manufactured in China

18 17 16 15 14 13 12 11 10 09
10 9 8 7 6 5 4 3 2 1

Edition number
This shows how many times the book has been reprinted, and when. The date and number elements of the code are deleted each time the book goes to press.

Verso
The left-hand page.

Gutter
Where the bound pages join.

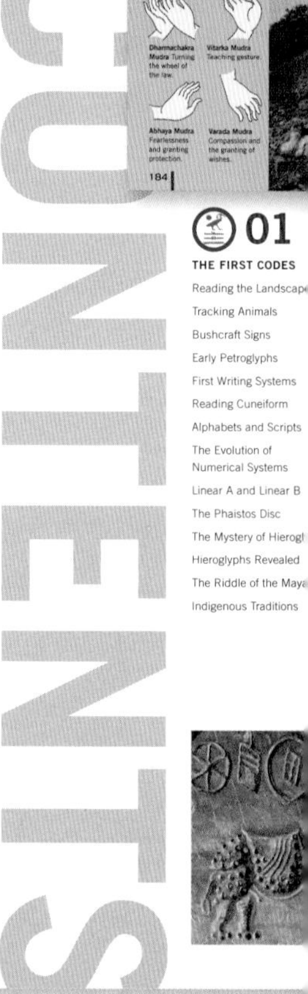

Navigating the book In a complex illustrated book such as this, there are a number of conventional core elements to be considered, many of which in turn have a 'language' of their own.

Typography
Various type faces and fonts will be assessed for impact, context, and different levels of information.

Layout
There is normally a grid or template that provides a structure for the designers to follow.

Running head
Normally indicates the chapter or section.

Italic face
Usually used for cross-referencing.

Leader lines
Indicate how an annotation relates to an illustration or diagram.

Folio number
The page number.

Recto
The right-hand page.

Contents This provides essential information about the structure of the book. Other navigational features may be listed in the Contents, such as an Index (a more refined look-up table of specific references), a Glossary (definitions of terms used in the book), an Appendix (which may include reference texts useful for the reader), and a Bibliography, which provides a list of sources and further reading.

Text navigation

There are a number of commonly used abbreviations and symbols that occur in texts, sometimes in more academic books, often deriving from a Latin root.

cf. *confer*, by way of comparison
e.g. *exempli gratia*, for example
et al. *et alia*, and others
etc. *et cetera*, and so forth
ff. and following
fl. *floruit*, flourished at the time given
ibid. *ibidem*, reference may be found in the same place as a previous reference
id./idem the same as something previously mentioned
i.e. *id est*, that is
loc. cit. *loco citato*, in the place previously cited
non obs. *non obstante*, notwithstanding
non seq. *non sequitur*, it does not follow
viz. *videlicet*, namely

Proofreading marks Raw text from almost any author requires editing. Editors and typesetters have developed a coded language of proofreading marks, which developed into a fine art when newspaper copy editors had to fit printable copy into confined templates for typesetters to adjust in a matter of minutes before going to press. Usually the copy editor would be provided with a 'galley' proof to mark up.

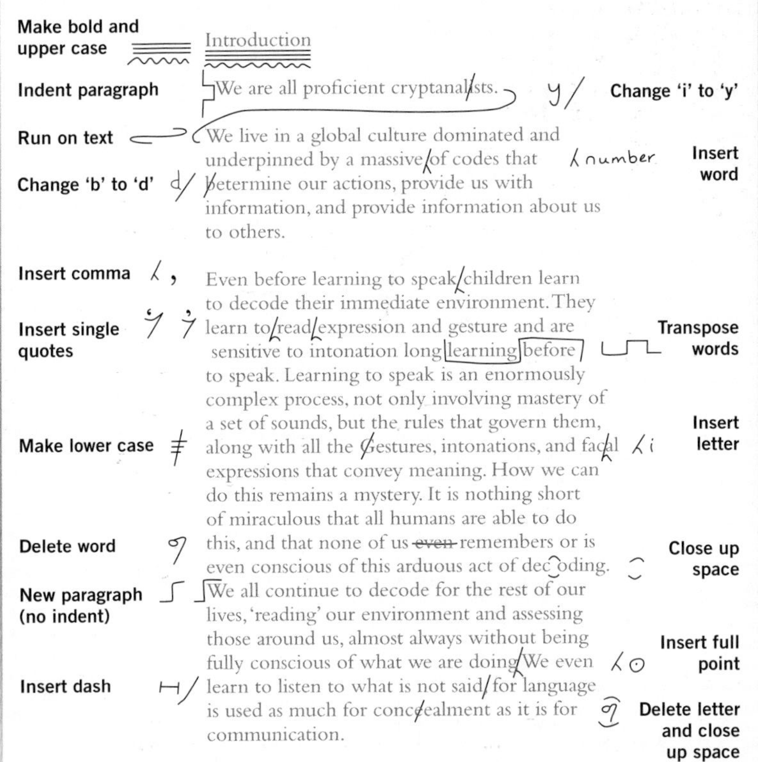

10

The smooth operation of a civil society depends upon an array of unspoken, undefined, and frequently unintelligible (but generally accepted) codes of behavior and deportment.

codes of human behavior

Beneath the panoply of social mores, traditions, and manners lie a welter of other signs and signals inherent in the human condition that fall all too frequently beyond our conscious control, and tell others much about ourselves that we might not care to reveal.

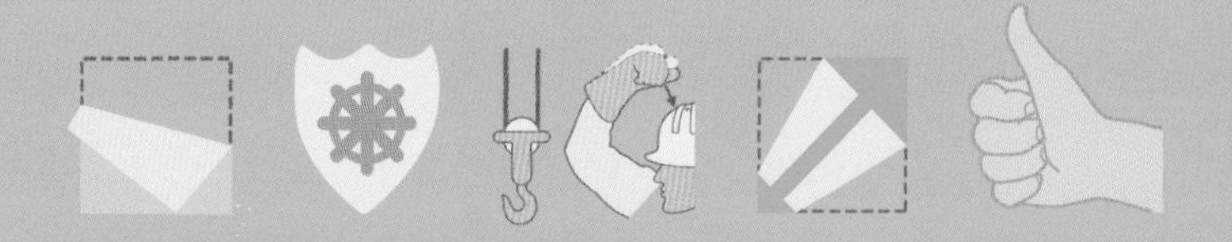

BODY LANGUAGE

Aside from verbal communication, the potential for determining or demonstrating moods and feelings by the manipulation of the face and body is huge. Alongside self-conscious use of body language such as winking, frowning, or waving, careful observation can reveal a complex of hidden – and often unintended – subconscious messages. An enormous amount of such information is instinctively understood – we can normally tell if someone is interested or bored by us, if they are embarrassed, or if they have 'something to hide.' The science of decoding what individuals reveal about themselves is now widely understood by psychiatrists and psychoanalysts, and this knowledge is used in personnel recruitment, interviews, and interrogation.

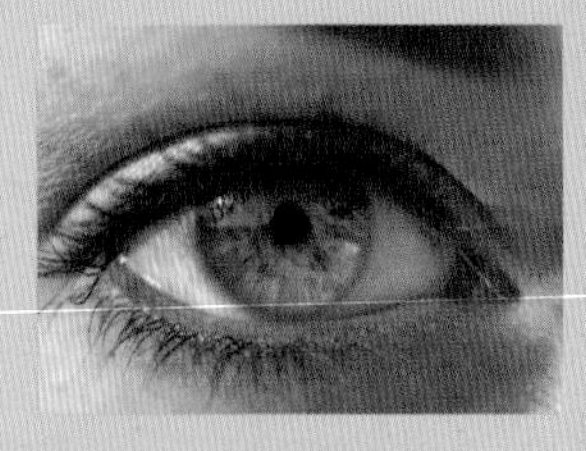

Body in control

While most of us strive to control the messages our bodies might be sending out, some reactions often prove uncontrollable. Blushing, perspiring, weeping, and reacting to pain often cannot be contained. The eye can give away many signals, the dilated pupil frequently indicating interest or attraction, while an inability to make or maintain eye contact usually means embarrassment or dishonesty.

Making faces

Artists from classical times onwards observed and sketched facial expressions and physical stances and poses as a means of expressing emotions in their work. However, the Austrian portrait sculptor Franz Xaver Messerschmidt (1736-83) was one of the first to attempt to catalog the range of human expression in a series of over 50 busts, based on studies made in lunatic asylums in Munich. Although often extreme, these studies reflect the Enlightenment's interest in every aspect of human behavior.

Conscious and subconscious communication

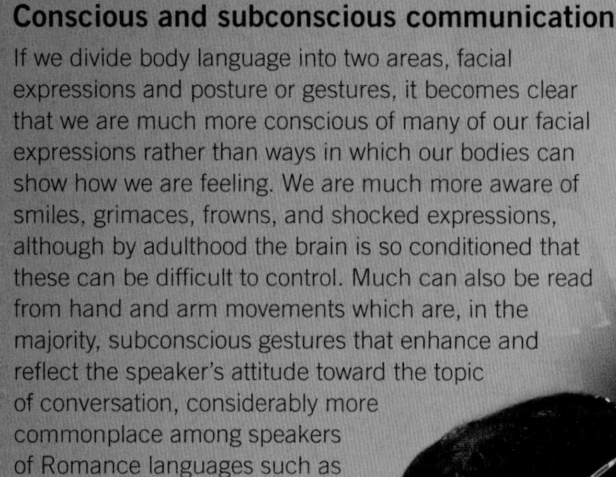

If we divide body language into two areas, facial expressions and posture or gestures, it becomes clear that we are much more conscious of many of our facial expressions rather than ways in which our bodies can show how we are feeling. We are much more aware of smiles, grimaces, frowns, and shocked expressions, although by adulthood the brain is so conditioned that these can be difficult to control. Much can also be read from hand and arm movements which are, in the majority, subconscious gestures that enhance and reflect the speaker's attitude toward the topic of conversation, considerably more commonplace among speakers of Romance languages such as Spanish or Italian.

Examples of more subconscious body language

Legs and arms firmly crossed Disinterest, annoyance, a defensive posture.

Leaning forward, hands to chin Attentive, interested, enthusiastic.

Playing with tie or hair (men) Nervous, uncertain.

Comfortably crossed legs, bouncing of the foot (women) Flirtatious invitation/sexual interest.

Eyes looking to left Obvious discomfort, often lying, bad in interviews.

Eyes looking to right Fact-finding, consideration, fine in interviews.

Head up, blank eyes Mild interest, perhaps thinking of something else.

Head tilted to one side, narrowed eyes Interest, positive consideration.

Tightening of the jaw/clenching teeth Frustration and anger.

Poker 'tells'

Poker is as much a game of skill as it is the chance fall of cards. Much of that skill resides in the ability to conceal one's own emotions during a game, and to 'read' what is going on in the mind of your opponents. Giveaway signs are known as 'tells.' In the movie *Casino Royale* (2006), James Bond uses his expert intuition to recognize when the criminal mastermind 'Le Chiffre' is bluffing – he blinks. When Le Chiffre knows his hand is awful, he really gives the game away by bleeding from one eye. Here are a few, more subtle, tells from the gaming tables.

Hand shakes
Look out for shaking hands when betting. Among new players this normally indicates they have a good hand, and are excited at the prospect of winning. Equally it may indicate a bluff.

Eyes down
Glancing at their chips just after the deal is complete (the 'flop') usually means a player has hit their hand. In contrast, staring at the 'flop' – searching for something – often means they missed. It may indicate a forthcoming bluff. Many professionals now wear sunglasses to conceal these tells.

Frozen time
Signs of increased tension: gum-chewers will often stop chewing when they bluff; similarly a person may momentarily stop breathing when making their play.

Talk the talk
With a strong hand players tend to be confident, talkative, and relaxed. Agitated behavior or forced conversation may indicate weakness.

I'm in
An eagerness to bet can reveal a lot. Players holding a strong hand are usually keen to get their bet in the pot. A key tell here is the player who usually waits, biding his time before calling, and then uncharacteristically bets quickly. However, taking some time to bet can conceal many ruses, and can unsettle the rest of the players.

Flirtatious fans

In 19th-century Spain, wealthy young ladies would always be accompanied by a chaperone outside the house. These chaperones were famously zealous, and were charged with overseeing the behavior of their young ladies, and ensuring that they were brought up in an honorable manner. Conversation with young men that strayed from virtuous subjects such as the weather, art, literature, and politics was forbidden, forcing the maidens to create their own means of communicating using their fans. A catalog of gestures developed, designed for covert courting and flirtation. Of course much of this was intuitive, but late 19th-century fan manufacturers began to publish 'guides' to fan language, partly perhaps to increase sales.

Moving the fan slowly over the chest I am single.
Moving the fan quickly in snappy movements over the chest I have a boyfriend or partner.
Opening and closing the fan, then touching the cheek I like you.
Touching the temple with the fan and looking skywards I think of you day and night.
Touching the tip of the nose with the fan Something doesn't smell good here (the man is displeasing her, perhaps by flirting with someone else).
Walking sideways, hitting the palm of the hand with the fan Careful, my chaperone is coming.
Opening and closing the fan then pointing with it Wait for me there, I'll be there soon.
Covering the mouth with the fan and looking suggestive Sending a kiss.
Carrying the fan closed and dangling from left hand I'm looking for a boyfriend.
Fanning very rapidly I'm not so sure about you ...
Closing the fan very rapidly Talk to my father ...
Placing the fan closed over the heart I love you very much.
Placing the fan open over the heart I want to marry you.
Giving the fan to the man My heart belongs to you.
Taking the fan from the man I want no more from you.
Covering part of the face with the open fan We've finished.
Letting the fan drop I'm suffering but I love you.
Hitting the left hand with the fan I like you.
Looking outside I'm considering it ...
Hitting right hand with fan I hate you.
Hitting dress with the fan I'm jealous.
Resting the fan closed on the left cheek I'm yours.

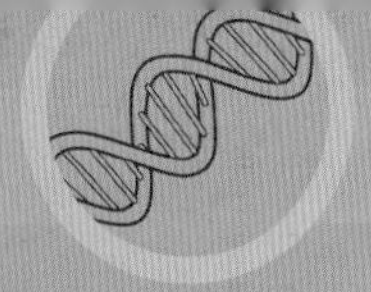

SURVIVAL SIGNALS

There are times when visual communication is more appropriate than verbal, or when manual signals clarify spoken communication. Since ancient times, and for obvious reasons, hunters have needed to convey information without making any sound; similarly, military personnel need to communicate, even across short distances, without being audible to adversaries (*see page 16*). Clear and simple signals are vital in these circumstances, particularly where lives may depend on the success of the communication. But there are many other areas in everyday life where signaling codes can be a matter of life – or death.

Help! Stranded

If stranded in a remote location due to shipwreck or an air crash, there is an internationally recognized ground-to-air signaling code designed to convey instant information to air search-and-rescue crews. Although easily conveyed as body signals or patterns, most aircraft and ships are equipped with flares for attracting attention and a colored blanket which can be laid out on the ground to send signals.

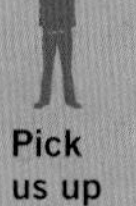

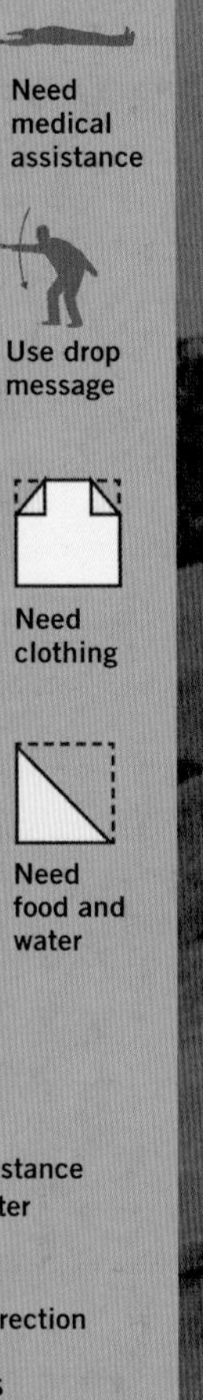

Pick us up | **Mechanical help needed** | **Need medical assistance**

Can proceed shortly | **Do not attempt to land here** | **Use drop message**

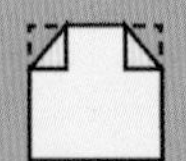
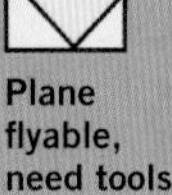

OK to land | **Plane flyable, need tools** | **Need clothing**

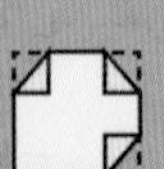
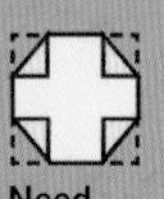
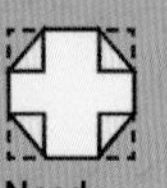

Need first aid supplies | **Need medical attention** | **Need food and water**

Ground-to-air patterns

- ┘ All well
- L Not understood
- V Require assistance
- X Require medical assistance
- F Require food and water
- N No
- Y Yes
- ↑ Proceeding in this direction

Air-to-ground responses

Message received, understood:
In daylight: tipping wings in a rocking motion from side to side.
At night: flashing green lights.

Message received, not understood:
In daylight: flying the plane in a right-handed circle.
At night: flashing red lights.

Mountain rescue

These signals are internationally recognized by all mountain rescue services and rely on sending sound or light messages.

Message	Signal	Sound or light signal
S O S	Red	**Three** short blasts/flashes, three long blasts/flashes; **repeat** after one minute interval.
Help needed	Red	**Six** blasts/flashes in quick succession; **repeat** after one minute interval.
Message understood	White	**Three** blasts/flashes in quick succession; **repeat** after one minute interval.
Return to base	Green	**Prolonged** succession of blasts/flashes.

Scuba signs

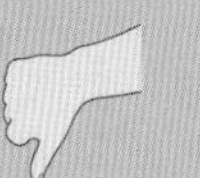
Descend

Ascend

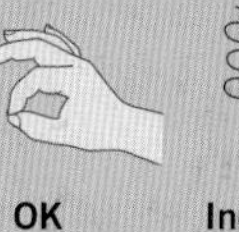
OK

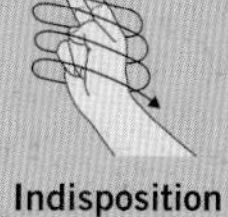
Indisposition

Something is wrong

Slowly

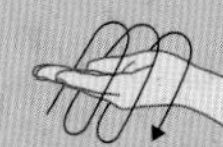
Fast

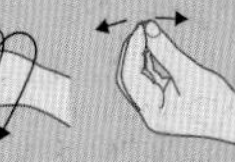
Did not understand

Speech is not possible underwater, but clear communication between divers is essential. A system of hand signals for most common messages has been devised.

On the road

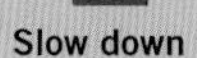
Slow down

Speed up

Staggered riding

Stop engine

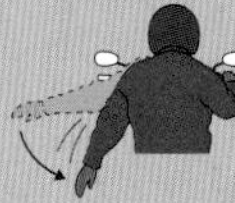
Pass

Hazard ahead

Hand signaling is rarely used today by automobile drivers, but is important for pedal cyclists, and motorcyclists in groups.

On site

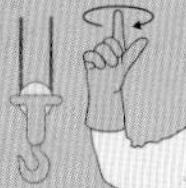
Hoist

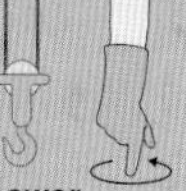
Lower

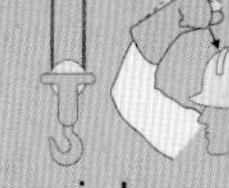
Use main boom

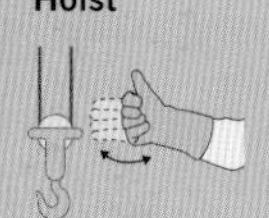
Raise boom, lower load

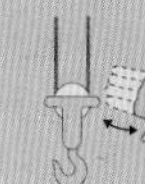
Lower boom, raise load

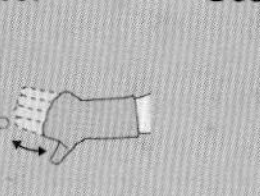
Move slowly

Transporting heavy loads safely above and across a busy building site relies on crane operators and ground staff understanding hand signals.

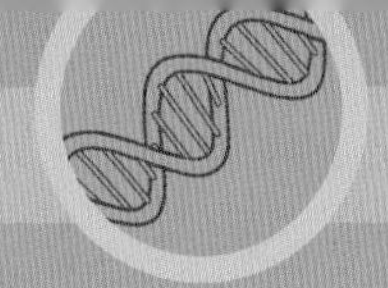

SPORTING CODES

In many sporting activities, visual codes are employed to convey messages to other participants, or to observers and scorers. In baseball, soccer, rugby football, cricket, and a host of other sports in which distance, or language, constitute barriers to immediate understanding, exaggerated gestures are used. Although modern technology can be of help, it is not always appropriate or efficient, and a system of clear manual signals is employed.

Signing the odds

Peculiar to Britain, and now infrequently used due to the increased use of mobile telephones, is a system of communicating betting odds known as 'Tic-tac.' Bookmakers taking bets need to be appraised of shifts and changes in betting patterns whilst embroiled in the clamor of ever-changing odds.

4-1

5-2

6-4

7-4

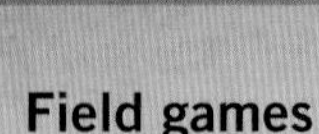

Field games

While the rules and signals of cricket remain inscrutable, other field games have developed a range of signals according to their needs. The globalization of many sports, most notably soccer, has meant that signaling by the umpire or referee now needs to be internationally endorsed. In games played by players who may not have a common spoken language, hand-signaled instructions or messages serve to clarify the official's judgments. They are also immensely important in situations where crowd noise would drown out the spoken word for all but those nearest to the official.

The rules of cricket are probably a mystery to many, and the umpire's signals almost certainly fail to make anything clearer to the uninitiated. However, they enable often finely judged decisions to be conveyed by the umpires from the center of the pitch to scorekeepers and spectators. The signal code has been long established, but there is still opportunity for umpires to introduce their own idiosyncratic mannerisms.

No ball

Six runs

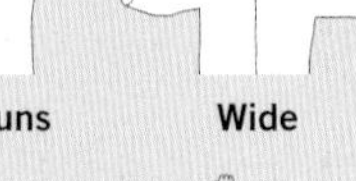

Wide

Out/wicket

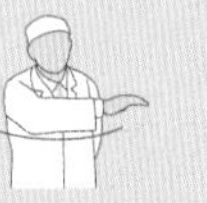

Four runs

Bye

Baseball signals
In a field game in which judgment on the spot by the umpire is of paramount importance to participating teams, the scorers, and the viewers (in the stadium or watching on television), baseball signals need to be clear and unambiguous. The fact that the umpires must stand in a 'direct line of fire' position behind the pitcher or catcher *(left)* emphasizes the importance of clear signaling.

Count

Play ball

Umpire strike

Strike or out

Safe

Time-out, foul, or dead ball

Soccer signs are less important than in many other sports – the laws of soccer are succinct and largely unambiguous. For the most part, both players and spectators are immediately aware of why a penalty or point has been awarded, so most soccer signs used by the referee simply maintain the rapid flow of action in the game.

Penalty

Free kick

Corner

No goal

Offside

Play on

Rugby Union rulings are often the result of an interaction between the referee and the linesmen. The referee needs to be able to rapidly tell both teams what the joint decision is, and convey this to the players, the scorekeepers, and the onlookers.

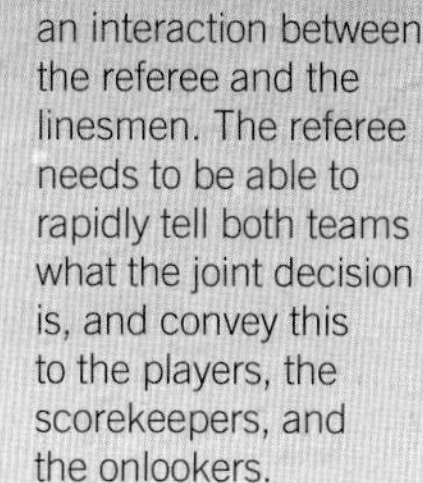

Try

Free kick

Penalty

Holding on

High tackle

Forward pass

Not straight

Advantage

ETIQUETTE

Profound cross-cultural misunderstandings can be generated by seemingly trivial customs such as greetings, gift-giving, table manners, dress, and general behavior. Etiquette is the codification of social behavior, an attempt to define the right way of behaving within society. While some etiquette simply emanates from respect for other people's comfort and feelings, more elaborate types of behavior develop in rarefied, hierarchical environments such as royal courts. Away from the refined air of the royal courts, social etiquette has evolved over the centuries as a generally accepted code of behavior, manifested in the concept of 'good manners.' While manners have become much less complex and formal over the last century, certain practices that ease social intercourse persist: 'please,' 'thank you,' and 'excuse me' should never be forgotten.

Court etiquette

Refined courtly etiquette reached its apogee at the palace of Versailles during the reign of Louis XIV of France (1638–1715). Correct behavior became a qualification for social advancement at court; elaborate etiquette also reinforced a rigid social hierarchy.

Entering No one was allowed to knock at the king's door. Instead, they were required to scratch at the door with the left little finger, and so many courtiers grew that fingernail longer than the others.
Contact A lady was never permitted to hold hands or link arms with a gentleman. She was required to place her hand on the gentleman's bent arm.
Sitting Ladies and gentlemen were not allowed to cross their legs in public; when a gentleman sat down, he slid his left foot in front of the other, placed his hands on the sides of the chair, and gently lowered himself onto the seat.
Greeting A gentleman was required to raise his hat high above his head when passing an acquaintance on the street.

Etiquette ordained the order of prominence at court, rigidly maintained complex customs of address, and determined who could sit or stand under what circumstances in the royal presence.

Offensive behavior

In much of the non-Western world – from the Pacific Islands to the Middle East – feet are considered offensive. Shoes are removed before entering homes, and wearing shoes in places of worship, such as mosques and temples, is considered deeply offensive. Feet are considered unclean, and there are many taboos about revealing the soles of the feet, which should never be pointed toward Mecca in mosques, or toward the shrine in Buddhist temples. In India the head is seen as the locus of the soul, and touching someone's head – especially a child's – is frowned upon. Koreans are appalled by nose-blowing, considered especially offensive at the dinner table.

Greetings

While Americans and Europeans can be reasonably sure that a handshake will not cause offense when meeting strangers, more elaborate customs prevail in Asia. Perhaps the most refined greeters of all are the Japanese, whose bow symbolizes respect and humility. Men bow with arms straight, palms flat and touching their legs; women bow with their hands slightly cupped, clasped in front of their thighs. The depth of the bow is significant, denoting minute gradations in social status. In Thailand the traditional greeting, or *wai*, is made by placing both hands together in a prayer position.

Gift-giving

Steering a route through the social minefield of gift-giving can be challenging. In some cultures, such as Japan and many Pacific islands, gifts are expected, and failure to give is considered offensive. In other countries, especially in northern Europe, it is not normal to give elaborate gifts, which are considered inappropriate. In China gifts should be given and received with both hands. A gift should be declined three times before accepting, as this shows a lack of greed. It should never be opened in the presence of the giver unless they insist. In much of Asia the elaborate wrapping of the gifts is as important as the contents. In China red or yellow paper should be used; black, white, and blue should be avoided. In South Asia green, red, or yellow wrapping is considered lucky; black and white should be avoided.

Say it with flowers?

Flowers may seem a safe choice of gift for a hostess the world over, but flowers have different connotations in different cultures. In the US lilies and gladioli are associated with funerals. In Japan camellias are considered unlucky, and yellow and white chrysanthemums are used in funeral arrangements, as they are in China. In France, too, chrysanthemums are considered a funereal flower, placed on graves on All Saints' Day (November 1), while in Switzerland white carnations are associated with mourning. In Japan flowers given in groups of four and nine are considered unlucky, whereas in China even numbers (except four) are considered inauspicious. In much of Europe bouquets of 13 flowers would be considered unlucky.

Nineteenth-century English social instruction handbooks and household manuals codified correct form in an increasingly mobile society, itemizing in minute detail the rules for attending and hosting dinner parties, including such information as how to word the invitations, ornament the table, order the courses, or seat the guests.

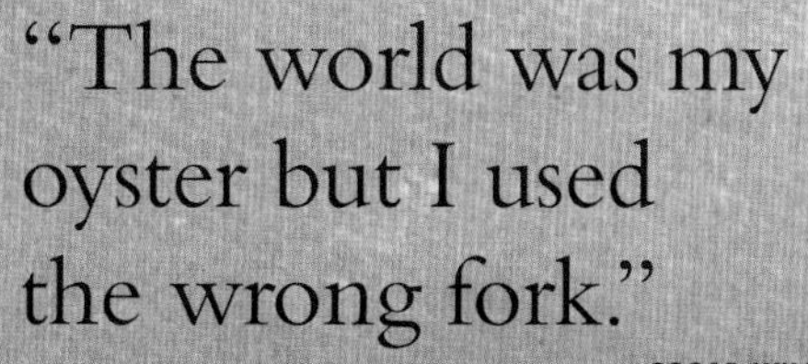

“The world was my oyster but I used the wrong fork.”

OSCAR WILDE.

Side dishes
Offer to those seated to your left or right before helping yourself.

The entrée dish
Wait until served, and do not begin to eat until all are served unless encouraged to do so by your hosts. In Europe an ‘entrée’ dish is not necessarily the main course.

Bread rolls
Should be broken with the hands, not cut with a knife.

Noblesse oblige

In 1954 Nancy Mitford published an essay entitled ‘The English Aristocracy,’ which spelled out the ways in which English class consciousness had permeated the very language, and use of an inappropriate term betrayed a lack of breeding: ‘u’ was acceptable (upper-class) usage; ‘non-u’ was the rest:

U	Non-U
Bike or bicycle	Cycle
Die	Pass on
Dinner jacket	Dress suit
Drawing room	Lounge
Good health	Cheers
House	Home
How d’you do?	Pleased to meet you
Lavatory or loo	Toilet
Looking glass	Mirror
Napkin	Serviette
Notepaper	Writing paper
Pudding	Sweet
Rich	Wealthy
Scent	Perfume
Sick	Ill
Sofa	Settee or couch
Spectacles	Glasses

Glasses
Beverages, usually wine, should be served in the correct glass with each course – do not help yourself.

Condiments
If not offered or passed around, do not stretch across your neighbor, but politely ask them to be passed.

Napkins
Never referred to as ‘serviettes,’ these should be placed on the lap, never tucked into the collar.

Using a fork
The fork should almost always be used with the prongs pointing downward, unless it is the only implement required (for instance, for a dessert).

Cutlery
Confronted by an array of knives, forks, spoons, and other implements, work your way in from the outside with each dish.

Soup spoons
Soup should always be scooped into the spoon in an action away from the body. Avoid tilting your soup plate, and do not dip your bread in it.

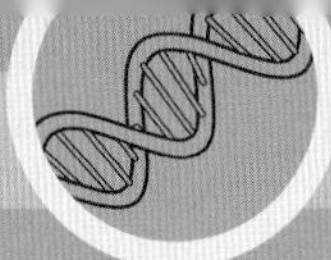

DRESSING YOUR MESSAGE

Today, in the West, we make very conscious decisions about what clothes we wear (and how and where we wear them) in order to send messages about who we are: it might be a Brooks Brothers ensemble, or tweeds and brogues, stiletto heels or court shoes, or simply T-shirt, jeans, and trainers. In many traditional societies, patterns, fabrics, and choice of apparel may form a strict grammar, a coded language establishing identity, status, and even where the wearer comes from. In others, there may be an overt meaning expressing a specific religious or cultural affiliation shown through choice of clothing or body decoration.

Uniforms

The suppression of the individual through uniform clothing has a number of meanings, often overlapping: identification of the wearer's function or rank (the military, the police, emergency services, the clergy); sanitary (health workers, culinary staff); to what institution the wearer belongs (schools, sports team); affiliation (team supporters, cult groups); or branding (we work for this particular company or product chain). The uniform sends a very specific message through clothing that the individual has a defined role.

Body decoration in the form of paint, tattoos, or scarification, has been used widely in separate cultures throughout history. The Roman historian Tacitus described the woad body paint used by British tribes (*above*). Body paint was also used by various indigenous peoples in North America and Africa. More permanent decoration in the form of tattoos was used in North America and most famously by the Maori peoples in New Zealand.

The Mayan tradition

The fusion of cultures in Latin America, with Catholic Iberian ideas intermixing with native Indian traditions, has produced a unique hybrid of signs and symbols, not least in the fabrics people wear. In Guatemala, the Latin American country with the greatest number of Maya descendants, a bustling market scene appears merely picturesque and colorful, but many people are conveying coded information about themselves through their choice of clothing.

Gender differentiation
Although Guatemalan men tend today to wear Western casual outfits, in certain communities both men and women wear woven blouses and skirts, but the style and cut of each is clearly differentiated for each sex.

White hats
Tend to indicate important members of the community, or that the wearer is the head of the household.

Changing allegiance
Upon marriage, a wife will adopt the patterns and styles of her husband's village.

Social status
Although traditional garments and textiles belie the fact that the wearer is probably from a rural community, within those societies the older and married women wear richer and more elaborate forms of dress.

Patterned shawls
Each village has its own unique pattern for woven shawls, blouses, and other garments. Those aware of this can identify exactly which community the wearer comes from.

Proscribed clothing

In addition to ritual clothing worn by priests, clothing and personal appearance among observers may be determined by religious laws. Male Sikhs are expected to wear turbans, and avoid cutting their hair or shaving. Muslim women are expected to cover their bodies and even their faces in public (*above*).

The Anabaptist Amish communities of the northeast USA live in effectively closed communities and reject electronic and other mechanical devices. They also have a particular dress code which eschews ornamentation or ostentation. Buttons are avoided in favor of hook-and-eye, snap, or pin fasteners. Printed patterns are not worn. Amish women usually wear blue, plain-cut, calf-length dresses, and often white aprons at home and dark capes and bonnets when outside. Single women wear a white cape. Men wear dark trousers, suspenders, vests, and hats, and after marriage tend to grow a beard (mustaches are not permitted). In summer, Amish children and some adults go barefoot.

The Amish use horse-drawn vehicles and do not like to be photographed.

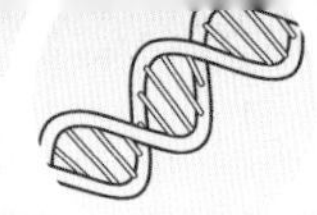

HERALDRY

Heraldic language

A strict heraldic language evolved to describe the colors, patterns, and devices that appear on the shield. This is known as 'blazonry,' and much of the language is derived from old French (or Norman French as used then in England). Most European nations have distinctive styles of heraldic form and design.

Heraldic colors (tinctures):

Argent	Silver (female)
Azure	Blue
Gules	Red
Murrey	Mulberry-crimson
Or	Gold (male)
Purpure	Purple
Sable	Black
Sanguine	Blood-red
Tenné/tawny	Orange
Vert	Green

Coats of arms were the status symbols of medieval Europe. They originally decorated the shields of soldiers on the battlefield, denoting the followers of individual feudal lords. Only people of the manor-holding classes and above had the right to bear arms. The distinguishing devices of medieval heraldry reflected a feudal society that was strictly stratified. Titles, which denoted complex degrees of standing within medieval society, were passed down to subsequent generations, and throughout Europe became synonymous with carefully delineated degrees of wealth, property, and power.

The tradition of heraldry

Coats of arms became a hereditary device from the mid-12th century in England and France, and from the 13th century arms appeared on family seals, and were displayed in family homes and on tombs. A coat of arms belongs to a family, not to a surname. Arms passed to younger sons as well as older, but were often altered in some way. If members of two arms-bearing families married, a composite shield of arms was created, with the husband's on the left, the wife's on the right. If a wife were her father's heir, the right to bear her family's arms would pass to her husband's family, the shield then being quartered. Thus coats of arms became a kind of pictorial family tree.

Heraldic designs

The basic design features on a heraldic shield (scutcheon or escutcheon) are called 'ordinaries.'

Bars A number of fesses.
Battled Horizontal partition in the form of battlements.
Bend A division running slantwise, from top left; one running from top right is 'bend sinister.'
Canton Rectangle in upper left quarter.
Charges Unique designs incorporated to fill the ordinary.
Checky Chequered.
Chevron Division like a roof gable.
Daunce A zigzagged feste.
Dexter The left side of the shield to the viewer.
Engrailed Scalloped.
Fesse Broad horizontal band.
Field The ground of a shield.
Fretty Thin interlaced bands.
Gobony Divided into rectangular segments, or gobets.
Gyron Triangular device.
Lozenge Diamond-shaped device.
Nebuly Device resembling the edge of a cloud.
Orle Band running parallel to the edge of the shield.
Pale Wide vertical band.
Party Vertical division.
Quarterly Divided into four by party and fesse.
Roundel Circular device.
Saltire Diagonal cross.
Sinister Right side of shield to the viewer.
Tressure A narrow orle, often flowered.
Vair Heraldic fur.
Voided Device with the centre cut out.

Blazoning animals

Animals are described using a number of terms:

Accosted	Side-by-side	**Dormant**	Sleeping
Addorsed	Back-to-back	**Salient**	Leaping
Attired	With horns	**Sejant**	Sitting
Couchant	Lying down	**Statant**	Standing
Courant	Running/galloping	**Vulned**	Wounded

The coats of arms of the various colleges at the University of Cambridge often incorporate heraldic devices of the founder or patron, or as with St. Catharine's, the attribute of its namesake.

The development of heraldry

Soon even towns, the clergy, and universities adopted heraldic devices, and coats of arms acquired a number of accessories: a crest (supposed to be worn on top of a helmet), mottoes, and supporters (figures or beasts supporting the shield). This is known as the complete 'achievement' of arms. Often rebuses were used to create a visual pun in a coat of arms (known as 'canting'), such as a hen on a hill for Henneberg in Saxony, or a conger eel, lion, and tun (barrel) for Congleton, UK.

The English aristocracy

England has the most complex and well-preserved aristocratic system in the world, partly because a hereditary peerage, created by writs of summons or by letters patent, is passed on to heirs of the first holder until the peerage becomes extinct. Until 1999 all hereditary peers were entitled to sit in the House of Lords. Life peers, nominated by the political parties or the House of Lords Appointments Commission, are not able to pass on their title. Their creation dates to the Appellate Jurisdiction Act of 1876. The ranks of the English peerage in order of precedence, with the correct form of address, are:

Duke	Referred to as His Grace, or Your Grace in direct conversation
Marquess	The Most Honourable
Earl	The Right Honourable
Viscount	The Right Honourable
Baron	The Right Honourable

All ranks below Duke are referred to as 'Lord' in conversation.

A baronetcy is a hereditary dignity; the holder is accorded the title 'Sir' and the suffix 'Bt' after his name. While peers once administered the lands associated with their title (e.g. the Duke of York, the Marquess of Anglesey), this has not been the case since the Middle Ages, with the exception of the Duchy of Cornwall and the Duchy of Lancaster, held by the Prince of Wales and monarch respectively, although a territorial designation is still added.

A heraldic bestiary

A selection of some of the curious mythological beasts that are used in heraldry:

Antelope Like a heraldic 'tyger' with serrated horns and a deer's legs. Fierceness, bravery.
Camelopard A giraffe.
Camelopardel A camelopard with two long, curving horns.
Centaur Half-man, half-horse, from classical mythology. Sensuality.
Dragon A horny head, barbed tongue, scaly back, armored rolls on the chest, bat-like wings, four legs with eagle's talons, and a pointed tail.
Enfield A fox's head and ears, a wolf's body, and an eagle's shanks and talons for front legs.
Griffin or Gryphon The head, breast, and claws of an eagle, with the hindquarters of a lion.
Harpy Vulture-like bird with the head and breasts of a woman.
Hippogriff A cross between a horse and a griffin, with the front of a female griffin and the back of a horse.
Hydra A seven-headed dragon.
Lucern A spotted cat with a short tail and tufts on its ears.
Musiman A cross between a ram and a goat with four horns.
Phoenix An eagle rising from the flames. Rebirth.
Python A winged serpent or snake. Wisdom.
Sea dog A dog with scales, webbed feet, and a dorsal fin on its back.
Sea lion A lion and a fish's tail. Similar hybrids produce sea wolves and sea horses. Maritime associations.
Tyger Like a lion, but with a long downcurving tusk on the end of his nose. A real tiger is blazoned as a 'Bengal Tiger.'
Unicorn A horse's body, a single long horn, a lion's tail, tufted hocks, cloven hooves, and a beard. The moon and lunar powers; often matched with the lion.
Wyvern A two-legged dragon. If 'proper' it is green with a red chest, belly, and underwings.

The imperial crest of the Russian Romanov dynasty, with its double-headed eagle, is particularly elaborate.

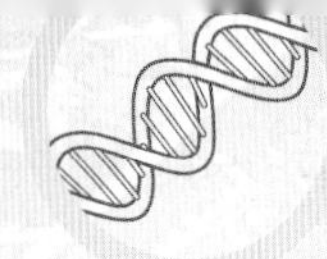

FORMAL DRESS CODES

Dress is an instantly recognizable way of signifying information about the wearers: their social class, wealth, occupation, or religious affiliation. Throughout history, it has also been used as a way of achieving distinction – from the senators of ancient Rome, who were uniquely allowed to wear purple-dyed togas, to the five-clawed dragon motif reserved for Chinese emperors and the ermine-trimmed robes of Tudor royalty. Clothes can also signify a sense of affiliation and belonging. A Scotsman's tartan indicates his clan, locale, and ancestry; a Muslim woman's all-enveloping burka makes it clear that she is submitting to the most conservative conventions of Islam. People's willingness to adhere to conventional codes of dress clearly signifies their readiness to conform to societal norms.

Prince Philip, Duke of Edinburgh The Queen's consort (not a formal title) wears the uniform of his most senior military rank, with full decorations.

Queen Elizabeth II The Queen wears a Parliamentary robe in crimson velvet, trimmed with gold lace, and lined with the finest coronation velvet. She wears the Imperial State Crown, made for the coronation of George VI in 1937, which travels in a separate carriage to the Palace of Westminster, and is placed on her head in the Robing Room.

Law lords On special ceremonial occasions judges and QCs wear long wigs, black breeches and silk stockings, and lace jabots. High court judges in addition have a scarlet and fur mantle, which is worn with his gold chain of office in the case of the Lord Chief Justice. The judges of the Court of Appeal have black silk and gold lace gowns.

Sumptuary laws and tradition

Sometimes dress codes are enshrined in law. In Tudor England, for example, sumptuary laws were introduced to curb immoderate excess. There were fears that, as England broke away from the codified simplicity of feudal society, people were beginning to squander money on clothes that were inappropriate to their station. According to the very detailed English Statutes of Apparel (1574), clothes were defined by social station; sable was only appropriate for high-ranking aristocrats, velvet was only permissible for wives of knights of the garter, and so on. The laws ensured that the rank of any member of society – from a duke to a working man or woman – was instantly recognizable by the fabric, color, and cut of the clothes they wore. These ideas are perpetuated in the traditional clothes worn at traditional events such as the State Opening of Parliament in the UK *(right)*.

Dress to impress

The English social season remains a bastion of formality. Invitations to weddings, dinners, and parties, and tickets to events such as horse racing at Ascot (*right*) and the Henley Regatta inevitably specify an arcane array of dress codes. Refusal to comply with, or ignorance of, the codes can cause acute social embarrassment.

Morning dress The traditional dress code for weddings and formal day events: men should wear a black or gray morning coat, matching gray or gray and black striped trousers, a plain white shirt, waistcoat, and a tie or cravat. Top hats are obligatory for the Royal Enclosure at Ascot. Women should wear dresses or skirts to the knee, or a well-tailored trouser suit. At Ascot, they are required to wear a hat that covers the crown of the head.

Ladies-in-waiting
The Queen's female attendants wear white full-length formal gowns.

The Serjeant-at-Arms
The bearer of the ceremonial mace during the Speaker's procession wears knee breeches and a long coat and sword.

Black Rod
The gentleman Usher of the Black Rod summons the House of Commons into the royal presence in the House of Lords. His costume dates to the mid-16th century – the period when the Commons were asserting their independence from the Lords. He wears a long black coat, cravat, garters, and pointed shoes.

Lords
Peers of the realm wear scarlet robes, trimmed with three-inch (7.5 cm) wide ermine bars and gold oak-leaf lace. The robes worn in the House of Lords date back to the 15th century. The number of bars of ermine and gold reveal the wearer's rank: dukes wear four rows of ermine and gold bars; marquesses wear three-and-a-half rows; earls three rows; viscounts two-and-a-half rows; and barons two rows.

Black tie Men should wear a black dinner jacket with silk lapels, black trousers, a white evening shirt, and a black bow tie. Women can wear dresses or skirts that skim the knees, or longer.

White tie The most formal, and rare, of dress codes. Men must wear a black tail coat with matching black trousers, a white shirt with a detachable wing collar, cufflinks and studs, a thin white bow tie, and evening waistcoat. Women should wear full evening dress.

Get ahead, get a hat! In the first half of the 20th century the streets of London and other capital cities were a sea of black bowler hats. Along with a pinstripe suit and rolled umbrella, bowler hats were an indispensable part of the respectable office worker's uniform. Today, such conformity has become outdated in many workplaces, and workers are encouraged to wear 'smart casual' clothes rather than business suits. Some offices have instituted the 'dress-down Friday,' a chance to discard more formal clothing on a regular basis.

DECODING THE UNCONSCIOUS

The mind in history

The concept of the 'unconscious' did not just appear fully formed in Vienna in 1900; for centuries many writers, artists, theologians, physicians, and philosophers had tried to define the human psyche. Shakespeare examines the unconscious mind in many of his plays (Freud later psychoanalyzed Shakespeare's *Hamlet*) and the unseen machinations of the human psyche have concerned philosophers from Socrates to Immanuel Kant and beyond. Ancient Greek and Roman physicians such as Galen (AD c.129-c.200) had tried to decode the characteristics of an individual's personality by the balance of the four supposed 'humors' in their body; later, in Renaissance Europe, there was a resurgence of interest in the humors.

The shadowy 'unconscious' is the mental activity that occurs somewhere beyond what we are consciously aware of thinking. Defining the unconscious remains a contentious issue among psychologists and psychiatrists. But it is now accepted that the conscious mind is merely a perceptual veneer and its thoughts therefore do not disappear: they are stored in the unconscious. Yet if we are not aware of these thoughts, how do we know they exist? The unconscious realizes itself – often obscurely – in our behavior, affects our conscious thought, and can exhibit physical symptoms as psychosomatic illness. The key 20th-century figures in the advancement of our understanding of the unconscious are Sigmund Freud (1856-1939) and Carl Jung (1875-1961).

Intimations of psychology

For several centuries in Europe the mentally ill were treated as curiosities, and London's Bethlehem Hospital (Bedlam) was part of the fashionable round of civilized entertainment. But, by the 19th century, the brain had become directly associated with the human mind; phrenology became a popular method of decoding the human consciousness, as well as an individual's character. Phrenologists used the bumps and fissures of the human skull to attempt to map or 'decode' the subject's personality. At around the same time Dr. Franz Anton Mesmer (1734-1815) had developed the technique of suppressing the conscious mind through hypnotism.

The humors

Yellow bile Choleric and quick to anger.

Black bile Depressed and melancholic.

Phlegm Sluggish and dull.

Blood Passionate and sanguine.

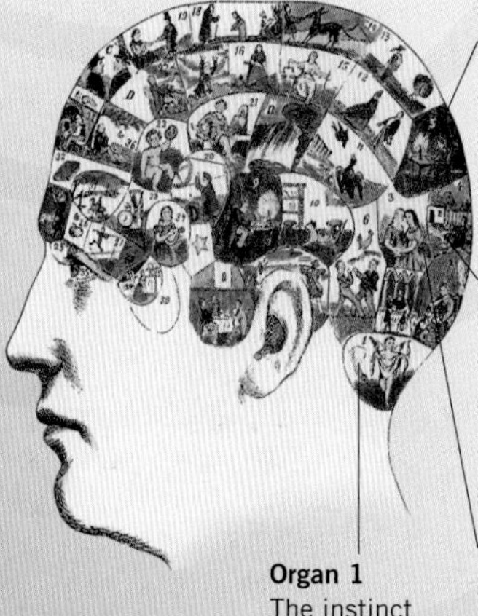

Organ 4 The instinct for self-defence and bravery; tendency to be aggressive.

Organ 5 The carnivorous instinct; murderous tendencies.

Organ 3 The capacity for fondness and friendship.

Organ 1 The instinct to procreate.

Franz Joseph Gall (1758-1828) developed phrenology, theorizing that the brain was divided into 27 'organs,' each with a specific function. Later phrenologists added more 'organs,' eventually identifying 43.

Freud and the 'talking cure'

Although Sigmund Freud popularized the theory of the unconscious (it was not his discovery, and the significance of the unconscious was widely acknowledged among Freud's 19th-century counterparts), his real contribution is the process of psychoanalysis – his system for deciphering the code of the unconscious mind. Freud formed many of his theories of the unconscious and developed the treatment of psychopathology in his practice in Vienna. Despite the debate surrounding his theories, the basis of the 'talking cure' (the term coined by one of Freud's patients known as 'Anna O.') remains a crucial tool in the treatment of mental illness and psychological disorders; even those who reject Freudian ideas often retain this groundbreaking approach (the concept of actually talking and listening to psychiatric patients was revolutionary at the time). By talking a patient through his or her problems, Freud hoped to bring the unconscious to the surface. The 'talking cure' began with Freud's contemporary Dr. Josef Brauer, and Freud used this as a starting point to develop his system of psychoanalysis.

Sigmund Freud in his London study. His accounts of his psychoanalytical cases remain enormously influential.

How Freud decoded the unconscious:

Amanesis Personal history as recollected by the patient in order to build up a profile for interpretation.
Free association Freud replaced hypnosis with this method. A patient freely relates a stream of consciousness, mentioning everything (however mundane or embarrassing). This creates a pathway through the patient's memory, revealing memories that have been hidden in the unconscious.
Interpretation of 'Freudian slips' Freud believed that it was not by chance that we mistake one word for another, forget a name etc. All these mistakes or Freudian slips can be analyzed and interpreted. For example, if a man were to mistakenly call his girlfriend by his ex-girlfriend's name, this might suggest unresolved feelings towards the previous girlfriend.
Interpretation of dreams Dreams are our encoded unconscious thoughts or desires according to Freud (*see page 234*); just like any code, he believed they can be cracked.

Psychic apparatus

Freud's final theory on the unconscious breaks down the boundaries between the conscious and unconscious components of the psyche. It divides the human psyche into three parts that Freud called 'psychic apparatus':
The ego This contains our conscious thought.
The id The mess of our primal unconscious that is driven by our most simple and basic urges.
The superego The second, more organized, part of the unconscious that often functions as our conscience. The ego acts as a broker for the id and superego, trying to balance the needs of both.

The final image from William Hogarth's moral cycle *A Rake's Progress* (1732) is set in Bethlehem Hospital (Bedlam), London, where the antics of the mentally disturbed are presented as an amusing distraction.

The visionary

Jung was initially a Freudian himself and corresponded with Freud for some time. However, Jung developed his own theories of the unconscious that differed from Freud in a key way: Jung separated the individual – personal – unconscious from what he termed 'collective unconscious.' Later called by Jung the 'objective psyche,' this is buried deep inside the psyche and contains the combined inherited experiences of a people. Springing from this idea Jung developed an interest in mysticism and spiritualism, becoming critical of the logic and science-dominated society of the 20th century; this interest led to his system of archetypes that symbolize the different facets of the collective and personal unconscious:

Animus The animus is the male personification of the unconscious in women.
Anima The female personification in men, often portrayed as negative in art.
Self 'The innermost nucleus of the psyche'; in a woman, it usually manifests as a 'superior female figure' such as a goddess or sorceress, and in a man as 'a masculine initiator or guardian,' such as a wise man or old wizard.
Shadow Represents attributes and aspects of our personality that we normally choose to ignore; it is a personification of the faults that we do not wish to acknowledge.

Carl Jung identified symbols in common human memory.

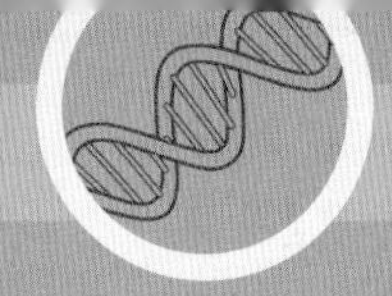

The Language of Dreams

Everybody dreams: dreaming is part of our sleep process, whether or not we actually remember them on waking. In ancient and modern cultures alike, we have always been fascinated by our dreams – or more precisely, fascinated by what our dreams could mean or tell us. Dreams as prophecy, dreams sent by a higher power, dreams as the key to the unconscious, dreams as a means of healing, dreams as completely random permutations of our thoughts; different people at different points in history have believed that dreams are all these things. This cacophony of different ideas means there can be no all-encompassing dictionary for the language of dreams, although there have been many attempts.

Dream temples

In ancient Greece, temples dedicated to Asclepius the god of medicine were called *asclepieia*; they were essentially places of healing where people went to be cured. To begin the healing process, a would-be patient spent a night in the temple, and the next day would tell a priest what he or she had dreamed; the priest would then interpret the dream and base his prescribed cure on what the dream revealed. The language of dreams for followers of Asclepius was an essential guide to treating illness. Such reliance on the interpretation of the symbolism of dreams is widespread in many cultures across the world.

"The interpretation of dreams is the royal road to a knowledge of the unconscious activities of the mind."

SIGMUND FREUD, *THE INTERPRETATION OF DREAMS*, 1900.

Down the rabbit hole

The significance of dreams is shown by how often dreams and the act of dreaming feature in works of art throughout history; dreams and their interpretation are important in both the Old and New Testament (Joseph as a dreamer and interpreter of dreams in Genesis, and Pilate's wife's dream in Matthew) and in classical works such as Homer's *Iliad* and Ovid's *Metamorphoses*, while the Roman emperor Constantine attributed his conversion to Christianity to a dream (*see page 43*). As a result, the dream poem was an extremely popular form in medieval Europe, notably in Geoffrey Chaucer's *Book of the Duchess*, Dante's *Divine Comedy*, and the *Hypnerotomachia Poliphili*, although such dream poems often served an allegorical purpose. Later writers have written in dream form, as, for example, Lewis Carroll's *Alice's Adventures in Wonderland* (1865).

Alice's adventures are presented as a dream narrative, combining surreal imagery with an underlying logic often derived from the author's experience as a mathematician.

The psychology of dreams

Freud theorized that dreams could be interpreted; each individual has their own 'key' to decoding the language of their dreams and therefore their unconscious. Freud believed that ultimately all dreams are unconscious wish-fulfillments, although their representation of an unfulfilled wish may be strange and obscure; he believed the same was true of any kind of dream – including daydreams. Jung attached greater significance to dreams and dreaming than Freud; like Freud, he saw them as outlets for our unconscious, but not merely as a key to the unconscious. He believed that dreams had their own internal language and logic and, for the more spiritual Jung, the unconscious world of dreams was as important as our waking life.

The Nightmare (1781) by Henry Fuseli shows a sleeping woman surmounted by an incubus, a male demon believed to visit and assault women when they sleep, with his steed, the 'nightmare.'

Living life through dreams

One particularly perplexing aspect of dreaming is our participation in unsettling activities, which both Freud and Jung recognized as the expression of suppressed desires or anxieties. Some dream activities or situations recur frequently enough to be tentatively identified:

Dancing Meant to signify good luck.
Flying To fly high forewarns of marital difficulties; to fly low symbolizes illness; falling forewarns of a downturn in luck, but waking before hitting the ground is a good sign.
Nudity Dreaming of being naked is a sign of looming scandal in your life.
Swimming Generally a positive omen; if you find yourself sinking, this forewarns of a struggle ahead; swimming underwater foresees worry and difficulties in your life.
Teeth Dreaming about loose teeth is an unlucky omen; if your teeth are knocked out, this forewarns of sudden disaster; examining your teeth is a warning that you must make sure your affairs are in order.

Animal dream symbols

The appearance of departed relatives or long-lost friends, often remembered in convincing detail, is not difficult to interpret, if unsettling upon waking. However, the appearance of animals has invited some speculation, although most cultures interestingly concur as to their significance.

Bees A positive premonition; bees symbolize a fertile, successful, and happy life for the dreamer.
Cats This can spell misfortune; black cats are associated with bad luck and dark forces; white cats indicate hard times ahead.
Crocodiles Indicate a hidden danger in the future of the dreamer.
Dogs A dead or dying dog may forewarn of the death of a good friend. Otherwise faithfulness.
Horses A black horse indicates mystery, and possibly the occult; a white horse represents prosperity and good fortune.
Lions Presage influential and prosperous friends who will assist you in the future.
Owls If you kill an owl or see a dead one in a dream, you will survive a dangerous experience.
Whales Symbols of good luck.

The science of Surrealism

There is no set 'text' for codifying the imagery that our brains conjure up when we are asleep, but dreams inspired visionary Romantic artists such as Henry Fuseli, Goya, and William Blake. In the 20th century the Surrealist movement produced artists Salvador Dalí (*above*), René Magritte, and Joan Miró, who appear to have instinctively identified a 'language' of dreams with which most of us can identify. The more literary Surrealists experimented with 'automatic writing' which often involved associative game playing, juxtaposing one word image with an automatically triggered response, supposedly revealing some sort of hidden unconscious 'truth' or 'meaning.' The American Beat writer William S. Burroughs meticulously recorded his (often drug-induced) dreams, and used his notes as a compositional device, notoriously in *The Naked Lunch* (1959), before exploring the 'cut-up' method of randomly assembling snippets of texts to provoke an 'automatic' but poetic reaction in the reader. The Surrealists were very impressed by cinema's dream-like editing, a technique used by film directors such as Luis Buñuel, Federico Fellini, and David Lynch to evoke convincing dreamscapes.

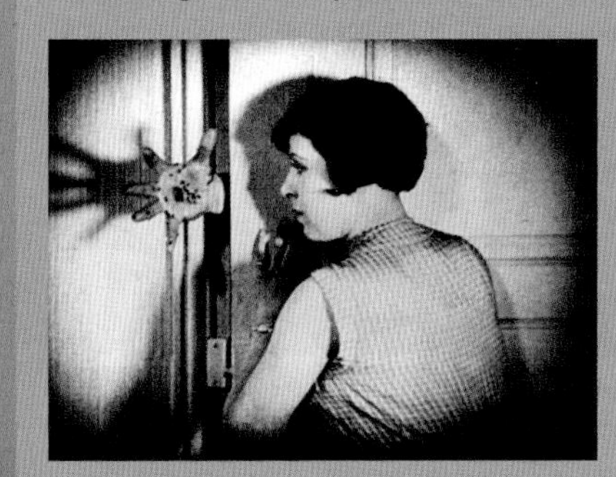

Un Chien Andalou (1928), an exercise in Surrealist filmmaking by Salvador Dalí and Luis Buñuel, paraded a number of nightmare dream sequences.

11

The development of ways of conveying information that transcend language has become increasingly important in our multi-national, globalized world. The most prevalent among these are visual and graphic codes.

visual codes

The problems to be met are many and various – whether they be overcoming barriers to education or communication with the physically challenged, or providing basic information in emergencies, on the highway, in transnational contexts, or simple instructions in the home. Animals too rely on a multiplicity of visual signals and other means of sending messages.

SIGNS AND SIGNAGE

It is only comparatively recently that the representation of human, natural, and other forms have been stylized to the same degree as the ancient Egyptian hieroglyphs to create a collection of symbols capable of communicating information in a graphic language as an alternative to written words. During the 20th century, two factors made the development of non-text based ways of communicating information of critical importance: globalization, and the rise of the automobile. The need for often lifesaving information to be conveyed swiftly and graphically created a burgeoning graphic design industry.

Aa Aa
Avant Garde Gothic Bauhaus

Aa Aa
Twentieth Century Gill Sans

Aa Aa
Rockwell Times New Roman

Graphic design

The first two decades of the 20th century, with the development of abstract fine art, and of minimalist architectural and product design, saw much experimentation with graphic design ideas. Effectively, the foundations of a transnational visual language were forged between 1907 and 1927. The Cubists' incorporation of printed ephemera into some of their collages suggested hidden messages, but it was the Russian Constructivist and Suprematist artists associated with the Bolshevik revolution who brought a new and politicized graphic meaning to the visual arts, especially in poster design. After 1919, with the foundation of the Bauhaus group in Germany, the idea of graphic codes for conveying specific information across the arts had become a central preoccupation. One important aspect of this was the development of a number of new typeface fonts (*above*), many of them sans serif, specifically designed for their unfussy visual clarity and impact when set large on posters and signs.

> "Art into Industry"
>
> BAUHAUS SLOGAN.

Graphic language: Isotype

In the early 20th century, the philosopher Otto Neurath (1882-1945) founded the Isotype Movement with the specific purpose of informing as widely as possible through the use of a non-verbal graphic language. Isotype (International System of Typographic Picture Education) was intended as an additional language, but one with minimal ambiguity, in order to convey vital information to the socially underprivileged. This was done through the creation of charts and tables and other graphic assemblages. Neurath saw the need to impose conventions in order to ensure the veracity of this language.

Reading Isotype The success of the Isotype code is in part its dependence on the user engaging in the recognition process. In this example of 'Moving Heavy Stones,' from an educational book on history, the reader has to learn what the shapes represent. Having absorbed this code, and the meaning of the relationships, the reader has a simple visual mnemonic, without the clutter of unnecessary peripheral illustrative material. Any one of the iconic shapes can then be used elsewhere with a clear definition.

Radio, Telephone, Automobiles

United States and Canada	Europe	Soviet Union
Latin-America	Southern Territories	Far East

Quantitative diagrams In particular, Neurath insisted that in charts an icon must represent a specific number of things – so more icons (of the same type) indicated more things. In this comparative chart, from 1939, each icon represents one million wirelesses, telephones, or automobiles. In the structure of the symbols, the Isotype Movement endeavored to produce clearly identifiable objects, and ruled that perspective could not be used. Between 1928 and 1940 a 'Symbol Dictionary' was compiled from the various icons devised for commissioned charts.

The Olympic Games

This was the first truly multinational sporting event, starting in the modern era at Athens in 1896. The five-ring symbol was introduced then, as was the establishment of French and English as the official Olympic languages. But with athletes from most of the world's 195 countries participating, and spectators attending, communication is an issue. Since 1964, for each Olympic Games there have been specially designed pictograms illustrating the classes of event. However, as long ago as the 1936 Berlin Olympics, pictograms were employed. The high profile event inspires the use of leading designers, each bringing their own visual treatment. For the 1964 Tokyo Olympics, graphic designer Yoshiro Yamashita working under the artistic direction of Masaru Katzumie produced a highly stylized and very powerful series.

Munich 1936	Tokyo 1964	Mexico 1968	Barcelona 1992
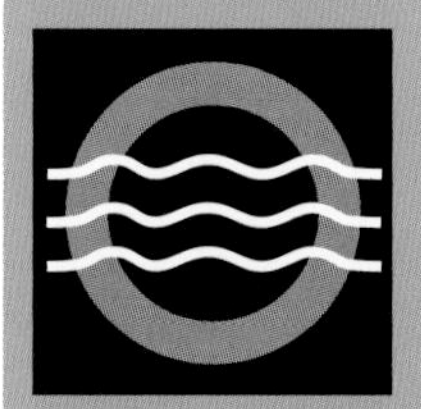 Swimming	 Swimming	 Swimming	 Swimming
 Gymnastics	 Gymnastics	 Gymnastics	 Gymnastics

Avoiding the language barrier

Visual communication without written language has become increasingly necessary with the massive growth in world trade and travel. Exported goods may need instructions for assembly, use, and transportation, in several or many languages. To avoid the cost of translation, and possible error and ambiguity, packing and content often carry non-verbal instructions in the form of immediately recognizable symbols and illustrations.

Assembly instructions for flat-pack furniture and other goods often exclude language, but are clearly labeled in a step-by-step manner with ingenious graphic instructions.

Washing labels

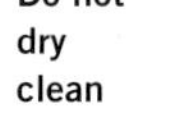
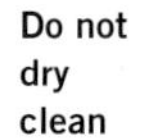

Warm machine wash · Non-chlorine bleach · Hot tumble dry · Warm iron · Do not dry clean

Garment retailers may have suppliers and outlets in many countries, so washing instructions are usually in symbol form.

Shipping labels

Hazardous waste · Infectious biohazard · This way up · Fragile · Handle with care

Goods in transit, whether by road, sea, or air often require extra information on their packaging to inform handlers.

On your PC

The revolution in use of the personal computer and mobile phones has seen pictograms further developed to provide icons designed to intuitively convey information or instructions on screen. This highly condensed art form was largely developed by Susan Kare who developed icons and fonts for Apple Computers, but has now provided increasingly complex, animated, colorful (and often humorous) icons for many best-selling PC firms. Her approach was inspired by road signs (*see page 240*), and the need to 'humanize' the computer interface. Her cursor hand, trash, and clock are universally recognized, but Apple's replacement of her alarm icon, the fizzing bomb, by the spinning 'beachball of death' is to be mourned.

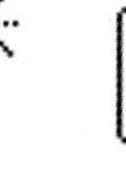
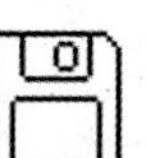

Welcome · Crash imminent · Disk · Painting Tool

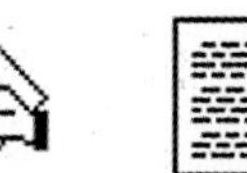

Notepad · Document · Question · Mail

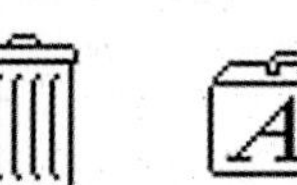
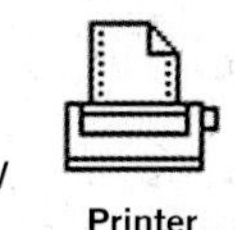

Trash · Font suitcase · Function timer/ Wait: loading · Printer

Icon language

The use of icons to form a language is now a possibility. Immediately recognizable (or easily learned) picture-word icons can be assembled to form sentences. Basic rules of visual grammar have been developed: certain icons are qualifiers indicating gender, type, or possessive pronoun, and appear in gray; black is used for emphasis; speed lines indicate a verb rather than a noun; arrows can indicate prepositions; and simple animation can emphasize both activities and emotions.

Officers are driving to court in New York

I look forward to hugging and kissing you

Dancing lady

A sophisticated icon language has been developed by Jochen Gros that combines minimalist icons with many witty 'emoticons' with a system of 'icon-handwriting' for showing emotions through animated pictograms.

HIGHWAY CODES

Road markings

Lines and other designs painted on the road surface provide essential and immediate information for the driver, and tend to be similar in different countries. Most are imperative or instructional.

Unbroken line, center of road	No passing
Broken lines, center of road	Beware when passing
Unbroken lines at side of road	Parking restricted
Double unbroken lines at side of road	Parking prohibited
Parallel lines crossing road	Pedestrian crossing
Hatchured lines	No entry zone
Crossed diagonal lines	No entry unless exit clear

Unsuccessful signs

The plethora of road signage, and ever increasing demands for new information graphics, has led to some unfortunate inventions. A suite of new toll-road information signs in France in 2008 included the strangely uninformative design above, apparently indicating a nearby woodland area.

As early as 1909 motoring organizations at an international meeting in Paris attempted to adopt a series of pictorial road signs. They agreed on only one sign – for dangerous crossings. This was to be erected 820 feet (250 m) before the hazard and was to be perpendicular to the roadway. Even now, despite the undoubtedly increasing need, there is no consistent international code, and each country tends to develop its own style. Different styles too have emerged as health and safety regulations in most countries require that signage for facilities, emergency exits, and so on, transcend the language barrier. Furthermore, issues such as cultural differences, gender distinctions, fashion, and developing technology mean that graphic communication continues to confront new challenges.

Road signs

Although some of this information is necessarily language-based – place names for example – other messages are more rapidly absorbed through the use of pictorial images, once the necessary code is understood. With road signs there need to be two systems of code. One represents the nature of the message – varying from useful information and warnings to instructions. For this, the geometric shape of the sign is used to indicate the function, whereas the pictorial content conveys the detail. Generally, circles or hexagons are used for prohibitory or restrictive signs, and diamonds or triangles are used for warnings. Color provides a different level of information. A red frame indicates an imperative or a hazard. Yellow or blue usually represent additional information. In some cases road signs need to be very specific to a region, as in the example of signs warning of kangaroos in Australia and reindeer in Norway. The meanings of these signs will seem obvious to most road users, but that is because they have been learned as a language while learning to drive. Many are graphically obvious, others need explanation. The British hump bridge sign, for example, was deemed to need a spelled out identification. However, over time the graphic device has been assimilated and now stands on its own. The 'Warning: Elderly People Crossing' sign has recently attracted understandable criticism in the UK on the grounds of ageism.

Instruction/imperative signs

No entry (France)

Stop (France)

No left turn (UK)

No bicycles (UK)

Hazard warning signs

Incline (Norway)

Incline (Japan)

Dangerous bend (Australia)

Dangerous bend (Ireland)

Dangerous bend (UK)

Dangerous bend (Norway)

Tunnel (Taiwan)

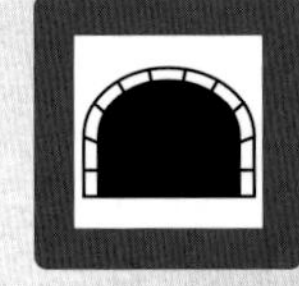

Tunnel (Germany)

The speed with which road signage needs to be deciphered by the driver requires simple images and simple concepts. An image struck through with a diagonal line is easily recognized as a negated or prohibited activity – for example the cycle prohibition sign. Others may prove more difficult to ascertain at a glance.

Information signs

Roundabout (UK)

Roundabout (US)

Pedestrian crossing (Poland)

Pedestrian crossing (US)

Pedestrian crossing (Sweden)

Elderly people crossing (UK)

Kangaroos (Australia)

Deer (Japan)

Reindeer (Norway)

Cattle (UK)

Deer (UK)

Toads (UK)

Emergency signs

The traffic through international airports, on public transport in general, and in many major tourist-destination cities has seen an explosion in specially commissioned suites of graphic codes. As important as road signage, the indication of emergency exits, fire escapes, and assembly points is a vital requirement in public and commercial buildings worldwide. Usually green or blue, sometimes red, these are prominently displayed and need to be unambiguous and immediately understandable.

For your convenience

Whereas generally the meaning ought to be clear from the image and context, pictorial symbols, like any other language, need to be learned and culturally recognizable. A door in a public place, such as a hotel or airport, with the image of a male or female is taken to indicate access to a public washroom. This assumes a cultural recognition of a particular style, generally Western, of dress – a male wears trousers, whereas a female wears a dress. Concealing the legs of the female goes some way to avoiding cultural upset. However, there needs to be sufficient differentiation between the gender images for the code to work.

Symbols also require some degree of longevity in order to inform. This selection of washroom signs displays some of the difficulties involved in creating a satisfactory multicultural design.Modern telephones bear little if any resemblance to their circular dialed two-piece forebears. However, as the old style of telephone is so unambiguous in appearance, images based on those models are still recognizable, and preferable to pictograms of modern telephone handsets.

Challenged Communication

Silence, please

Many medieval monastic communities, notably the Trappists and the Benedictines, were encouraged to adopt a vow of total or partial silence, and many developed signing languages which are still in use today. In the arts, gestures were particularly important. The Spanish painter de Navarrete (El Mudo, 1526-79), was a deaf mute. He was educated by Benedictines, and developed a sufficiently successful signing system for him to become court painter to King Philip II. His paintings (*below*) are notable for the pronounced use of gestures.

Code systems to aid communication for people with either aural or visual impairment have been in existence for some time. Gesturing, using mimicry of recognizable actions, will have been a natural recourse of those with speech and hearing impediments. Indeed these methods are used in formally structured signing systems such as Makaton. However, in order to extend the potential for communication, a formalized system that could be commonly understood, at least by those 'speaking' the same language, needed to be developed. Verbal communication may not be an issue for the visually challenged, but the written word was obviously not accessible until the creation, and acceptance of, the touch-reading system Braille in the 19th century.

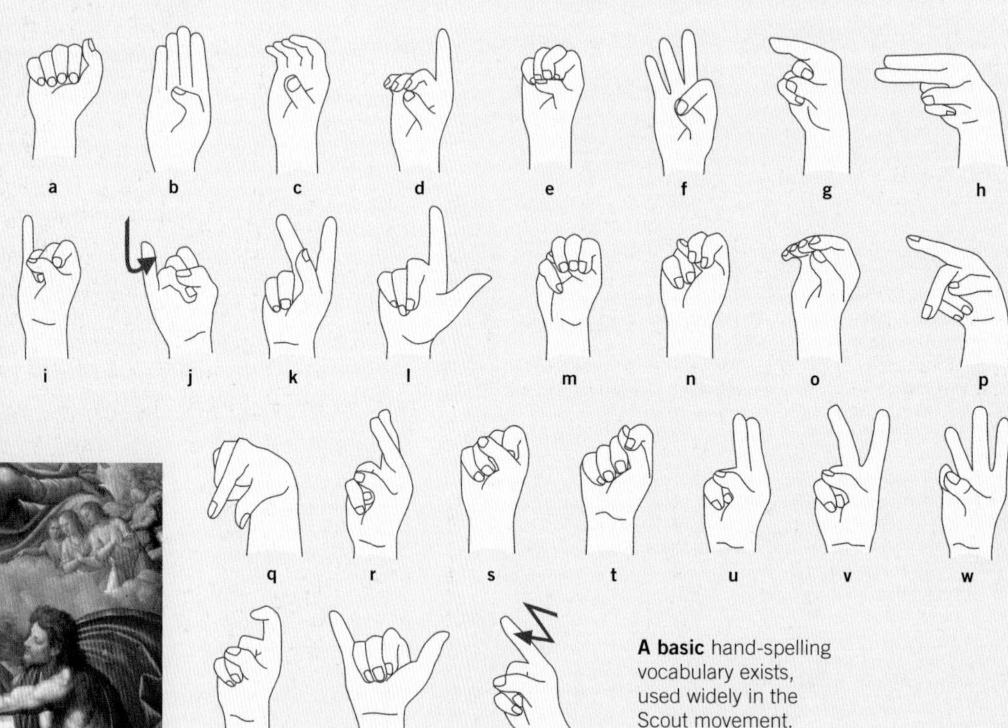

A basic hand-spelling vocabulary exists, used widely in the Scout movement.

Signing for the deaf

Signing systems are based on the user's localized language – signs represent letters or specific words. Even with English, there are variations. There is a US signing system and a British signing system, and although there may be similarities, the differences impede ease of communication between the two. Most modern signing systems are based not only on hand gestures but on mouthing and the use of other 'body' language and the space around the body. They also rely on a vocabulary of signs rather than a basic alphabet, although the alphabet remains important for conveying words (such as names) not covered by the vocabulary. Linguistic, cultural, and pedagogical issues have meant that a host of national systems have grown up; although a 'universal' signing system – Gestuno – has been invented, its adoption has been limited (rather like the 'international' language, Esperanto).

Writing for the visually impaired

The Braille system was devised by Frenchman Louis Braille in 1821 after he tried a system of 'night writing' developed by Charles Barbier. Originally commissioned by Napoleon for silent communication by the military at night (and also adapted for semaphore), it relied on a grid of 12 dots in two rows of six. Combinations of dots, pressed up through cardboard, represented letters and were read by touch, but the system was judged too difficult to learn (especially among barely literate soldiers in the field) and abandoned. Braille (1809-52, *left*) created a simplified version based on just six dots, each letter easily recognized. Braille's system revolutionized written communication for the visually impaired.

Modern Braille (*above*) has been extended to represent shorthand, mathematical symbols, and musical notation. It has also been expanded to an eight-dot display for use with computers.

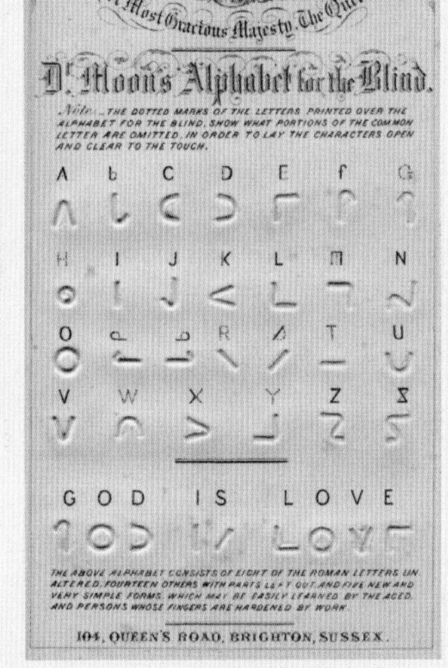

Moon type An alternative to Braille was developed by Englishman Dr. William Moon in 1845. Moon replaced Braille's dot matrix with curved and slanted characters more closely resembling alphabetic forms.

mummy

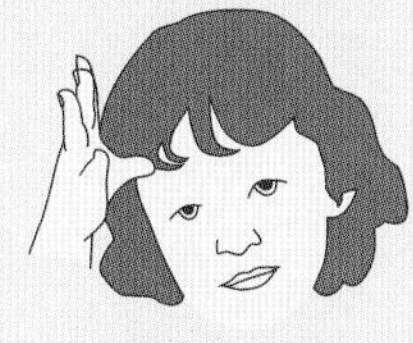

daddy

bird

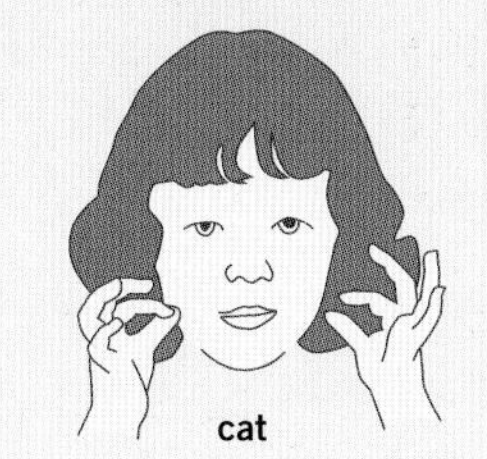

cat

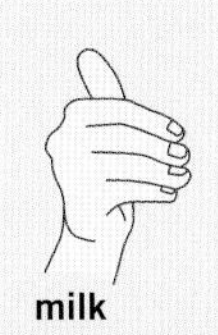

milk

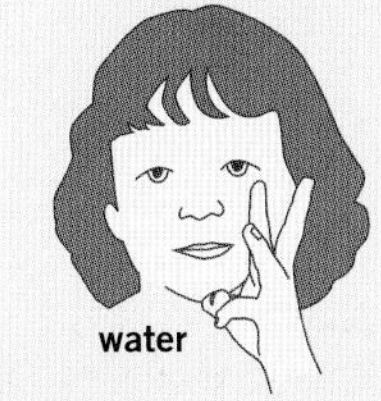

water

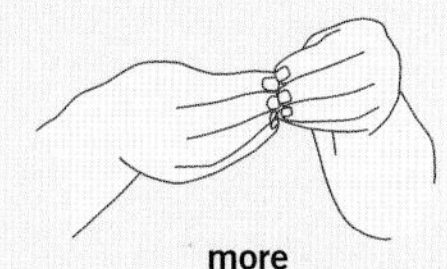

more

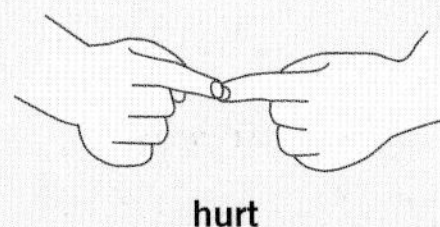

hurt

Signing for speech

Other beneficiaries of signing systems include those with speech development difficulties. Muscle tone in people with Down's Syndrome can impede verbal delivery even when the person knows exactly what they wish to say. Infants who are still too young to command language will also know what they want but are unable to express it in words. Makaton is a specialized signing system, based on British sign language, specifically designed to facilitate communication in these cases. It also includes recognizable mimicry to express things such as 'drink,' 'milk,' 'bread,' 'diaper,' and so on. These give the system a limited degree of cross-language usage.

The Braille system

Louis Braille realized that Barbier's system, which involved 12 dots, was too difficult to 'read' by fingertip, and would involve moving the finger in order to read a single letter. Braille concentrated on producing an alphabet and numerology based on cells of up to only six dots.

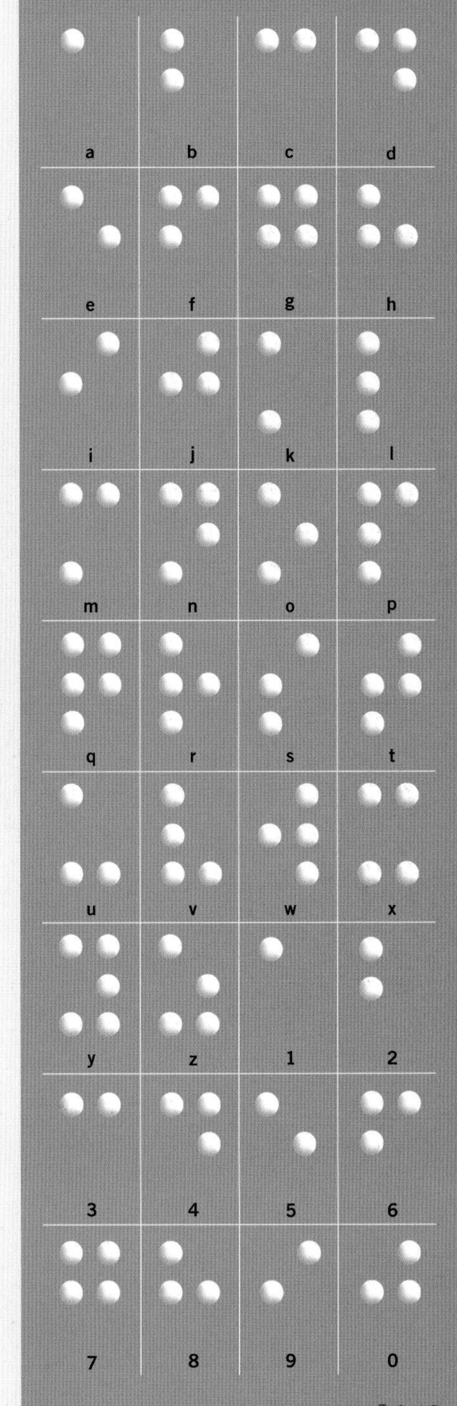

Describing Music

Musical notation today is a largely standardized language that instructs musicians how to interpret written symbols corresponding to sounds and silences in order to produce music. But it was not always so. The problem was how to find a language to accurately describe a sound or a series of sounds. Particular elements of notation and variations upon them have evolved at different times and in different places over the centuries, although by around 1350 the fundamentals of the now familiar bar lines and time signature structure of the musical score had been established.

Development of musical notation

The first musical notations were 'neumes' (*above*), devised over 1,000 years ago.

800-1200 Music notated with 'neumes,' small squiggles and dots representing notes or note groupings. By 1100 neumes arranged vertically to show relative pitch.
c.1020 Guido of Arezzo (c. 995–1050) invents basic staff (stave) with spaces, lines to indicate pitch, and hints at clefs, accidentals.
1260 Modern notation evolves with attempts to formalize the relationship between a note's shape and value.
1350 Time signatures and bar lines.
1400s The natural sign in use with sharp and flat signs.
1500s Tempo and dynamic markings.
1520s Ledger lines, ties, and slurs.
1600s G (treble) clef in harpsichord music.
1700s Bowing marks, fingering indications.
1770s Pedaling signs for pianoforte.
1780s G (treble) and F (bass) clefs generally accepted.
1800s Tempo, pulse, dynamic levels become more extreme; phrasing, articulation, expression receive more attention; ornamentation becomes absorbed within style, or fully written out.

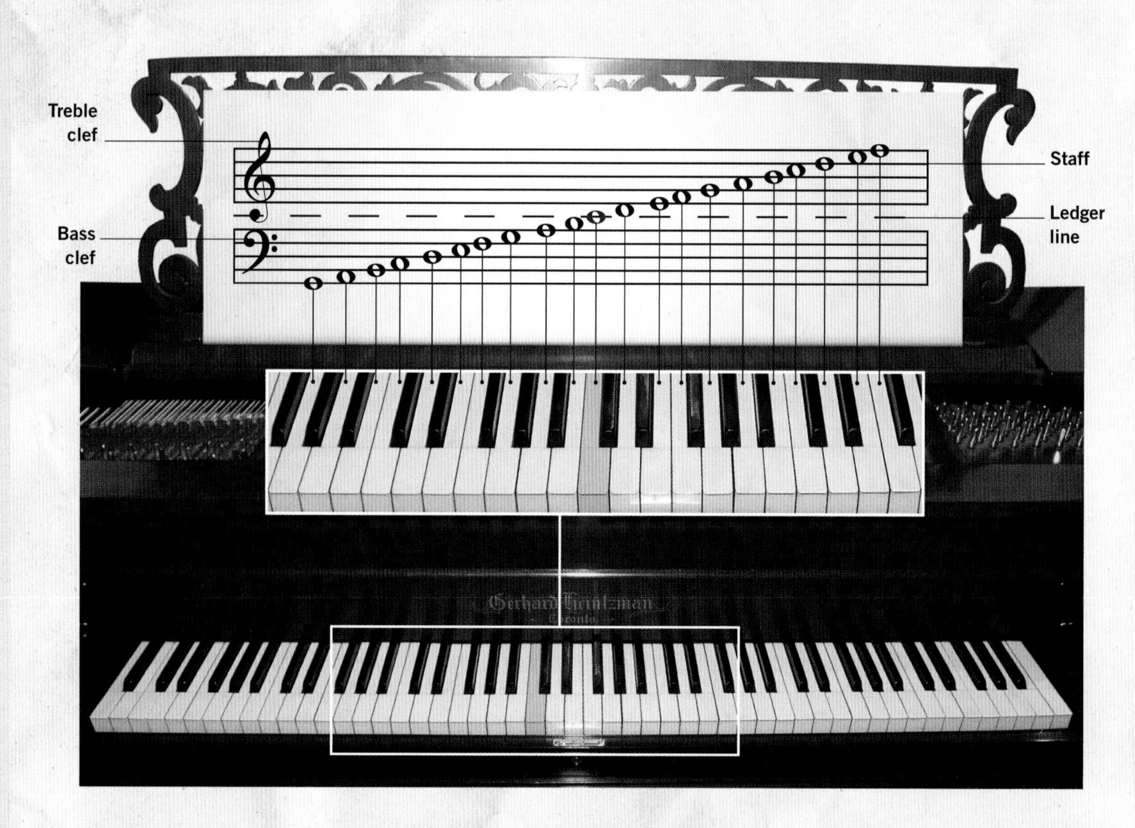

Structure of notation

Notation in Western music indicates two of the principal properties of musical sound, namely, its pitch and duration. A sign called a 'clef' appears at the left of each staff to indicate which line or space signifies a particular note. There are two main clefs, the G (treble) and F (bass) clef. The G clef is so-called because the loop at the bottom wraps around the line on the staff corresponding to the note G above middle C. The F clef is so-called because the dots are placed above and below the line marking F below middle C. The G clef is used for higher pitches, the F clef for lower pitches. A flautist, violinist, or soprano reads from the treble clef, a bassoonist or cellist from the bass clef, and a pianist from both clefs. A very high or very low note that of necessity has to extend above or below a clef is indicated by a 'ledger line.' A note in a musical 'score' (a manuscript or its printed copy, *see page 246*) indicates pitch according to its position on the 'staff' and its respective 'clef.' Duration is expressed by the shape of a note, whether it has a black or hollow head and whether it has an attached stem or not. The letters A through G are assigned to the various pitches, which are placed as individual notes on, between, above, or below five horizontal lines known as a 'staff,' or 'stave.'

Pitch and notation

On the piano, all eight white keys from middle C up to the C above (C, D, E, F, G, A, B, C), or down to the C below, correspond to an 'octave' (eight notes). Within this octave, the 'diatonic scale' comprises the seven different pitches. Black keys separate some of the white keys. Combined, the white and black keys add up to a total of 12 possible notes, which correspond to the 'chromatic scale.' A gap between two notes is an 'interval.' The black keys on a piano are 'sharps' (#) or 'flats' (♭). A sharp is used to raise (sharpen) a note and a flat to lower (flatten) a note by one semitone.

For example, the interval between F and G is a whole tone, whereas that between F and F# (sharp) is a semitone. Scales can be in a major or minor key. A major scale will have a corresponding 'relative minor' scale. In any scale there will be a combination of whole tones and semitones (half tones), depending on the scale. The intervals between the notes remain the same for all scales of the same type (e.g. major scales, minor scales). The relative scales major to minor are comprised of the same tones, but starting and ending on a different tone. Here are some examples.

Diatonic scale of C major – starts and ends on C

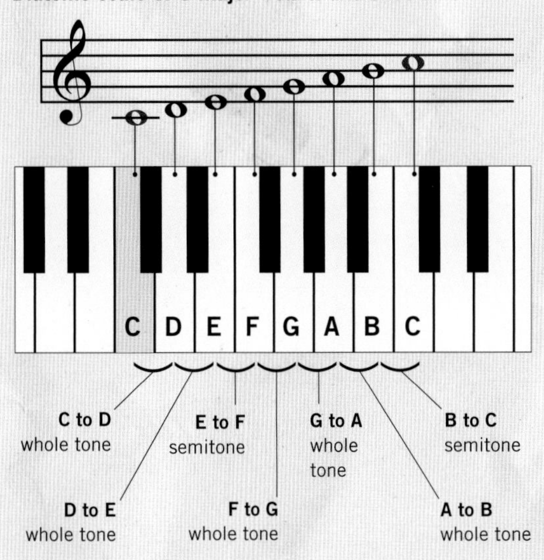

Diatonic scale of A major – starts and ends on A

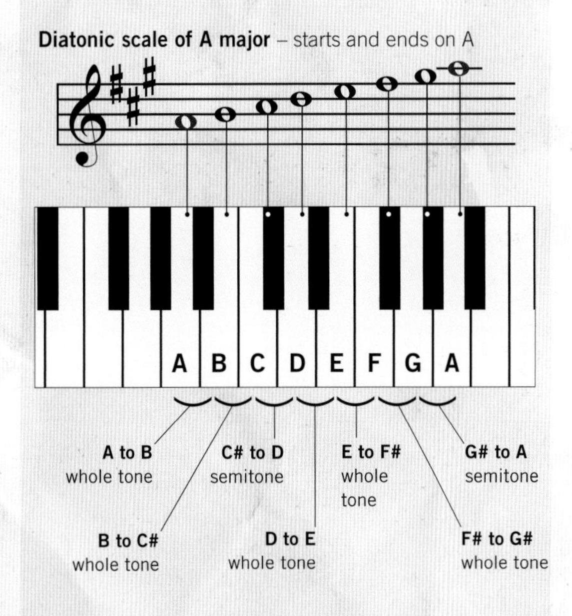

Diatonic scale of A minor – starts and ends on A

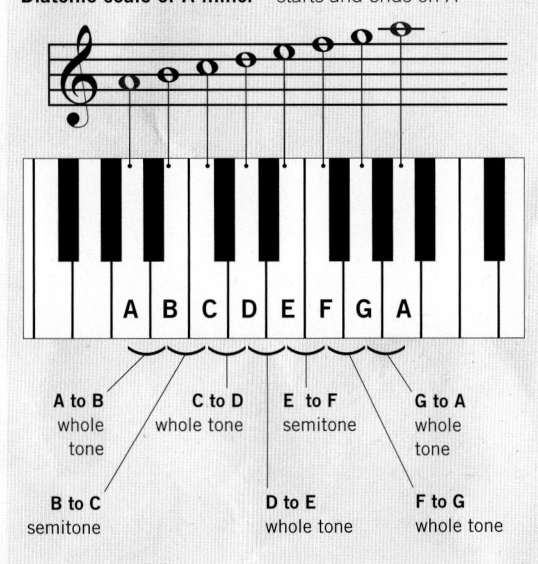

Diatonic scale of F# minor – starts and ends on F#

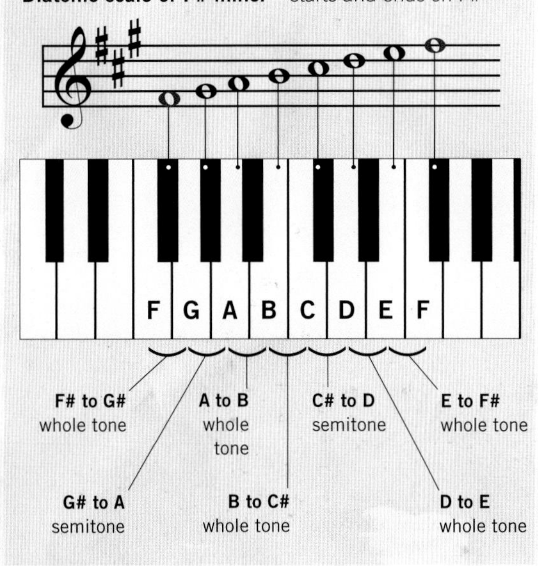

Rhythm and tempo

A number of Italian expressions are used to describe the beat or pulse of the music, and the tempo, which is the speed of the beat.

Grave	Very slowly
Lento	Slowly
Largo	Broadly
Larghetto	Rather broadly
Adagio	Leisurely
Andante	Walking pace
Moderato	Moderately
Allegretto	Fairly fast
Allegro	Fast
Vivace	Lively
Presto	Very fast
Prestissimo	As fast as possible

Dynamics

Other Italian terms are used to describe the intensity or volume of sound required. These are often abbreviated.

Pianissimo (pp)	Very quietly
Piano (p)	Quietly
Mezzo piano (mp)	Moderately quietly
Mezzo forte (mf)	Moderately loudly
Forte (f)	Loudly
Fortissimo (ff)	Very loudly
Crescendo <	Increasing in volume
Diminuendo >	Decreasing in volume

The length of an organ's pipes relate directly to the notes they produce.

Musical Scores

For orchestral or ensemble music, composers create a manuscript musical score, which carefully details the parts to be played by each instrument. This sheet of manuscript annotated by Wolfgang Amadeus Mozart (1756-91) shows part of his final orchestral dance *Il Trionfo delle Donne* (K607). Mozart designated the part to be played by each instrument, and the key and tempo in which it should be played.

In his short life Mozart composed 655 pieces of music, including 59 symphonies, 176 chamber pieces, and 23 operas.

Musical notation

Manuscript scores such as this show the complexity of abstract thought required to communicate accurately a detailed musical concept to ensemble musicians. A prodigiously talented composer, and an outstanding pianist, Mozart was accomplished in all styles and genres of music (his greatest love was opera). He was one of the first composers to become an independent, international superstar.

Key signature

This is a group of flat or sharp signs placed after the clef symbol on the staff to indicate the key in which the composition has been written. Their positions indicate which notes are to be consistently flattened or sharpened throughout, unless indicated otherwise, thus laying down the prevailing 'tonality' of the piece. This piece is in E♭ major, but other key signatures are marked as follows:

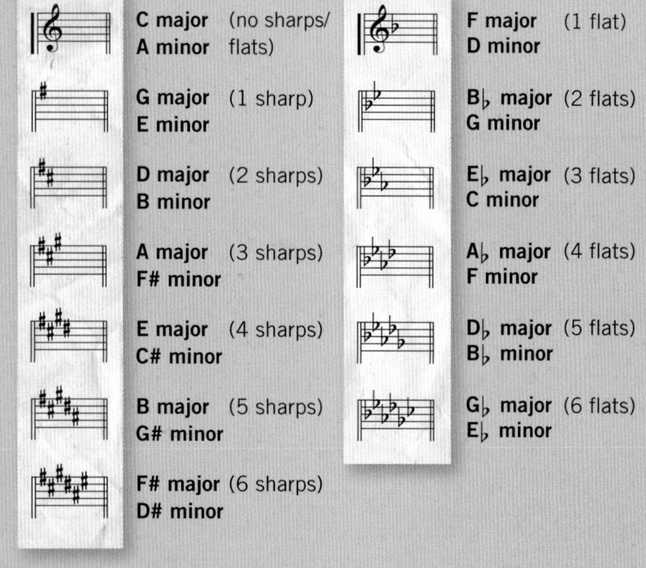

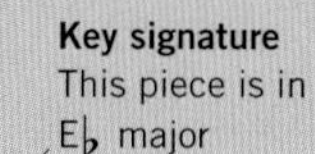

Meter

A musical composition is organized by a pattern of regular pulses that together are referred to as the 'meter.' One completed pattern is contained within a 'bar.' The overall meter is indicated after the clef and key signature by a stacked 'time signature' in the form of a fraction such as 2/2, 2/4, 3/4, 4/4, 3/8, 6/8, 9/8, and so on, positioned after the symbols for treble or bass clef. The top number indicates the number of beats in the bar, the bottom number the note value of each beat in the bar. Thus, 2/4 denotes two beats to a bar, each beat being a crotchet (or quarter note). The sign 'c' at the start of the first staff (stave) means the same as 4/4, and the sign '¢' means 2/2.

Note values and rests

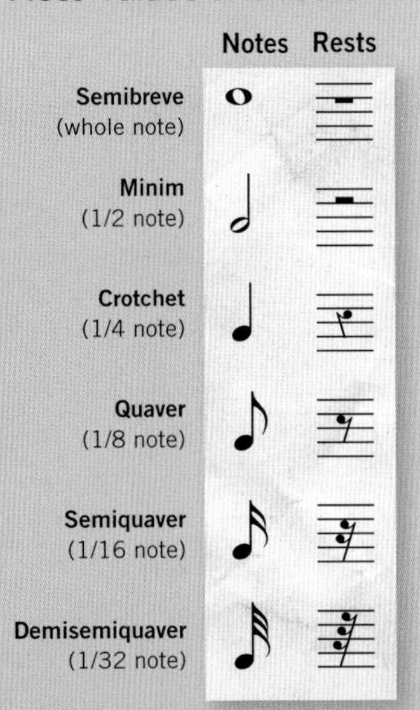

Each of these symbols displays how long a note lasts: one note is equal to a whole bar, so if the time signature is 4/4, a semibreve will last four beats, a crotchet one beat, and so on.

Additional symbols of notation

Accidentals are sharp, flat, or neutral notes that vary from the key signature (*opposite*).

Ties are used to demonstrate an undescribable note value. Here, the two notes together will last for 5/16 of a bar (1/4 + 1/16), and will be played as one note.

Slurs are used to 'blend' the notes into a continuous sound – there should be no audible gap.

Staccato is essentially the opposite of a slur: each note is short, clear-cut, and punchy.

Tabs

Although the modern rock composer Frank Zappa (1940-93) insisted that his innumerable band members should be able to 'read' music, he was an exception. Many modern popular musicians are self-taught, and tend to play by ear or example. The prevalence of the guitar as the main harmonic instrument, and the advent of the Internet, has made many popular music scores available by the development of 'tabs' in the form of fingering patterns.

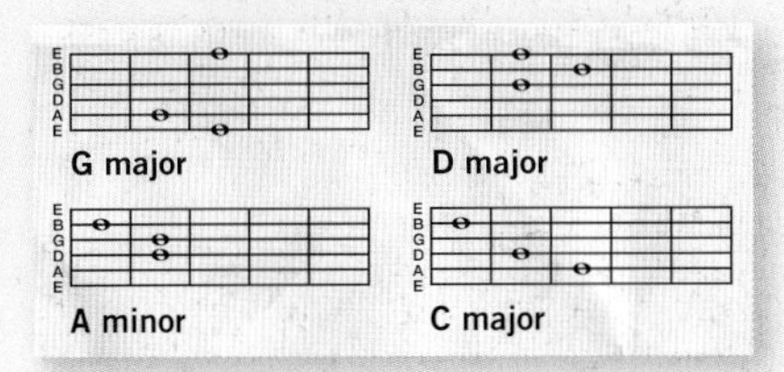

Modern challenges

Technical variations in notation also occur based on the performance medium (vocal, orchestral, electronic, and so on); the musical genre (string quartet, symphony, concerto, cadenza); and modern experimental techniques in vocal, instrumental, or multimedia performances, in which instruments can be 'played' or used in unconventional ways – for example, in John Cage's piano music.

John Cage's scores include *Composition for a Tape Recorder* (*above*) and complete silence in *4' 33"*.

ANIMAL TALK

Take a walk in a rain forest and the cacophony that surrounds you is overwhelming. Frogs surround a pond in a temperate zone and create a terrible din. Wherever there are animals there is noise, color, movement, and scent. Communication is taking place at every level, in every form. If we could fully decode this symphony of signals our understanding of the natural world would be immeasurably increased. Most of the many patterns of communication animals use are inherited, but they are often complex, finely tuned, surprising, and strangely beautiful.

Human to animal

Humans have a long history of training and domesticating animals dating back around 20,000 years. Interspecies communication is vital in this, but how much we can actually 'talk to the animals' is debated. Dogs and cats have an uncanny ability to read human moods. The ability of primates and cetaceans to communicate with us is well documented but, more surprisingly, birds can learn to communicate to a high degree as well. Dr. Irene Pepperberg at the University of Arizona trained an African gray parrot named Alex for 30 years. Alex was a good mimic and he learned the names of 50 objects plus attributes that included seven colors, five shapes, relative size (bigger/smaller), the material the objects were made from, and numbers up to six. He also understood sameness, difference, and absence. He would speak short sentences putting words in their correct order, and make up novel, syntactically correct sentences. He even made up new words combining syllables of other words to name new objects, coining the word 'banerry,' a combination of 'banana' and 'cherry,' for a juicy red apple he was given. Perhaps birds are not so bird-brained after all. Alex died suddenly in 2007, but Dr. Pepperberg's work continues with other African grays.

Sound

Humans are frequently only aware of other animals because they make a noise. Many animals, however, use sound below or above the register of our ears. Elephants were once thought to have a form of extrasensory perception: related groups maintain parallel tracks, well beyond their range of vision, through all types of terrain, and yet turn simultaneously to come together – all with no apparent communication. We now know that they communicate with rumbles below the register of our ears (infrasound). This low frequency sound travels further and is less impeded by obstacles than higher registers, enabling elephants to hear each other miles apart. Hippos produce infrasound calls above and below the waterline – infrasound travels further and faster underwater. Other animals known to use infrasound include rhinos, giraffes, alligators, lions, tigers, okapi, and some birds. Sensitivity to low-frequency sound may also explain the ability of animals to detect natural disasters such as earthquakes and tsunamis.

At the other end of the spectrum, many animals, including dolphins, bats, birds, and insects, can emit and detect sounds of a higher frequency (ultrasound). This is often associated with forms of echolocation.

The flat, fatty pads that form the elephant's feet are also thought to be sensitive receptors, capable of detecting infrasound vibrations.

Good vibrations

Messages may also be sent by vibration. Male stone flies tap out rhythms on a branch which are answered by receptive females with a complementary rhythm. Certain male web spiders pluck a soothing love song on a female's web to initiate mating (and avoid being eaten). Jumping spiders and wolf spiders use vibration alone, or add a leg-waving semaphore to attract a mate. The vibrating 'timbals' of the male cicada produce sounds with a volume of up to 120 decibels at close range – the pain threshold for human ears. Crickets rub their wings together and mate only with other crickets of the same species who produce the correct 'sound.' Some species are so finely attuned to other members that they will more readily mate with fellow hybrids than either of their parent's stock. Chirping often attracts predators: the parasitic ormia fly has evolved a sensitive eardrum and homes in to lay its larva on a singing male cricket.

In the oceans cetaceans produce a wide range of sounds. Salt water transmits infrasound better than freshwater. Blue whale calls travel thousands of miles through the open ocean. Dolphins have signature whistles that may function as names. Male humpback whales (*below*) sing long and complex songs in their breeding grounds.

Scents and pheromones

Many animals have highly adapted sensory organs that allow them to communicate using chemical messages. Ants leave trails of pheromones to guide fellow ants to food sources. Alarm pheromones provoke attack behavior, helping them to work together to overcome predators, while some emit pheromones that provoke rival ants to fight among themselves. Among mammals, scents are often left to mark territorial boundaries, often in the form of urine or feces. Other information communicated by scent may concern an individual's social status, sex, breeding condition, and state of health. A pungent message that reads 'keep away' is powerfully communicated by skunks and stink badgers – they are rarely attacked.

Birds of paradise are among the most spectacular birds to display brightly colored markings to attract a mate. Songbirds have a 'learned' component to their communication: baby birds raised in isolation never learn to sing properly, while females are attracted to males that more faithfully reproduce the songs of their fathers. Regional dialects have even been noted, as have food calls and alarm calls.

Visual communication

Most animals convey information visually. A dog with its teeth bared and hackles raised is hard to mistake as friendly. Primates use combinations of gestures, sounds, and facial expressions, and male fiddler crabs wave their claws to entice a female. Squid and cuttlefish have highly developed eyes and can communicate by flashing fast-changing patterns across their skin, when not blending into the background.

Poisonous animals are often brightly colored to warn off predators. These colors can be mimicked by other species that are not poisonous who thereby 'borrow' the defense mechanism, such as wasps and hoverflies, or the venomous coral snake (*top*) and the harmless milk snake (*bottom*).

The honeybees dance

Naturalists from Aristotle to recent times have puzzled over the honeybee's ability to transmit information about food sources to other bees in the hive. The mystery was finally unraveled by Karl von Frisch in the early 20th century. He observed that bee scouts, on returning to the hive from a source of pollen or nectar, dance a straight line shimmy. They then circle back to the beginning, first to the left, then to the right, forming a figure eight. Other bees crowd in to 'watch,' sensing air currents with their antennae. The straight line portion of the dance conveys information about the direction of the food source relative to the sun. The speed of the shimmy and the number of cycles per minute indicate distance from the hive. These results have been confirmed more recently using man-made, active, bee replicas placed in hives that have successfully conveyed information about target food sources by dancing the amazing dance of the honeybee.

Direction
The scout bee shimmies in the direction of food source.

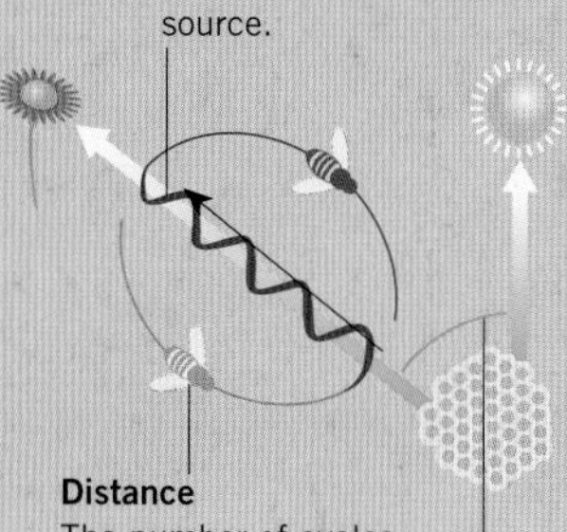

Distance
The number of cycles per minute indicates the distance from the hive.

Location
The angle specifies the degree to left or right of the sun.

EXTRATERRESTRIALS

While Steven Spielberg's E.T. found he could communicate with his young earthling companions quite easily, the very real challenge of anticipating ways in which extraterrestrial intelligences might communicate with us, or *vice versa*, has preoccupied scientists, astronomers, and science fiction writers for the last half century – a preoccupation that became more pronounced as humans took their first steps into space.

Quasars and pulsars

After the development of radio astronomy in the 1950s, the phenomenon of radio broadcast messages emanating from far beyond our solar system fascinated many scientists. Were these messages sent by an alien intelligence? For many years it was thought that they were, and many attempted to decrypt the signals.

By 1960 it had become clear (although the sources were still invisible) that these radio waves, known as pulsars, were caused by redshift sources of electromagnetic energy broadcast by supermassive black holes formed as galaxies are created, known as quasars, a contraction of 'quasi-stellar radio source.' Among the first to be accurately identified was 3C 273, some 2.44 billion light years away. By the time the radio messages are detectable, the galaxy may well have formed and already become extinct.

Messages to outer space

A number of attempts to summarize, in a 'universally' understandable coded form, our terrestrial existence, our position in, and our knowledge of the universe have been developed and propelled into space. Each has been limited by the available technology at the time of the launch.

The Pioneer plaques

Both unmanned space probes designed to travel to and beyond the edge of our solar system, *Pioneer 10* (1972) and *Pioneer 11* (1973) carried a plaque designed by astrophysicists Carl Sagan and Frank Drake. Engraved on gold-anodized aluminum, the plaques measured 9 inches (23 cm) by 6 inches (15 cm) and were mounted on the external struts of the craft.

"This is a present from a small, distant world... "

MESSAGE FOM US PRESIDENT JIMMY CARTER ON THE *VOYAGER* GOLDEN RECORD.

Hydrogen
Schematic representation of the hyperfine transition of hydrogen, thought to be the most abundant element in the universe. The binary digit 1 is used to indicate the spinflip transition of a hydrogen atom from its electron state, indicating both a unit of measure and of time.

Spacecraft
The outline of the Pioneer spacecraft was included to indicate the relative size of the human figures.

Human beings
The human male and female were originally to have held hands, but it was thought they may be interpreted as a single being. Racial stereotypes were avoided, but the humans appear to be Caucasian. The man is shown with a gesture of greeting, which also indicates his opposable thumb. The woman's genitals are not indicated, a last-minute modification of the design to appease NASA.

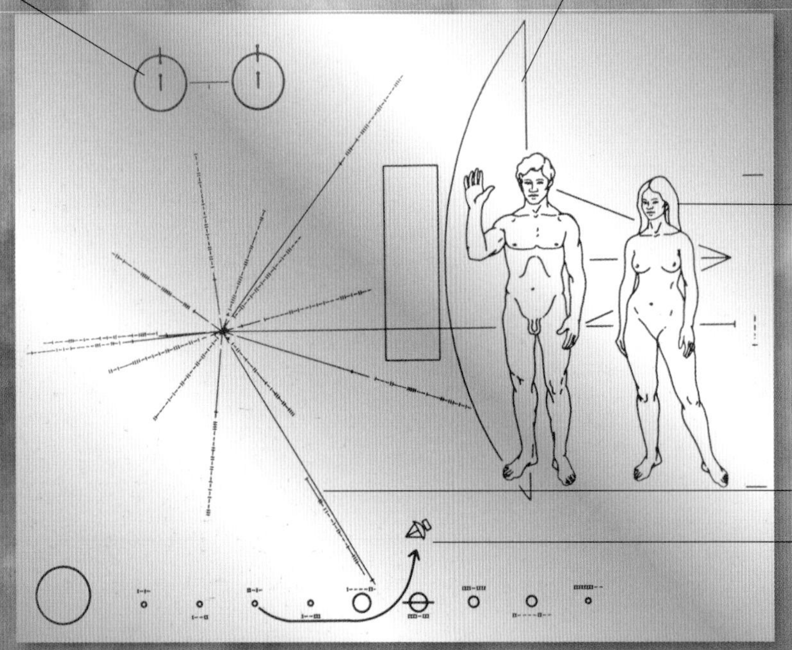

Galaxy
The plan indicates the position of our Sun in relation to the center of the galaxy, with 14 pulsars and their frequency periods indicated. From this, an alien intelligence might calculate the position of our solar system.

Solar system
The solar system is laid out in linear form, showing the order of the planets orbiting the Sun. The trajectory of *Pioneer* is also shown.

The Arecibo message

A radio broadcast was transmitted upon the reopening of the Arecibo radio telescope in 1974, aimed at star cluster M13, about 25,000 light years away. It was the most powerful man-made signal ever transmitted. It comprised 1,679 binary digits, a number chosen because it is the product of two prime numbers, 73 and 23 (a semiprime), and therefore could only be broken down into a grid of 73 by 23, or *vice versa*. In fact 73 rows in 23 columns was the correct algorithm for decoding the message. In turn, the broadcast lasted exactly 1,679 seconds, and was not repeated.

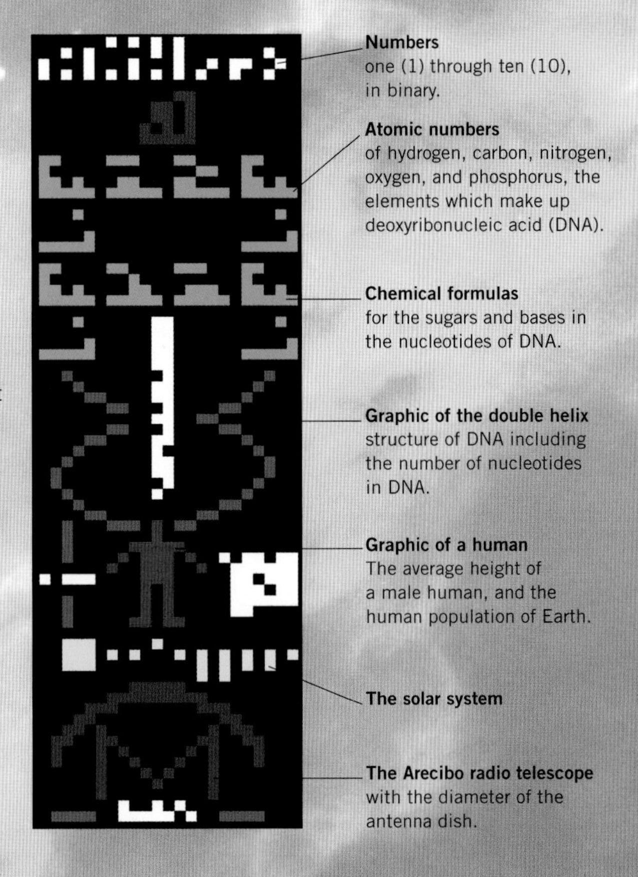

Numbers one (1) through ten (10), in binary.

Atomic numbers of hydrogen, carbon, nitrogen, oxygen, and phosphorus, the elements which make up deoxyribonucleic acid (DNA).

Chemical formulas for the sugars and bases in the nucleotides of DNA.

Graphic of the double helix structure of DNA including the number of nucleotides in DNA.

Graphic of a human The average height of a male human, and the human population of Earth.

The solar system

The Arecibo radio telescope with the diameter of the antenna dish.

Cosmic Calls

A series of nine Cosmic Call messages were broadcast in 1999, and in 2003, aimed at various star clusters. These incorporated the original Arecibo message, with a number of other digital text, video, and image files, including the 'Rosetta Stone,' developed by Stephane Dumas and Yvan Dutil covering a wide range of mathematical functions, formulas, and processes.

EXPLANATION OF RECORDING COVER DIAGRAM

THE DIAGRAMS BELOW DEFINE THE VIDEO PORTION OF THE RECORDING

Voyager Golden Record

An engraved phonographic record disk was developed by the same team who designed the *Pioneer* plaques to be placed on the *Voyager 1* and *Voyager 2* spacecraft launched in 1977. It contained 115 analog visual images, samples of 55 of the world's languages, 90 minutes of music (unfortunately EMI opposed the inclusion of The Beatles' *Here Comes the Sun*), some radio broadcasts, and a message in Morse code. It was coated in uranium-238 to allow it to be dated, and had a sleeve with various playing instructions on it.

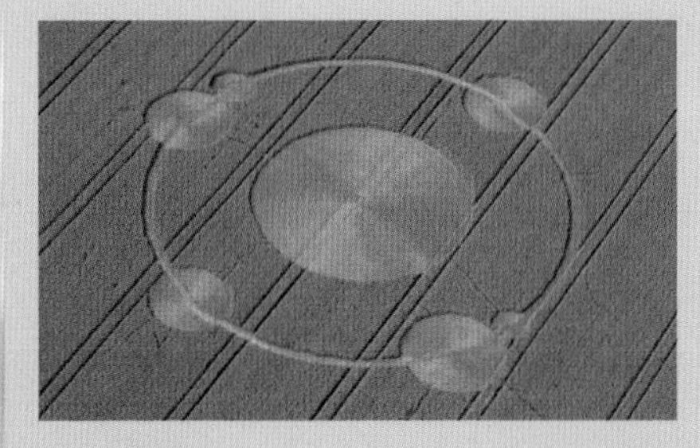

Crop circles

While reports of alien encounters and abductions mainly seemed to occur in the American Southwest, the crop circle phenomenon was originally British. From the mid-1970s, bizarre patterned imprints in flattened cereal crops appeared in spring and early summer. By the 1990s the patterns were becoming more elaborate and widespread, appearing in Russia, Japan, and North America. Considerable speculation ensued: were these etched by passing (or landing) extraterrestrial craft? Did they embody some secret, coded message for the farmers of Earth? The theories neatly tied in to New Age ideas concerning ley lines, stone circles, and other 'inexplicable' phenomena, until some British perpetrators (usually using planks attached to their feet to compress the plants) held their hands up. It was all a hoax.

Star words

Direct interaction with aliens is likely to prove tricky. H.G. Wells identified the problem in his novel *The War of the Worlds* (1898). The protagonists in the 1960s cult TV series *Star Trek* had access to a universal translator system, an idea often toyed with by science fiction writers. Several languages have been developed by scientists, including the mathematically-based systems Astraglossa (1953), Lincos (Lingua cosmica, 1960), and algorithmic messages, sets of mathematical and logic symbols which can form a basic programming language, designed as a two-way communication device between 'us' and 'them.'

12

The impulse to find hidden meanings or values in everything from historical texts to everyday events has long been the province of soothsayers and conspiracy theorists.

imaginary codes

The desire to invent alternative languages and embedded messages has become a prime preoccupation of mystery and fantasy writers as well as con artists. The appeal of a concealed significance is undeniable, as is the difficulty of distinguishing the genuine from the imaginary.

Modern Magic and Mayhem

In spite of religious censure and the general rationalism of the industrial age, in the late 19th century necromancy (*see page 56*) exploded into a vogue phenomenon known as Spiritualism – a spurious cult of séances (*right*), mediums, and ghostly apparitions which fascinated the Victorians. Despite repeated exposures of fraud, Spiritualism was embraced as a surrogate religion by doctors, university professors, clergymen, housewives, and the intelligentsia, such as Sir Arthur Conan Doyle and the Nobel Prize-winning physicist, Lord Rayleigh. As with the alchemists and necromancers of an earlier age, an obsession with arcane alphabets and codes containing magical properties was shared by its enthusiasts. Possibly this was a spiritual reaction to the threat of a rising secular, scientific culture, but the allure of arcane rituals, hidden meanings, and magical coded languages has persisted to present times in various forms.

Spiritualism and Theosophy

Spiritualism began in 1848 in Hydesville, N.Y. when the teenage Fox sisters demonstrated the purported ability to commune with the spirits of the dead. The mysterious Russian aristocrat and medium, Madame Blavatsky (1831-91), was a world traveler and fearless champion of the modern occult revival. Despite her dubious reputation, in the 1870s Blavatsky founded the Theosophical Society and anticipated many a New Age religion.

The Golden Dawn

At the turn of the 19th century, the revival in magical practices was exemplified by The Hermetic Order of the Golden Dawn, which began in England but spread rapidly. They advocated study of the Rosicrucians (*see page 58*), the Tarot, astrology, and geomancy as a basic tenet of admission, and among their members were senior civil servants and the poet W.B.Yeats. The so-called 'Cipher Manuscripts,' a collection of some 60 folios describing magical initiation rites, written in a mixture of Agrippa's Theban alphabet (*see page 57*) and Hebrew, containing crude drawings of symbols and ritual equipment, became a touchstone for the group. The background of the manuscripts are hotly debated, but seem likely to be a mid- to late-19th-century confection of intriguingly coded nonsense.

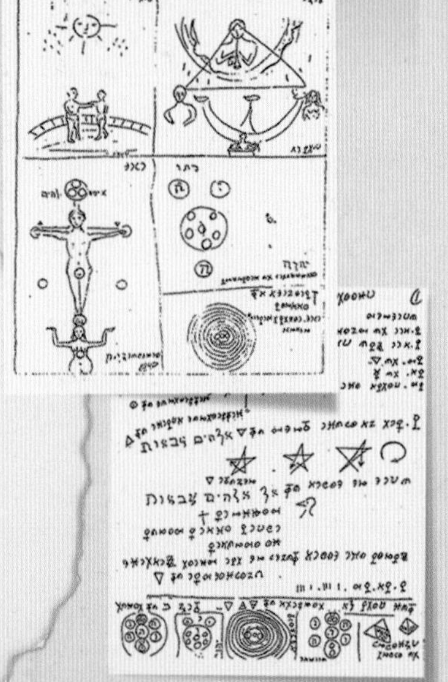

The 'Cipher Manuscripts' appear to contain magic formulae described in an arcane ciphertext.

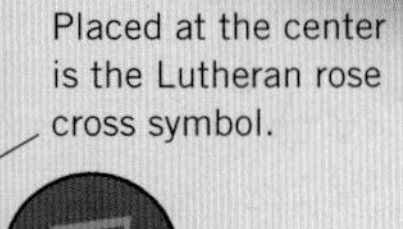

The rose cross
Placed at the center is the Lutheran rose cross symbol.

Magic writing
A heady mixture of Hebrew and other alphabetic letters are strategically located around the design.

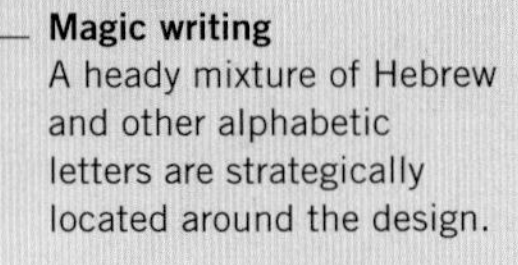

The pentangle
Variations on the pentangle and the Star of David appear alongside symbols for masculine and feminine.

The members of the Golden Dawn were keen advocates of Rosicrucian thought, and developed their own potpourri of mysterious Hermetic symbols to embed 'meaning' into their publications.

007 and all that

A surprising number of apparently quite rational people have been associated with bizarre Hermetic practices. English thriller writer Dennis Wheatley, a contemporary associate and then adversary of the mystic Aleister Crowley, and a member of Britain's Secret Intelligence Service (SIS) during World War II, clearly believed that occult practices had value. *The Devil Rides Out* and *The Ka of Gifford Hillary* are just two of many of his novels in which satanic or necromantic rites are seen to work. In later books, such as *They Used Dark Forces*, he even portrayed Hitler as a Satanist.

Some have argued that Ian Fleming, also a secret serviceman, and the creator of James Bond, seeded his novels with coded references to arcane and magical practices: the agent's code number 007 has magical significance, as does his upgrade number 7777 in *You Only Live Twice*. The secret code Bond is sent to Japan to access in the same novel is called 'Magic 44.' One of the possible models for Bond's spymaster boss 'M' was Maxwell Knight of MI5, also a known associate of Aleister Crowley. During his time at Naval Intelligence, Fleming proposed that Crowley feed falsified horoscopes to Nazi henchman Rudolf Hess, and that the Enochian alphabet (*see page 57*) be used as a coding device.

Best-selling thriller writers Dennis Wheatley (1897-1977) and Ian Fleming (1908-64, *below*) were both involved in intelligence activity during World War II, and each hovered on the fringe of what was often seen as a 'fashionable' fascination with the occult.

The 'Great Beast'

Self-described, but also known by the press as 'The Wickedest Man in the World,' Aleister Crowley (1875-1947) came from a privileged, but religiously strict, background and drifted into magic and opium addiction via the Golden Dawn group after attending Cambridge University. He soon split from them to follow his own path within such organizations as the A ∴ A, and the Ordo Templi Orientis (O.T.O.). Crowley published many occult texts, including his *Book of Thelema* and his own 'translation' of the 'Cipher Manuscripts,' (*opposite*) and devised his own hexagram (*right*).

A	B	C	D	E	F	G
H	I	J	K	L	M	N
O	P	Q	R	S	T	U
V	W	X	Y	Z		

Crowley designed his own 'dagger' alphabet to be used to invoke spirits in magical ceremonies, reminiscent of similar magical alphabets devised by Agrippa and John Dee (*see page 57*).

The Bible Code

The renewed left-field interest in a Bible code (also known as Torah code) and its decipherment is a modern interpretation of a classic obsession; from medieval Kabbalists (*see page 54*) to British scientist Isaac Newton and the French theologian and philosopher Blaise Pascal (1623-62), both Jewish and Christian Bible scholars have posited that the Old Testament may contain coded messages. In modern academia, this has proved a contentious issue among theologians, scientists, and mathematicians, as well as being the subject of media hype.

Origins of modern Bible code

In 1988 three mathematicians at the Hebrew University, Jerusalem published a paper in *The Journal of the Royal Statistical Society*, and again in 1994 in the academic journal, *Statistical Science*, called 'Equidistant Letter Sequences in the Book of Genesis.' Doron Witztum, Yoav Rosenberg, and Eliyahu Rips used computer programs to extract 'meaningful' messages encoded in the first book of the Jewish Torah; the Torah (also known as the Five Books of Moses, or the Pentateuch) contains Genesis, Exodus, Leviticus, Numbers, and Deuteronomy, and is written in Hebrew. The paper (often known as the WRR paper) derived from the work of Rabbi Michael Dov Weissmandl (better known for his attempts to save the Jews of Slovakia from the Nazis during the Holocaust) who as a student spent time manually working out Bible codes in the Torah. According to its creators, Equidistant Letter Sequence method is only applicable to the Torah and they strongly believe that any attempt to apply this method to English translations of any parts of the Jewish or Christian Bible is invalid.

One example of many prophecies found in the Torah using ELS foresaw the US atom bomb attack on Hiroshima in 1945, dated American-style.

Shape	Word	Translation	Verse	Position	Skip
	יפנ	Japan	Numbers 25:13	230779	6266
	שואהאטומית	Atomic holocaust	Numbers 29:9	237020	-3133
	יפנ	Japan	Numbers 29:9	237042	3
	רשת	8/6/1945	Deuteronomy 8:19	265216	-1

Equidistant Letter Sequence

In simple terms, to find an ELS, pick somewhere to start (any letter will do) and a skip number (a skip number is the number of letters skipped between selected letters). Then choose letters from the text at the equal intervals designated by the skip number. If this is applied to an entire book (Genesis was the original book used in the WRR paper), the result is a string of letters; infinite strings of letters can be found by changing the starting point, and the value of the skip number. By reading the string horizontally, vertically, backwards, or diagonally, it is possible to find names or dates or whatever it is you are looking for; this is made much simpler by running a computer program on the text.

In Darren Aronofsky's 1998 movie *Pi*, mathematician Maximillian Cohen, the movie's central character, meets a Hasidic Jew who conducts mathematical research on the Torah. He tells Max that the Torah is composed of a string of numbers that are in fact a code sent by God.

The Bible Code books

Michael Drosnin, an American journalist, began investigating Bible code in 1992; this culminated in his notorious and best-selling book, *The Bible Code*, published in 1997, and was followed by his equally profitable *The Bible Code II* in 2002. Drosnin claims to have used Equidistant Letter Sequence to decode what he believes are predictions and prophecies of important world events that affect not just the Jews but society as a whole. He alleges that the Bible predicted the assassination of Yitzak Rabin, as well as the Kennedy brothers. His claims have been strongly refuted by one of the authors of the original ELS paper, Eliyahu Rips, as well as Harold Gans, a retired Defense Department cryptologist, Professor Menachem Cohen, an Israeli Bible expert, and Professor Brendan McKay in the Department of Computer Science at the Australian National University. McKay used ELS methodology on the Hebrew translation of *War and Peace* to link his own name and date of birth in his paper 'Tolstoy Loves Me.' He states that Drosnin used no methods that can be classified as scientific, nor did he find anything that could not be found in any book, English or Hebrew, if you have the motivation to look hard enough.

> "When my critics find a message about the assassination of a prime minister encrypted in *Moby-Dick*, I'll believe them."
>
> MICHAEL DROSNIN, *NEWSWEEK*, 1997.

Assassinations predicted in *Moby-Dick*

Whereas Weissmandl and the authors of the WRR paper (*see opposite*) insisted that the ELS technique was only valid when applied to the Torah in Hebrew, the fact is that it can reveal 'hidden' messages in any text. Drosnin's methodology, when applied by McKay to Herman Melville's *Moby-Dick* (1851) revealed predictions of the following killings:

- **Prime Minister Indira Gandhi**
- **Soviet exile Leon Trotsky**
- **The Reverend Martin Luther King**
- **President John F. Kennedy**
- **President Abraham Lincoln**
- **Prime Minister Yitzak Rabin**
- **The death of Princess Diana**

This does not mean that *Moby-Dick* really does predict these events, it just proves that using Drosnin's 'methodology' you can find whatever you are looking to find without contradicting the laws of probability!

The Beale Papers

The legend

A buffalo-hunting expedition in the American Southwest in 1817: one night whilst supper was being cooked one of the party accidentally discovered a rich gold deposit. Noting the location, the party returned and mined for 18 months, accumulating a vast amount of gold and silver bullion. Fearing for the safety of their hoard in this lawless territory, the party decided that it needed to be hidden somewhere secure. Several of their number, including one Thomas J. Beale, were sent east to find a suitable location, finding their way to Lynchburg, Virginia.

In 1862 an elderly man, Robert Morriss, former proprietor of the Washington Hotel in Lynchburg, Virginia, entrusted a sheaf of papers to a friend. One was a letter, describing the discovery of gold in the American Southwest, and its subsequent burial in a safe place in Virginia. The other three sheets were covered in numbers, numerical substitution ciphers. In 1885, a 23-page booklet was printed by James B. Ward, an 'agent' for Morriss' anonymous 'friend,' and published in Lynchburg. It purported to describe a mystery dating back over half a century involving hidden treasure and a convoluted plot. The booklet is the only source for both the background to the Beale story and for the ciphers that lie at the heart of the mystery. To this day, the booklet remains an enigma. Nobody has detected the key to the first and third ciphers, although many have unsuccessfully attempted to locate the treasure based on the scraps of information in the 'decrypted' second cipher. Theories abound: or was it simply a huge hoax by the printer of the booklet?

The Beale ciphers

Only the second of the three sheets of ciphertext was decrypted, apparently by Morriss's anonymous friend. He assumed that the numbers on all three sheets represented individual letters, however the numbers were too numerous and varied to be a straightforward alphabetic substitution cipher. Maybe it was a book cipher (*see page 79*)? The problem was what book, or which text? The author of the pamphlet claims that he spent years and a fortune accumulating and testing popular texts, and hit upon one which seemed to work: the Declaration of Independence. By numbering every word in the document, then assuming the numbers in the ciphertext indicated the initial letter of each word, a coherent message emerged.

317, 8, 92, 73, 112, 89, 67, 318, 28, 96, 107, 41, 631, 78, 146, 397, 118,
98, 114, 246, 348, 116, 74, 88, 12, 65, 32, 14, 81, 19, 76, 121, 216, 85,
33, 66, 15, 108, 68, 77, 43, 24, 122, 96, 117, 36, 211, 301, 15, 44, 11,
46, 89, 18, 136, 68, 317, 28, 90, 82, 304, 71, 43, 221, 198, 176, 310,
319, 81, 99, 264, 380, 56, 37, 319, 2, 44, 53, 28, 44, 75, 98, 102, 37,
85, 107, 117, 64, 88, 136, 48, 151, 99, 175, 89, 315, 326, 78, 96, 214,
218, 311, 43, 89, 51, 90, 75, 128, 96, 33, 28, 103, 84, 65, 26, 41, 246,
84, 270, 98, 116, 32, 59, 74, 66, 69, 240, 15, 8, 121, 20, 77, 89, 31,
11, 106, 81, 191, 224, 328, 18, 75, 52, 82, 117, 201, 39, 23, 217, 27,
21, 84, 35, 54, 109, 128, 49, 77, 88, 1, 81, 217, 64, 55, 83, 116, 251,
269, 311, 96, 54, 32, 120, 18, 132, 102, 219, 211, 84, 150, 219, 275,
312, 64, 10, 106, 87, 75, 47, 21, 29, 37, 81, 44, 18, 126, 115, 132, 160,
181, 203, 76, 81, 299, 314, 337, 351, 96, 11, 28, 97, 318, 238, 106, 24,
93, 3, 19, 17, 26, 60, 73, 88, 14, 126, 138, 234, 286, 297, 321, 365,
264, 19, 22, 84, 56, 107, 98, 123, 111, 214, 136, 7, 33, 45, 40, 13, 28,
46, 42, 107, 196, 227, 344, 198, 203, 247, 116, 19, 8, 212, 230, 31, 6,
328, 65, 48, 52, 59, 41, 122, 33, 117, 11, 18, 25, 71, 36, 45, 83, 76,
89, 92, 31, 65, 70, 83, 96, 27, 33, 44, 50, 61, 24, 112, 136, 149, 176,
180, 194, 143, 171, 205, 296, 87, 12, 44, 51, 89, 98, 34, 41, 208, 173,
66, 9, 35, 16, 95, 8, 113, 175, 90, 56, 203, 19, 177, 183, 206, 157,
200, 218, 260, 291, 305, 618, 951, 320, 18, 124, 78, 65, 19, 32, 124,
48, 53, 57, 84, 96, 207, 244, 66, 82, 119, 71, 11, 86, 77, 213, 54, 82,
316, 245, 303, 86, 97, 106, 212, 18, 37, 15, 81, 89, 16, 7, 81, 39, 96,
14, 43, 216, 118, 29, 55, 109, 136, 172, 213, 64, 8, 227, 304, 611, 221,
364, 819, 375, 128, 296, 1, 18, 53, 76, 10, 15, 23, 19, 71, 84, 120,
134, 66, 73, 89, 96, 230, 48, 77, 26, 101, 127, 936, 218, 439, 178,
171, 61, 226, 313, 215, 102, 18, 167, 262, 114, 218, 66, 59, 48, 27,
19, 13, 82, 48, 162, 119, 34, 127, 139, 34, 128, 129, 74, 63, 120, 11,
54, 61, 73, 92, 180, 66, 75, 101, 124, 265, 89, 96, 126, 274, 896, 917,
434, 461, 235, 890, 312, 413, 328, 381, 96, 105, 217, 66, 118, 22, 77,
64, 42, 12, 7, 55, 24, 83, 67, 97, 109, 121, 135, 181, 203, 219, 228,
256, 21, 34, 77, 319, 374, 382, 675, 684, 717, 864, 203, 4, 18, 92, 16,
63, 82, 22, 46, 55, 69, 74, 112, 134, 186, 175, 119, 213, 416, 312, 343,
264, 119, 186, 218, 343, 417, 845, 951, 124, 209, 49, 617, 856, 924,
936, 72, 19, 28, 11, 35, 42, 40, 66, 85, 94, 112, 65, 82, 115, 119, 236,
244, 186, 172, 112, 85, 6, 56, 38, 44, 85, 72, 32, 47, 63, 96, 124, 217,
314, 319, 221, 644, 817, 821, 934, 922, 416, 975, 10, 22, 18, 46, 137,
181, 101, 39, 86, 103, 116, 138, 164, 212, 218, 296, 815, 380, 412,
460, 495, 675, 820, 952

The first cipher has attracted the most attention as it is meant to provide the location of the buried treasure (*left*).

The third cipher claimed to identify the shareholders and their relatives, who should be reimbursed should the trove be found (*right*).

71, 194, 38, 1701, 89, 76, 11, 83, 1629, 48, 94, 63, 132, 16, 111,
95, 84, 341, 975, 14, 40, 64, 27, 81, 139, 213, 63, 90, 1120, 8, 15,
3, 126, 2018, 40, 74, 758, 485, 604, 230, 436, 664, 582, 150, 251,
284, 308, 231, 124, 211, 486, 225, 401, 370, 11, 101, 305, 139, 189,
17, 33, 88, 208, 193, 145, 1, 94, 73, 416, 918, 263, 28, 500, 538,
356, 117, 136, 219, 27, 176, 130, 10, 460, 25, 485, 18, 436, 65, 84,
200, 283, 118, 320, 138, 36, 416, 280, 15, 71, 224, 961, 44, 16, 401,
39, 88, 61, 304, 12, 21, 24, 283, 134, 92, 63, 246, 486, 682, 7, 219,
184, 360, 780, 18, 64, 463, 474, 131, 160, 79, 73, 440, 95, 18, 64,
581, 34, 69, 128, 367, 460, 17, 81, 12, 103, 820, 62, 116, 97, 103,
862, 70, 60, 1317, 471, 540, 208, 121, 890, 346, 36, 150, 59, 568,
614, 13, 120, 63, 219, 812, 2160, 1780, 99, 35, 18, 21, 136, 872, 15,
28, 170, 88, 4, 30, 44, 112, 18, 147, 436, 195, 320, 37, 122, 113, 6,
140, 8, 120, 305, 42, 58, 461, 44, 106, 301, 13, 408, 680, 93, 86,
116, 530, 82, 568, 9, 102, 38, 416, 89, 71, 216, 728, 965, 818, 2, 38,
121, 195, 14, 326, 148, 234, 18, 55, 131, 234, 361, 824, 5, 81, 623,
48, 961, 19, 26, 33, 10, 1101, 365, 92, 88, 181, 275, 346, 201, 206,
86, 36, 219, 324, 829, 840, 64, 326, 19, 48, 122, 85, 216, 284, 919,
861, 326, 985, 233, 64, 68, 232, 431, 960, 50, 29, 81, 216, 321, 603,
14, 612, 81, 360, 36, 51, 62, 194, 78, 60, 200, 314, 676, 112, 4, 28,
18, 61, 136, 247, 819, 921, 1060, 464, 895, 10, 6, 66, 119, 38, 41,
49, 602, 423, 962, 302, 294, 875, 78, 14, 23, 111, 109, 62, 31, 501,
823, 216, 280, 34, 24, 150, 1000, 162, 286, 19, 21, 17, 340, 19, 242,
31, 86, 234, 140, 607, 115, 33, 191, 67, 104, 86, 52, 88, 16, 80, 121,
67, 95, 122, 216, 548, 96, 11, 201, 77, 364, 218, 65, 667, 890, 236,
154, 211, 10, 98, 34, 119, 56, 216, 119, 71, 218, 1164, 1496, 1817, 51,
39, 210, 36, 3, 19, 540, 232, 22, 141, 617, 84, 290, 80, 46, 207, 411,
150, 29, 38, 46, 172, 85, 194, 39, 261, 543, 897, 624, 18, 212, 416,
127, 931, 19, 4, 63, 96, 12, 101, 418, 16, 140, 230, 460, 538, 19, 27,
88, 612, 1431, 90, 716, 275, 74, 83, 11, 426, 89, 72, 84, 1300, 1706,
814, 221, 132, 40, 102, 34, 868, 975, 1101, 84, 16, 79, 23, 16, 81,
122, 324, 403, 912, 227, 936, 447, 55, 86, 34, 43, 212, 107, 96, 314,
264, 1065, 323, 428, 601, 203, 124, 95, 216, 814, 2906, 654, 820, 2,
301, 112, 176, 213, 71, 87, 96, 202, 35, 10, 2, 41, 17, 84, 221, 736,
820, 214, 11, 60, 760

115, 73, 24, 807, 37, 52, 49, 17, 31, 62, 647, 22, 7, 15, 140, 47, 29, 107, 79,
84, 56, 239, 10, 26, 811, 5, 196, 308, 85, 52, 160, 136, 59, 211, 36, 9, 46,
316, 554, 122, 106, 95, 53, 58, 2, 42, 7, 35, 122, 53, 31, 82, 77, 250, 196,
56, 96, 118, 71, 140, 287, 28, 353, 37, 1005, 65, 147, 807, 24, 3, 8, 12, 47,
43, 59, 807, 45, 316, 101, 41, 78, 154, 1005, 122, 138, 191, 16, 77, 49, 102,
57, 72, 34, 73, 85, 35, 371, 59, 196, 81, 92, 191, 106, 273, 60, 394, 620,
270, 220, 106, 388, 287, 63, 3, 6, 191, 122, 43, 234, 400, 106, 290, 314, 47,
48, 81, 96, 26, 115, 92, 158, 191, 110, 77, 85, 197, 46, 10, 113, 140, 353, 48,
120, 106, 2, 607, 61, 420, 811, 29, 125, 14, 20, 37, 105, 28, 248, 16, 159,
7, 35, 19, 301, 125, 110, 486, 287, 98, 117, 511, 62, 51, 220, 37, 113, 140,
807, 138, 540, 8, 44, 287, 388, 117, 18, 79, 344, 34, 20, 59, 511, 548, 107,
603, 220, 7, 66, 154, 41, 20, 50, 6, 575, 122, 154, 248, 110, 61, 52, 33, 30,
5, 38, 8, 14, 84, 57, 540, 217, 115, 71, 29, 84, 63, 43, 131, 29, 138, 47, 73,
239, 540, 52, 53, 79, 118, 51, 44, 63, 196, 12, 239, 112, 3, 49, 79, 353, 105,
56, 371, 557, 211, 505, 125, 360, 133, 143, 101, 15, 284, 540, 252, 14, 205,
140, 344, 26, 811, 138, 115, 48, 73, 34, 205, 316, 607, 63, 220, 7, 52, 150,
44, 52, 16, 40, 37, 158, 807, 37, 121, 12, 95, 10, 15, 35, 12, 131, 62, 115,
102, 807, 49, 53, 135, 138, 30, 31, 62, 67, 41, 85, 63, 10, 106, 807, 138, 8,
113, 20, 32, 33, 37, 353, 287, 140, 47, 85, 50, 37, 49, 47, 64, 6, 7, 71, 33,
4, 43, 47, 63, 1, 27, 600, 208, 230, 15, 191, 246, 85, 94, 511, 2, 270, 20, 39,
7, 33, 44, 22, 40, 7, 10, 3, 811, 106, 44, 486, 230, 353, 211, 200, 31, 10, 38,
140, 297, 61, 603, 320, 302, 666, 287, 2, 44, 33, 32, 511, 548, 10, 6, 250,
557, 246, 53, 37, 52, 83, 47, 320, 38, 33, 807, 7, 44, 30, 31, 250, 10, 15,
35, 106, 160, 113, 31, 102, 406, 230, 540, 320, 29, 66, 33, 101, 807, 138,
301, 316, 353, 320, 220, 37, 52, 28, 540, 320, 33, 8, 48, 107, 50, 811, 7, 2,
113, 73, 16, 125, 11, 110, 67, 102, 807, 33, 59, 81, 158, 38, 43, 581, 138, 19,
85, 400, 38, 43, 77, 14, 27, 8, 47, 138, 63, 140, 44, 35, 22, 177, 106, 250,
314, 217, 2, 10, 7, 1005, 4, 20, 25, 44, 48, 7, 26, 46, 110, 230, 807, 191, 34,
112, 147, 44, 110, 121, 125, 96, 41, 51, 50, 140, 56, 47, 152, 540, 63, 807,
28, 42, 250, 138, 582, 98, 643, 32, 107, 140, 112, 26, 85, 138, 540, 53, 20,
125, 371, 38, 36, 10, 52, 118, 136, 102, 420, 150, 112, 71, 14, 20, 7, 24, 18,
12, 807, 37, 67, 110, 62, 33, 21, 95, 220, 511, 102, 811, 30, 83, 84, 305, 620,
15, 2, 10, 8, 220, 106, 353, 105, 106, 60, 275, 72, 8, 50, 205, 185, 112, 125,
540, 65, 106, 807, 138, 96, 110, 16, 73, 33, 807, 150, 409, 400, 50, 154,
285, 96, 106, 316, 270, 205, 101, 811, 400, 8, 44, 37, 52, 40, 241, 34, 205,
38, 16, 46, 47, 85, 24, 44, 15, 64, 73, 138, 807, 85, 78, 110, 33, 420, 505,
53, 37, 38, 22, 31, 10, 110, 106, 101, 140, 15, 38, 3, 5, 44, 7, 98, 287, 135,
150, 96, 33, 84, 125, 807, 191, 96, 511, 118, 40, 370, 643, 466, 106, 41, 107,
603, 220, 275, 30, 150, 105, 49, 53, 287, 250, 208, 134, 7, 53, 12, 47, 85,
63, 138, 110, 21, 112, 140, 485, 486, 505, 14, 73, 84, 575, 1005, 150, 200,
16, 42, 5, 4, 25, 42, 8, 16, 811, 125, 160, 32, 205, 603, 807, 81, 96, 405, 41,
600, 136, 14, 20, 28, 26, 353, 302, 246, 8, 131, 160, 140, 84, 440, 42, 16,
811, 40, 67, 101, 102, 194, 138, 205, 51, 63, 241, 540, 122, 8, 10, 63, 140,
47, 48, 140, 288

The second cipher (*above*), apparently decrypted by the author of the pamphlet, tantalizingly reveals the existence of the treasure, but little else (*opposite*).

The mysterious Mr. Beale

According to the booklet, Morriss described how a tall, dark, swarthy man named Thomas J. Beale arrived at his hotel in 1820. He wintered there, acquainting himself with various local characters and befriending the proprietor. In the spring he left, reappearing two years later, and wintering once again at the hotel. In the spring he left to hunt buffalo and grizzlies in the West, but not before entrusting Morriss with an iron strongbox, with the instruction that, should he fail to return within the next ten years, Morriss should break the lock. "You will find, in addition to the papers addressed to you, other papers which will be unintelligible without the aid of a key to assist you. Such a key I have left in the hand of a friend in this place, sealed and addressed to yourself, and endorsed not to be delivered until June 1832." Beale never returned.

For years Morriss resisted the temptation to open the box, and the promised 'key' never materialized. Finally, in 1845, Morriss succumbed to his curiosity, broke the lock, and discovered four sheets of paper. The first, in Beale's handwriting, described the discovery of the gold and silver deposits in the Southwest by Beale and his associates, and outlined their plan to hide the bullion in Virginia for safekeeping. The other three sheets were simply covered in numbers, evidently a cipher of some sort.

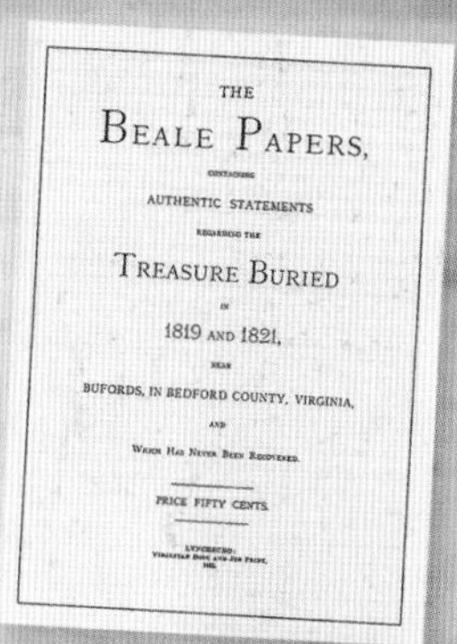

THE

BEALE PAPERS,

CONTAINING

AUTHENTIC STATEMENTS

REGARDING THE

TREASURE BURIED

IN

1819 AND 1821,

NEAR

BUFORDS, IN BEDFORD COUNTY, VIRGINIA,

AND

WHICH HAS NEVER BEEN RECOVERED.

PRICE FIFTY CENTS

The title page of the booklet printed in 1885.

Secrets of the Declaration of Independence

An initial test on the first paragraph of the Declaration illustrates the principle:

"When, in the course of human events, it becomes necessary for one
1 2 3 4 5 6 7 8 9 10 11 12

people to dissolve the political bands which have connected them with
13 14 15 16 17 18 19 20 21 22 23

another, and to assume among the powers of the earth, the separate and equal
24 25 26 27 28 29 30 31 32 33 34 35 36 37

station to which the laws of nature and of nature's God entitle them, a decent
38 39 40 41 42 43 44 45 46 47 48 49 50 51 52

respect to the opinions of mankind requires that they should declare the causes
53 54 55 56 57 58 59 60 61 62 63 64 65

which impel them to the separation."
66 67 68 69 70 71

Taking the first few numbers of the ciphertext, a pattern emerges:

115, 73 24, 807, 37, 52, 49, 17, 31, 62, 647, 22, 7, 15, 140, 47,
- - a - e d e p o s - t e d - n

29, 107, 79, 84, 56, 239, 10, 26, 811, 5,
t - - - o - n t - o

Even in this sample the beginnings of a message emerge with the word 'deposited.'

The second cipher decrypted

"I have deposited in the county of Bedford, about four miles from Buford's, in an excavation or vault, six feet below the surface of the ground, the following articles, belonging jointly to the parties whose names are given in [cipher] number 3, herewith:

The first deposit consisted of one thousand and fourteen pounds of gold, and three thousand eight hundred and twelve pounds of silver, deposited November, 1819. The second was made December, 1821, and consisted of nineteen hundred and seven pounds of gold, and twelve hundred and eighty-eight pounds of silver; also jewels, obtained in St. Louis in exchange for silver to save transportation, and valued at $13,000.

The above is securely packed in iron pots, with iron covers. The vault is roughly lined with stone, and the vessels rest on solid stone, and are covered with others. Paper number 1 describes the exact locality of the vault so that no difficulty will be had in finding it".

MYSTERY AND IMAGINATION

The concept of a hidden meaning – especially in a literary text – is almost irresistible. The 19th century saw a number of very popular fictions whose resolution depended on the cracking of a cipher or code. It was also the century that saw a huge growth in the popular press, newspapers often setting puzzles for their readers to solve. The principal exponents of the mystery genre, Edgar Allan Poe and Sir Arthur Conan Doyle, were both fascinated by codes, ciphers, and hidden messages, and featured them in their work.

Treasure maps

The search for a long-lost treasure, and the mysterious clues to finding it, provided the most common motif for many writers. Robert Louis Stevenson (1850-94, *above*) perfected the art in *Treasure Island* (1883), even providing the reader with a treasure map drawn up by Bartholomew, the Edinburgh Geographical Institute (*below*), including a number of riddles which had to be solved by the protagonists in order to track down the location of the treasure. This was an early example of an entirely imaginary map of a fictional location. Since then Tolkien's Middle Earth, C.S. Lewis' Narnia, Pratchett's Discworld, and many others have been mapped.

Mystery or imagination?

The American writer Edgar Allan Poe (1809-49) is regarded as the father of several literary genres – the 'tale,' the horror story, the detective or mystery story, and science fiction. Fascinated by the scientific developments of his day, Poe wove such themes into many of his tales, but cryptography was a particular obsession, so much so that he challenged readers of Philadelphia's *Alexander's Weekly Messenger* to send him cipher messages to decrypt. He never failed to meet the challenge, although his abilities were limited to substitution cryptograms. However, after solving 100 ciphers, he retreated, going on to publish two ciphertexts which he claimed had been sent in by a reader (but were probably conceived by Poe himself), offering the public the chance to win a competition if solved. They remained unsolved for 150 years until an Internet challenge produced the solutions.

Edgar Allan Poe counted cryptography as only one of his obsessions.

The *Gold-Bug*

Poe's tale, *The Gold-Bug* (1843, revised 1845), is a story of missing treasure, madness, and obsession; the key to the location of Captain Kidd's lost trove is a mixture of bizarre geographical locations and a message in the form of a seemingly complex cryptogram written with invisible ink on a piece of vellum found on a beach by the melancholy protagonist Legrand. The message read:

"53‡‡†305)) 6*;4826)4‡.)4‡) ;806*;48†8¶60))85;1‡(;:‡*8†83(88)
5*†;46(;88*96*?;8)*‡(;485) ; 5*†2 : *‡(;4956*2 (5*—4) 8¶8*;40692 85)
;) 6†8) 4‡‡;1 (‡9;48081 ;8:8‡1 ;48†85;4) 485† 528806*81 (‡9;48;(88
;4(‡?34;48)4‡;161 ; :188;‡?;"

The narrator sees no sense in it, and Poe's use of numbers and symbols in place of letters, arranged in a single string, seems to make it more perplexing to the reader. However, Legrand explains, step by step, how he decrypted the message using straightforward frequency analysis (*see page 68*) to identify a tentative alphabet, whilst also looking for doubled symbols and repeated strings in the cryptogram, finally unraveling the message as:

"A good glass in the bishop's hostel in the devil's seat forty-one degrees and thirteen minutes northeast and by north main branch seventh limb east side shoot from the left eye of the death's-head a bee line from the tree through the shot fifty feet out."

At first almost as mystifying as the cryptogram, the decrypted message is worked through by Legrand – 'glass' meaning telescope; 'bishop's hostel' a reference to a local rocky outcrop, on which 'the devil's seat' forms a look-out point; from here, following the bearings given, a distant tree can be identified with a skull lodged in the branches; a weighted line dropped through the left eye of the skull marks a point on the ground 50 feet (15 m) from the location of Kidd's treasure.

Enter Sherlock Holmes

Sir Arthur Conan Doyle's most enduring creation was, unsurprisingly, a gifted code breaker – after all, he solved most of the mysteries in his casebook using a cryptanalytical approach to the most seemingly intractable problems.

The Adventure of the Dancing Men

In 'The Adventure of the Dancing Men,' Holmes once again astonishes his companion Dr. Watson with his ineffable logic. But, despite being the author of "a trifling monograph upon the subject, in which I analyse one hundred and sixty separate ciphers," Holmes in fact relies on straightforward frequency analysis to reach his conclusion, although unfortunately he does so too late to save the hapless victims of the cryptographer. Possibly, this is because the coded message is delivered in seemingly harmless drawings, but also Holmes is fed the code in small batches, each not enough to analyze the overall code effectively. Holmes' client passes a series of cryptograms in the form of dancing matchstick men.

Sir Arthur Conan Doyle (1859-1930). His world-famous hero, Sherlock Holmes, used the forensic techniques of a cryptanalyst to solve most of his cases, although few tales involve actual ciphers.

1 The first message

Holmes immediately assumes that the 'dancing men' are ciphers for letters, and goes on to identify that the most common letter will be 'e,' and also assumes that the flag-waving figures signify something, maybe word-endings.

Having established which cipher represents 'e,' Holmes goes on to make assumptions about the possible missing letters, filling in some blanks:

A M/ H E R E/ A .. E/ .. .. A .. E ..

He assumes that the last two words are the signature or name of the cryptographer, whose first name might be 'Abe' – possibly American.

2 The second message

This starts with a series of cipher figures which make little sense, spelling out only **A .. / E .. R .. .. E ..**, which Holmes assumes (correctly) might indicate where the cryptographer is staying, followed by:

Knowing his client's American wife's name was Elsie, Holmes fills in the 'I,' 'L,' and 'S,' then focuses on the first word. Assuming that the cryptographer is trying to engage Elsie in a meeting of some sort, Holmes assumes the first three letters are 'C,' 'O,' and 'M,' leading to:

C O M E/ E L S I E

At this point, Holmes can fill in more of the missing letters from the first message:

A M/ H E R E/ A B E/ S L A .. E ..

Holmes contacts the US police to find out if anyone with a name like 'Abe Slaney' is on their records. He is on the right track.

3 The third message

This appears to be from Elsie in reply, and provides Holmes with a tentative 'v.'

4 The final message

Holmes pieces this together from his progress above, but it has arrived at Baker Street too late.

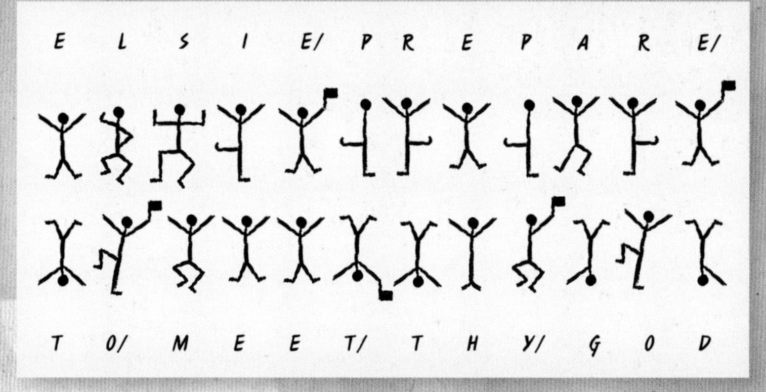

Only when Holmes has seen this message, guessing at some of the letters, does he have enough material to solve the problem, but unfortunately the threatening sender of the coded messages has already struck. Elsie is suffering from an apparently self-inflicted wound, and Holmes' client is dead, shot by an American gangster named Abe Slaney, a former associate of Elsie's who came out of the past to try to blackmail her, but the supersleuth's deductions mean that he is tracked down by the law. The code Slaney used was thought to have been in use in the Chicago underworld.

Fantasy Codes

In the 20th century the popularity of fantasy and science fiction was accompanied by many authors, filmmakers and computer game developers creating parallel worlds furnished with invented alphabets, calendars, numerical systems, and encoded mysteries. The Middle Earth novels by J.R.R. Tolkien (1892–1973) include a number of invented languages, and represent the first of several high points in this richly inventive genre.

The runes spelling Arne Saknussemm, the Icelandic explorer whose coded message sparks off the action in *Journey to the Center of the Earth*.

Early science fiction

Although Edgar Allan Poe may have been the first writer to invest his mystery writing with coded messages (*see page 260*), other 19th-century examples may be found. In Frenchman Jules Verne's 1864 novel *Journey to the Center of the Earth*, the action opens with the discovery of an ancient message in Icelandic runes, the unscrambling of which dominates the first four chapters. Firstly, Professor Lidenbrock translates the runic characters, but the translation merely makes a jumble of letters, clearly a cryptogram. By arranging the characters in vertical columns, then reading across the rows, again little emerges, until the Professor takes the first letter of each group of characters, then the second, and so on. Still no obvious pattern emerges, until it is realized that, read backwards, a passage in Latin appears. It provides the adventurers with a route map to the center of the Earth. Twenty years later, in Verne's 1885 novel *Mathias Sandorf* the plot hinges around a grille cipher (*see page 80*).

Tolkien's runes

A veteran of the Somme battlefields, and Professor of Anglo-Saxon and then English Literature at Oxford, J.R.R. Tolkien invested his *Lord of the Rings* trilogy (1954-55) with a great credibility by drawing on his academic skills. His vision of the ultimate struggle between humble good and world-destroying evil was enriched by his coherent vision of Middle Earth (he provided maps of the realm) and a detailed mythological history, manifested through various runic languages and scripts he invented and detailed in related works, *The Hobbit* (1937) and *The Silmarillion* (1977). He invented alphabets and languages from an early age, and claimed that his vision of Middle Earth grew out of the languages, rather than the other way round. He invented his first language, Quenya (which evolved into 'High Elven'), in 1915. The examples below spell out 'The Hobbit.'

Cirth

Directly modeled on Norse and Anglo-Saxon runes, and used for inscriptions for various Middle-Earth languages such as Quenya, Sindarin (based on Welsh), and Dwarvish, Cirth is written left to right, with dots dividing words.

Tengwar

Reminiscent of Tibetan or Brahmin script, Tolkien invested most effort in this imagined writing system: it has a variety of 'modes' depending on which language is being written (mainly Sindarin or Quenya).

Sarati

Used for a handful of inscriptions, Sarati is written in vertical columns arranged left to right. This alphabet comprises consonants with vowel diacritics placed either before or after the consonant.

Fantasy languages

St. Thomas More (1478-1535), the English statesman and philosopher, invented an ideal alphabet in his political allegory *Utopia* (1516, *see right*). Over the last half century, a host of languages and alphabets have been created for fantasy settings.

Thomas More's ideal script is used here to spell 'Utopia.'

A	B	C	D		
				Ancients' Alphabet	Created for the TV series *Stargate SG-1*.
				Ath Alphabet	Created for the novel *Crest of the Stars* (1996), the first of the *Seikai* trilogy by Hiroyuki Morioka.
				Aurek-Besh	Debuting in *Return of the Jedi*, this alphabet by Stephen Crane appears in other *Star Wars* films.
				Gnommish	Featured in the Artemis Fowl series of children's books by Eion Colfer.
				Hylian	Invented for the *Legend of Zelda* and other Nintendo fantasy games.
				Kryptonian	Introduced to DC's *Superman* comics in the 1970s, E. Nelson Bridwell's invention also features in the TV series *Smallville*.
				Marain	Iain M. Banks created this language in his *Culture* novel cycle.
				SGA	The Standard Galactic Alphabet was invented by Tom Hall for the *Commander Keen* series of computer games.
				Tenctonese	Possibly derived by Joe Hawthorne from Pitman shorthand; created for the *Alien Nation* films.

The Star Trek franchise has seen several alphabets to express alien languages, including Vulcan, Klingon, and Romulan, the latter being written in Kzhad (*above*).

Cyberpunk

The potential power of computers to dominate and eventually overwhelm human life has been richly explored, most notably in the 'cyberpunk' novels of William Gibson (b.1948), such as *Neuromancer* (1984) and *Count Zero* (1986), in which the exchange between coded data and human beings is seen as a new paradigm, combining digital and organic codes in a seamless and interchangeable network. In *The Matrix* cycle of films by the Wachowski brothers a future world is imagined in which humans merely play a part in an entirely computer-generated 'reality.' Humans have become small strings of code in a coded universe.

The American writer Neal Stephenson (b.1959) has created a suite of cult novels which interweave detailed historical research (and real characters) with theories concerning the mathematical science and cultural history of cryptography (with fictional characters), forming a dense historical narrative covering much of the last 400 years. Beginning with *Cryptonomicon* (1999), which looked at Alan Turing's development of computing concepts at Cambridge and Bletchley Park (*see page 118*) and at the creation of a modern data haven, he expanded his themes in an elaborate swashbuckling prequel, the *Baroque Trilogy* (2003-04). Many major mathematicians, natural scientists, and philosophers are featured in the loose cycle (including Hooke, Newton, and Leibniz), as are ideas concerning cryptography, mechanical computers ('logick mills'), alchemy and science, libertarianism, and capitalism.

DOOMSDAY CODES

It has many names: Doomsday, Armageddon, the Apocalypse, Judgment Day. Throughout history people of many religions, from many societies, have believed that they were living at the 'end of days,' that there were only 'minutes to midnight.' In the last 50 years, more than ever we believe the end of days is looming: nuclear holocaust, environmental disaster, global pandemic, World War III – these are popular ways to end the world in the media. Many fringe groups and paranoid individuals believe the Apocalypse is nigh, and moreover that it has been predicted already; they believe that if we can crack the right code, we will see it for ourselves.

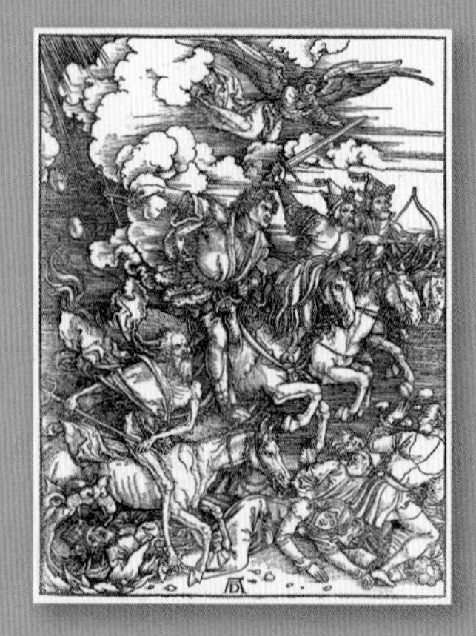

The Four Horsemen
The medieval fear of War, Conquest, Pestilence, and Death (as in the Albrecht Dürer print, *above*) remains with us. A strange mixture of astronomical, astrological, and other calculations have produced an enormous number of 'doomsdays.' Nostradamus postulated that the most inauspicious dates for the human future would appear when:
Good Friday fell on April 23, St. George's Day;
Easter Day fell on April 25, St. Mark's Day;
Corpus Christi fell on June 24, St. John the Baptist's Day.
This has occurred in the following years, and some calamitous events have been attributed to the 'coded' calculations of Nostradamus: AD 45, 140, 387, 482, 577, 672, 919, 1014, 1109, 1204, 1421,1451, 1546, 1666, 1734, 1886, 1945. The next dates to beware of are 2012 and 2096.

Nostradamus

Born Michel de Nostredame, the Frenchman Nostradamus (1503-66) was one of the leading astrologers and physicians of the Renaissance. During his lifetime, Nostradamus made some 6,338 predictions in his best-selling annual publications of 'almanachs,' 'presages,' and 'prognostications'; in recent years, people have been most interested in his 'perceptual prophecies' that are thought to foretell world history well into the fourth millennium. People have linked these prophecies to events as diverse as the rise of Hitler and the Kennedy family in the USA, and have used these apparently 'fulfilled' prophecies as evidence that Nostradamus was a genuine prophet. Many myths and rumors surround Nostradamus – for instance that he was buried upright with a medallion round his neck predicting when he would be dug up – and there are many problems with his prophecies. Some believe they were written in code, but in fact the order of books such as *Les Centuries* have just become scrambled texts, corrupted over the centuries; there have also been problems with copied texts differing from early facsimiles and the problem of interpretative translation.

ESTANT aſsis de nuit ſe-
cret eſtude,
Seul repouſé ſus la ſelle d'æ
rain,
Flambe exigue ſortant de
ſolitude,
Fait proferer qui n'eſt à croire vain.

The first 'Century' quatrain
Being seated by night in secret study,
Alone resting on the brass stool:
A slight flame coming forth from the solitude,
That which is not believed in vain is uttered.

Archaic spelling
Probably originally written in Low Latin, the quatrains have been translated into French; about five percent of the terms are not recognizably French, and another five percent are Old French, Greek, or Latin.

As was typical in 16th-century literature and writing of all kinds – even in what we would term scientific writings – Nostradamus wrote in quatrains (four-line verse), using flowery and poetic language, and deliberately used obscure Greek and Latin vocabulary; although to the uneducated this might seem like 'code,' in fact it is just metaphor, presumably to keep his predictions ambiguous enough not to upset any people of influence.

Among the dangerous dates predicted by Nostradamus (*left*) was 1666, when the Great Fire of London destroyed the city (*right*). It was synchronicities such as this (which nevertheless meant little to a peasant in France or China) that have provided an element of credence for millennialists.

Minutes to midnight?

The Doomsday Clock is the creation of the *Bulletin of the Atomic Scientists* at the University of Chicago; since its inception in 1947, it has been regularly maintained by its creators. Its purpose is mainly symbolic as it is supposed to represent the changes and developments in science and technology that are pushing civilization closer to the End; the positioning of the hands of the clock's face represents how close to 'midnight' civilization currently is. The clock's maintainers take into consideration the potential of political, economic, and environmental influences on impending doom such as nuclear war, global warming, and the development of biotechnology: the clock stood at two minutes to midnight in 1953 (the USA and the Soviet Union had tested nuclear weaponry within nine months of each other), and in 1984, in the midst of the Cold War. As of January 17, 2007, the clock stood at five minutes to midnight.

Apocalypse averted

Y2K, the 'year 2000 problem,' or the 'millennium bug,' was the fear that the timing systems programmed into computers around the world would not be able to handle the rollover into the third millennium AD, and that the effects of this would be potentially disastrous on a world that relies so heavily on computer technology. Would we lose control of our nuclear power stations...our hospitals...our weaponry? The problem stemmed from the programming design of early computers that, it was believed, would cause date-related processing to become defective between and after December 31, 1999 and January 1, 2000. Many governments and private companies invested hugely in upgrading computers to make sure they were 'Y2K-safe.' In fact, not much happened on December 31,1999, either in countries that spent a lot of time and money to insure their computers were Y2K compatible, or in countries that did not. In Australia, the worst did happen: in two states, bus-ticket validation machines stopped working, but luckily nobody died.

2012: 'The End of Days'?

December 21, 2012 has been cited by many New Age followers as the date for a cataclysmic event that will either end or change human civilization forever. Some base their theory on the Mesoamerican Long Count Calendar (*above*) used by the Maya that dates back to 1800 BC; this measures dates over long periods of time (anything longer than 52 years) and the winter of 2012 is at the end of the 5,125-year cycle the calendar covers. This date coincides with the 'Galactic Alignment,' the alignment of the solstice Sun with the equator of our galaxy, the Milky Way, and is further used by some New Age believers to support their theories. Oddly, it coincides with one of Nostradamus' apocalyptic dates (*opposite*). However, academic Mayan specialists dismiss these ideas, saying there is no reason to believe the end of the Long Count Calendar signifies the end of the world (the world certainly existed before the beginning of the calendar) or even to believe that the Maya intended it to.

Cryptography entered a new age with the advent of computers. Not only were these extraordinary tools to transform the ways in which codes could be generated and attacked, but the only way of communicating with them was through coded languages.

the digital age

The ubiquity of binary digitized systems in modern life means that innumerable aspects of our existence are dependent upon combinations of zeros and ones, from the operation of states, government, security, and finance, to the domestic – how we communicate, move around, acquire or spend money, entertain ourselves, and look after our health and well-being. We have all, by necessity, become cryptographers.

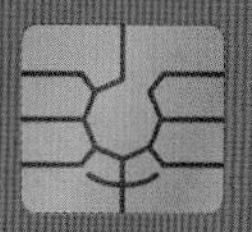

THE FIRST COMPUTERS

Charles Babbage (1791-1871) was a pioneering English mathematician and gifted engineer who designed and built the first ever machines used to solve mathematical problems. Whilst studying mathematics at Cambridge University he became increasingly frustrated by the human error that existed in the mathematical tables available at the time. In the absence of calculators, the values for functions like sine, tangent, and log had to be looked up from specially prepared books of tables – which took many weeks of painstaking calculation by hand to prepare. Babbage conceived of a machine that would be able to calculate the values of these functions to a high degree of precision, quickly and without the scope for human error – the very first computer.

Babbage the inventor
Born in London, Babbage studied at Trinity College then Peterhouse, Cambridge. He went on to form the Analytical Society in 1812, with John Herschel and George Peacock, and was Lucasian Professor of Mathematics at Cambridge (1828-39). Although he is most closely associated with the famous Difference Engine, he developed a number of other important inventions.

The printer Babbage designed a version of the Difference Engine that would be able to print its output to paper – the first printer.

The Analytical Engine Although never built, this was designed to be a more powerful version of the Difference Engine. The operator would have been able to program what the engine would calculate by using a series of punch cards. Almost a century later, the first electronic computers used exactly this method to read instructions and store data.

Occulting lights Used in lighthouses and lightships that rhythmically go from light to dark, establishing a regular signal, enabling mariners to tell what lighthouse they are close to (*see page 166*).

The cowcatcher Fixed to the front of railway engines for clearing obstacles from the track.

The Difference Engine

Charles Babbage realized that by using a special mathematical process called 'The Method of Finite Differences' his machine could do all of the long-winded calculations needed essentially by doing large numbers of subtractions – no multiplication would be necessary. This was achieved by a densely integrated series of cogs and ratchets. The Difference Engine was Babbage's first design, although engineering limitations of the time meant that the Engine was never fully completed. Babbage later designed an improved version of the Difference Engine, a working version of which was built from his designs 20 years after his death.

A gear from the results part of the machine, showing one of the four sets of numerals 0-9 printed on them. The row of gears would show the 31-digit result of the calculation the machine had been set to solve.

A full-scale reconstruction of Babbage's Difference Engine was completed in 1991 by London's Science Museum. It includes some 25,000 working parts, weighs 15 tons and is eight feet (2.4 m) tall. The first trial calculation returned a figure of 31 digits.

Two of Babbage's original gear cogs, with a 'carry-forward lever'. As the main shaft of the engine rotated, the columns of cogs would rotate and, depending on the position of the other cogs in the column, cause them to rotate as well. Each cog was attached to a wheel of numbers (31 in each column) that would show the current state of the calculation. The carry-forward lever triggered the transmission of the result of one set of columns to the next.

Early punch cards

Punch cards were first developed in the late 18th century to 'program' patterns in mechanical weaving looms. They were also used for the player-piano (pianola). Babbage adapted the idea for his Analytical Engine, inventing the first computer program.

Babbage's punch cards, designed for use with the Analytical Engine, would allow the user to program the machine to carry out various calculations. Although the Analytical Engine was never built, punch cards like these were used to program computers until the late 1970s.

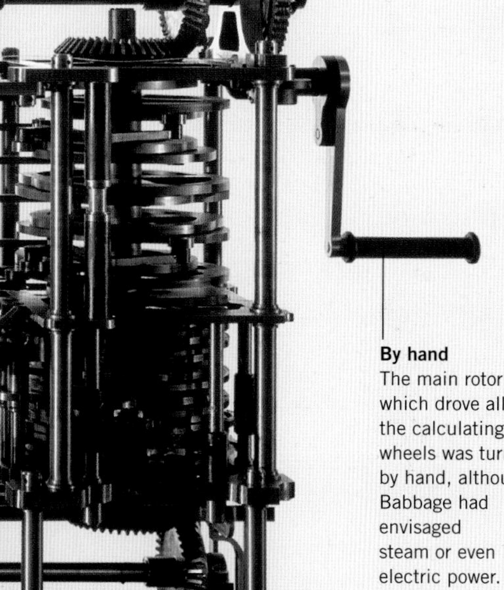

By hand
The main rotor which drove all the calculating wheels was turned by hand, although Babbage had envisaged steam or even electric power.

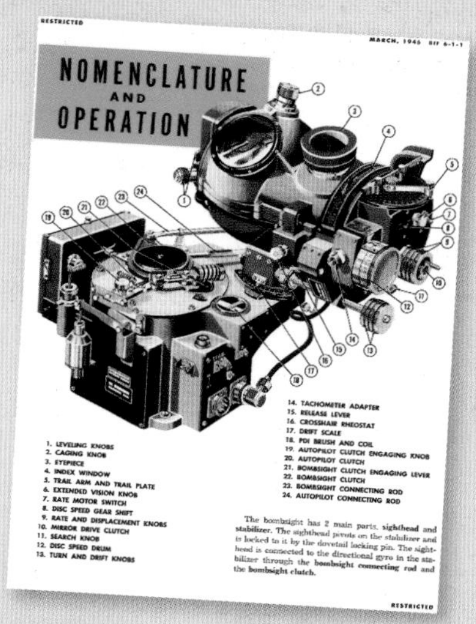

RESTRICTED

NOMENCLATURE AND OPERATION

1. LEVELING KNOBS
2. CAGING KNOB
3. EYEPIECE
4. INDEX WINDOW
5. TRAIL ARM AND TRAIL PLATE
6. EXTENDED VISION KNOB
7. RATE MOTOR SWITCH
8. DISC SPEED GEAR SHIFT
9. RATE AND DISPLACEMENT KNOBS
10. MIRROR DRIVE CLUTCH
11. SEARCH KNOB
12. DISC SPEED DRUM
13. TURN AND DRIFT KNOBS
14. TACHOMETER ADAPTER
15. RELEASE LEVER
16. CROSSHAIR RHEOSTAT
17. DRIFT SCALE
18. PDI BRUSH AND COIL
19. AUTOPILOT CLUTCH ENGAGING KNOB
20. AUTOPILOT CLUTCH
21. BOMBSIGHT CLUTCH ENGAGING LEVER
22. BOMBSIGHT CLUTCH
23. BOMBSIGHT CONNECTING ROD
24. AUTOPILOT CONNECTING ROD

The bombsight has 2 main parts, sighthead and stabilizer. The sighthead pivots on the stabilizer and is locked to it by the dovetail locking pin. The sighthead is connected to the directional gyro in the stabilizer through the bombsight connecting rod and the bombsight clutch.

RESTRICTED

Mechanical computers

The modern computer is a digital system, made out of electronic circuitry rather than cogs and gears – but mechanical computers similar to Babbage's were developed, such as the Norden Bombsight (*above*) fitted to US planes during WWII, which could compute airspeed, altitude, windspeed and drift, and groundspeed, and would link with an early form of autopilot to lock planes on target during a bombing run. Today, mechanical computers are being developed for use in situations where conventional electronics might fail – such as post-conflict situations, or environments with extremely high temperatures or radiation levels.

Supercomputers

In 1954, when IBM produced its first commercially available computer, the IBM 704 – almost a century after Charles Babbage invented his Difference Engine, but a mere decade after Alan Turing developed his 'bombes' (*see page 120*) – the company's marketing department predicted a market of only six machines. Half a century later, almost two billion computers have been sold, and of these there are currently over a billion in use today around the world. Computers are machines that subsist on code: coded languages constitute the way that we humans, for the time being their masters, talk to them (*see page 272*), and provide the way computers operate, think, and interconnect. Soon these increasingly portable and compressed devices may become their own code masters – possibly all but eliminating the need for human intervention in the foreseeable future.

The NEC Earth Simulator in Yokohama, Japan was created in 1997 and was one of the first and fastest supercomputers. It runs simulations of global climate models, and monitors climate change and geophysical processes. The Japanese government relies on it for accurate weather forecasting. It is capable of 35.86 trillion floating-point calculations per second. The development of an even larger machine was announced in 2008.

Your life in their hands

Advanced 'supercomputers' have been developed over the last decade that are now being entrusted with highly complex tasks that impact on the security of us all. The ASC Purple supercomputer at Terascale Simulation Facility at the Lawrence Livermore National Laboratory in Livermore, California is one of the world's fastest supercomputers. It is linked to another state-of-the-art system, Blue Gene/L. The ASC Purple's job is to ensure that the US nuclear weapons stockpile is secure and reliable without the need for testing, and it does this by running continual simulations and other checks. Access to the supercomputer is enshrouded with bureaucracy and, critically, top-security codes and firewalls to keep hackers at bay. Nevertheless, in 2007 reports were published of concentrated hacker attacks originating from Asia on this and other top-level systems.

ASC Purple comprises a ring of 196 IBM Power5 SMP servers. The system contains 12,544 microprocessors with 50 terabytes of memory and 2 petabytes of disk storage. It runs on IBM's AIX 5L operating system and uses 7.5 megawatts of electricity (enough to power 7,500 domestic homes), and requires a purpose-built cooling system. It has a processing speed of 100 teraflops (or one hundred trillion floating-point operations per second).

Memory development

One of the major challenges in the development of modern computers was finding an efficient means of storing coded data. Both programs and data files were initially inputted and outputted on punch cards (*see page 272*), but delay-line technology, developed during radar research in World War II, was to become one of the first electric memory systems. In parallel, the Williams-Kilburn cathode ray tube was introduced in 1947, and became a rival binary memory system. Magnetic core memory, which paved the way for modern desktop and laptop computers, was invented in 1949, but not widely applied until the late 1950s.

Cathode ray tubes stored binary data using secondary emission, a point drawn on the tube becoming positively charged, the area around it negatively charged, creating a 'charge well' which preserved the data.

Electric delay-line memory units worked in series, and were typically made up of electric enameled copper lines wrapped around a metal tube.

The ferrite rings Spaced about 0.04 inches (1 mm) apart, each store one bit (a '0' or a '1') of memory.

A magnetic core plane in close-up. Planes are arranged in stacks, with alternating electromagnetic charges passing through them.

Wiring The red wires are X or Y wires, the green sense or inhibit wires.

The IBM Series/1 minicomputer was launched in 1976, aimed at experienced programmers rather than the domestic market.

Split cultures

The computer world was traditionally divided into those who made hardware (the computers) and those who made software (the code). Today, integrated systems have blurred the distinctions, allowing a corporation like IBM enormous power, especially when undertaking governmental or defense contracts. On the other hand, the stylish, innovative, and user-friendly Apple consumer products (*below*) have become fashion icons. Meanwhile, computer programming code has become the modern equivalent of oil or gold – a high-value commodity. Indeed, possession of core code saw Bill Gates, founder of Microsoft, become in a matter of a few short years one of the wealthiest men on the planet.

TALKING TO COMPUTERS

In some ways, computers are very much like people. In order to communicate one needs to share a common language; to program computers to perform tasks one must 'talk' to them through a language that they understand. The vast majority of computers operate in binary code (essentially combinations of '1's and '0's). This inevitably makes direct communication extremely difficult for human beings, who are unused to this type of language. For this reason, as digital computers have proliferated and have been adapted to many new kinds of usage, new ways of programming them have been developed.

Moore's Law

The number of transistors on a chip (which determines processing power) doubles every two years, a phenomenon first outlined in 1965 by Gordon Moore, co-founder of Intel.

1971
First Intel chip
23000.74MHz

1993
First Pentium3
100,000,300MHz

2006
First Core 2 Duo
291,000,000,320MHz

Programming languages

Programming languages are classified by their 'level.' The syntax of low-level languages generally reflects the underlying processing unit's architecture and operation. A good example is an 'assembly language,' which directly feeds native instructions to the microprocessor. This means that while highly efficient programs can be written, they are difficult to use, and are not 'portable,' being processor-specific. High-level languages such as C and Java more closely resemble the English language, and allow for large programs to be developed relatively easily: most Web browsers, office productivity applications, and image-editing programs are written in high-level language, which allows applications to be portable but, as more 'translation' has to be done, run more slowly.

'Compiled' and 'interpreted' programs

Programming languages can be divided depending on the method that is used to 'translate' the commands into processor instructions: 'compiled' or 'interpreted.' Compiled languages such as C, C++, and COBOL involve a program that takes the original high-level source code and translates it into machine code, which can then be executed by the processor. Interpreted languages rely on a program that executes the original source code directly without translation, as and when a particular command is required. Interpreted-language programs such as Java are extremely portable, and can be executed on many systems, with programs written in this language running on mobile phones to supercomputers without needing much modification. However, interpreted languages generally execute more slowly than compiled ones.

Media

For decades, Charles Babbage's concept of punched cards was used to program computers (*see page 269*), relying on a system of communication not unlike Morse code or Braille. Punched cards were also used for data storage and retrieval until the 1970s, although the last three decades have seen a revolution in programming and data storage solutions.

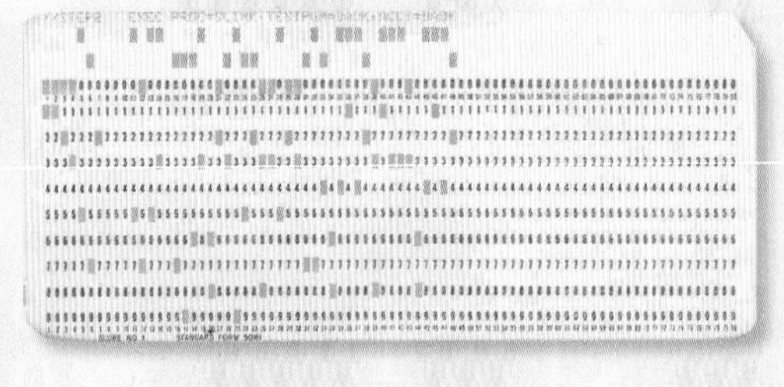

From 1928, punch cards became fairly standardized in format, with 80 columns of potential bits, and ten rows for data storage, with a control row at the top.

1846
Punched paper tape
Paper

1956
FORTRAN developed by IBM; the first direct way of addressing the computer via syntax.

1963
Compact cassette tape
20KB plus
Magnetic

Late 1960s
SGML, the first text markup interface.

1976
5.25" Floppy disk
256KB
Magnetic

1850 1900 1950 1960 1970 1980

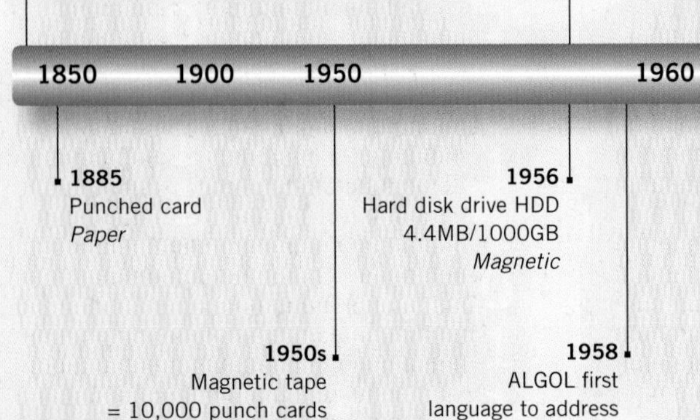

1885
Punched card
Paper

1950s
Magnetic tape
= 10,000 punch cards
Magnetic

1956
Hard disk drive HDD
4.4MB/1000GB
Magnetic

1958
ALGOL first language to address algorithms.

1963
ASCII developed to communicate to the computer via language-based commands in binary instructions.

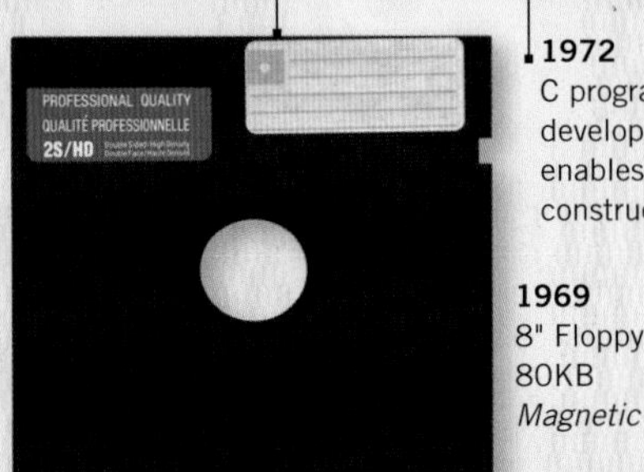

1969
8" Floppy disk
80KB
Magnetic

1972
C programming language developed by Bell Telephone enables users to alter or construct programs.

1979
Compact disk
CD-ROM
700MB
Optical

"I am a HAL 9000 computer, production number 3."

2001: A SPACE ODYSSEY **(1968).**

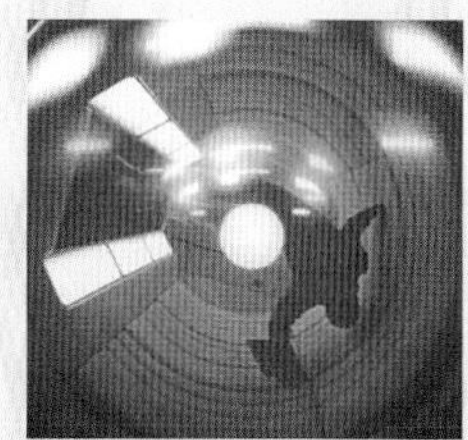

In Kubrick's film *2001*, the spaceship crew hold friendly conversations with Hal, the onboard computer. Programmed to ensure the success of the mission, Hal identifies the crew as a threat, and starts to eliminate them.

ASCII	Character	ASCII	Character	ASCII	Character
32	(space)	64	@	96	`
33	!	65	A	97	a
34	"	66	B	98	b
35	#	67	C	99	c
36	$	68	D	100	d
37	%	69	E	101	e
38	&	70	F	102	f
39	'	71	G	103	g
40	(	72	H	104	h
41	)	73	I	105	i
42	*	74	J	106	j
43	+	75	K	107	k
44	,	76	L	108	l
45	-	77	M	109	m
46	.	78	N	110	n
47	/	79	O	111	o
48	0	80	P	112	p
49	1	81	Q	113	q
50	2	82	R	114	r
51	3	83	S	115	s
52	4	84	T	116	t
53	5	85	U	117	u
54	6	86	V	118	v
55	7	87	W	119	w
56	8	88	X	120	x
57	9	89	Y	121	y
58	:	90	Z	122	z
59	;	91	[	123	{
60	<	92	\	124	|
61	=	93	]	125	}
62	>	94	^	126	~
63	?	95	_	127	DEL

ASCII code

The most popular and long-lived encoding language is called ASCII (American Standard Code for Information Interchange). With this system all of the letters of the alphabet, all of the numerals, and many punctuation characters can be described in a form that computers can work with. An ASCII character is stored in a single byte, which is eight bits. Traditionally the eighth bit of the character was reserved as an error-checking bit, so altogether there were seven bits to work with. Binary is base two, so there are two to the power seven values that can be represented – this is why ASCII codes run from zero to 128. Left are the 'printable' ASCII codes. All of the other ASCII codes are special and are reserved for controlling the systems that use ASCII. The bulk of these 'control codes' are now obsolete – many date from the days when computers didn't have screens, but printed all output to paper; for example, there are codes for carriage return, as on an old-fashioned typewriter.

How ASCII code works The numbers 39 72 101 108 108 111 44 32 67 111 109 112 117 116 101 114 33 39 translates to 'Hello, Computer!' – including punctuation.

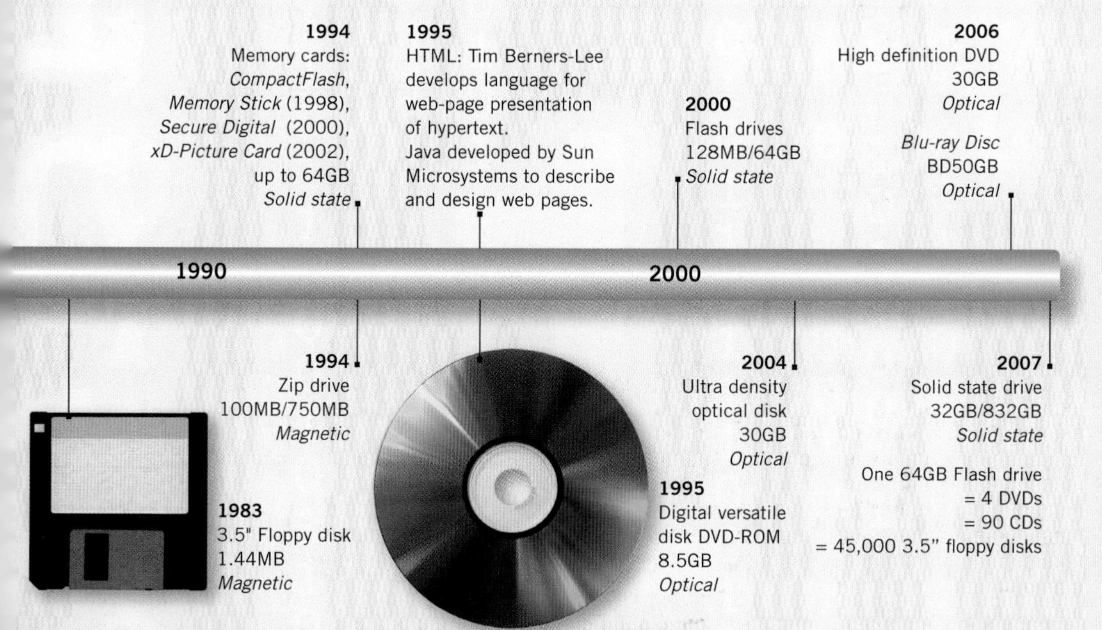

HTML and the Internet

Almost every Web site you will find on the Internet is written in HTML (Hypertext Markup Language). This is used to describe elements of a Web page. The developer writes the text of the page and encloses each section in 'tags' of HTML, which describe whether the text is plain paragraph text, a hyperlink, part of a list, styled as bold or italic, and so on. This information is sent to your Web browser when you open the Web site, and your Web browser decides how to display these elements of the Web page accordingly. As computer languages go, HTML is very simple to read and write, which was crucial in the speed at which the Internet developed – it took very little technical knowledge to write a simple Web site.

Scripts

These are languages that do not talk directly to the machine architecture but interface with programs, adapting them to different uses; some of the most prolific scripting languages are used on the Internet. PHP and ASP facilitate Web servers to display dynamic content that responds to the user (such as e-commerce pages) complementing HTML. CSS (Cascading Style Sheets) is a scripting language for Web browsers, which can determine how Web pages are rendered in the choice of fonts and colors.

Unicode and Multibyte

Unicode and Multibyte character sets are the 'sequels' to ASCII. ASCII always suffered from the limitation that there are no characters with accents – needed in many European languages (let alone the thousands of characters needed to encode Mandarin and many other non-European languages). Unicode solved this, allowing one to have accents (cedillas, acutes, graves etc.), and also to represent the syllable sounds of syllabaries, permitting the encoding of Mandarin or Japanese; it will in the next few years contain Egyptian hieroglyphs and many other arcane characters, such as Linear A, Linear B, and the Phaistos disc characters (*see pages 28, 30*).

Alice, Bob, and Eve

Perhaps the most famous people in modern day cryptography are two fictional characters – Alice and Bob. In the field of computer sciences, quantum physics, and cryptography, when two parties need to communicate, these are the names that they are given. In the past, it would have been 'Person A,' 'Person B,' 'Person C,' and so on, but in the late 1970s 'A' became 'Alice,' 'B' became 'Bob,' 'C' became 'Carol,' and a small cast of other names joined the group. Nowadays most cryptographic systems are described in terms of Alice, Bob, and their associates. The most common addition to the group is 'Eve' – the 'eavesdropper' who is trying to intercept communications.

Public and private keys

In most cryptographic systems, information is encrypted with a 'key.' For centuries the system was 'symmetric' – both the sender and receiver needed to know both the algorithm and the key, and combining the two pieces of information meant that the receiver merely worked backwards through the encryption using the key (*see pages 64-87*). This remained relatively secure as long as the key was a secret shared only by the sender and receiver. In modern digital cryptography a key is typically an extremely long number that is used to encrypt each character of the data. Different types of keys exist, and there are many ways to generate them. In addition, most modern systems also use 'asymmetric' keys, often called 'public' and 'private' keys (known as the RSA). The public and private keys have the property that anything encrypted by the private key can only be decrypted by its public key counterpart, and vice versa. The first step in creating these key pairs is to generate a very large random number – often a prime number (*left*).

PGP

The most widely used asymmetric public key/private key digital encryption system is called 'Pretty Good Privacy' (PGP), and was first released by Philip Zimmermann (*left*) in 1991. It was designed to provide 'digital signatures' and to securely encrypt plaintext, especially for use on the Internet. The system soon ran into trouble with the security services, as it was so impregnable that it could be used by criminals and terrorists. Any encryption systems involving more than 40 bits were technically classified as 'munitions/weapons' by US export regulations, and PGP never used less than 128 bits. Criminal proceedings began, but Zimmermann simply published PGP in book form, sidestepping export regulations via the First Amendment.

Alice

Starting with a very large random number (*see right*), Alice generates two keys – a public key and a private key. She sends a copy of the public key to both Bob and Carol. If Alice wants to send a message to Bob and Carol, she encrypts the message with her private key. As she is the only one who owns a copy of it, anything that can be decrypted using the corresponding public key must have come from Alice. With this system, Alice is the most trusted member of the group, as she holds the one and only private key.

Bob

Bob receives a copy of Alice's public key. Bob and Carol decode Alice's message with their public keys. If Bob wants to send a message to Alice, he encrypts it with his public key. Only Alice, with the corresponding private key, can decode the message.

Primal dream

Prime numbers are of particular importance in modern cryptography for generating unique keys. A prime number is a natural number that can only be divided completely into one or itself. Many numbers that look prime (prime numbers are always odd numbers) can take you by surprise:

3 prime Divides only into 1 (giving 3) and 3 (giving 1).

5 prime Divides only into 1 and 5.

7 prime Divides only into 1 and 7.

9 not prime Divides into 1 (giving 9) and 3 (giving 3).

11 prime Divides only into 1 and 11.

13 prime Divides only into 1 and 13.

15 not prime Divides into 1 (giving 15), 3 (giving 5), and 5 (giving 3).

19 prime Divides into 1 (giving 19) and 19 (giving 1).

21 not prime Divides into 1 (giving 21), 3 (giving 7), and 7 (giving 3).

27 not prime Divides into 1 (giving 27), 3 (giving 9), and 9 (giving 3).

37 prime Divides only into 1 and 37.

49 not prime Divides into 1 (giving 49) and 7 (giving 7).

Prime numbers have bewildered and fascinated mathematicians since the ancient Egyptians. They occur seemingly without pattern all the way through the set of natural numbers. It takes a lot of computing power to test whether a very large number is prime, and because prime numbers are used to generate keys in modern day cryptography, identifying large prime numbers has become exceptionally important.

"Prime numbers grow like weeds among the natural numbers, seeming to obey no other law than that of chance."

DON ZAGIER, NUMBERS THEORIST, 1975.

The RSA algorithm was first described by Clifford Cocks, a mathematician working at the British security center GCHQ in 1973. The concept was later developed independently by Ron Rivest, Adi Shamir, and Leonard Adleman (the name deriving from the initial letters of their surnames) at MIT, and published in 1977.

Generating the keys

Alice generates two prime numbers, p (e.g. 223) and q (199) – although in reality these would be much larger prime numbers.

From these prime numbers more numbers are generated, n (44377 – the product of $p \times q$), and ϕ, which is $(p-1) \times (q-1) = 43956$.

Now a special integer e is chosen (e.g. 5), which relates n and ϕ.

Finally a secret number d (35165), is chosen that relates e to ϕ. At this stage the public key is (n, e) and the private key is (n, d).

Alice sends the public key to Bob and Carol. Now they have a copy of (n, e), which allows them to decrypt messages from Alice.

Alice encrypts a message and sends it to Bob. The ciphertext is the plaintext converted into numbers which are raised to the power e and multiplied by the modulus of n (modulus is a wraparound number, like a 24-hour clock). This text is now completely encrypted, here divided into five-number sequences, and is sent to Bob. 'Hello Bob' becomes '26946 09392 37665 23986 12461.'

Bob decrypts the message. Bob has the message, '26946 09392 37665 23986 12461' and uses his key, (n, e) to decrypt it. The plaintext is the ciphertext raised to the power d and multiplied by the modulus of n. If the key is correct then: 26946 09392 37665 23986 12461 will read: 'Hello Bob.' Otherwise a garbled message will be generated.

Eve the 'eavesdropper' is excluded from the system. If Eve obtains a copy of the public key, she can send messages masquerading as Bob or Carol, but will not be able to decrypt messages from Bob or Carol. This is a limitation of the system – although in reality public keys are regularly regenerated and redistributed to the trusted parties.

Eve

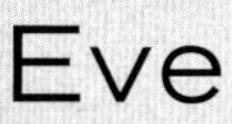

Carol

Carol receives a copy of Alice's public key. If Carol wants to send a message to Alice, she encrypts it with her public key. Only Alice, with the corresponding private key, can decode the message.

If Bob sends the same message to Carol using the public key, she won't be able to decrypt it, and vice versa, as Carol does not have the private key.

The problem of randomness

Asking a computer to generate a random number is extremely difficult. By their very nature computer processors have to follow instructions to carry out any task – you cannot ask a computer to pick out a random number. With most modern PCs, this problem is sidestepped by using the 'clock cycle counter.' Every computer processor contains a crystal that oscillates at a certain frequency. The time taken for one cycle is the same as the lengthiest single operation a processor can perform. The oscillation of the crystal is used to keep the processor in time, and all operations are synchronized and start at the beginning of every crystal oscillation. Every time that the processor 'ticks' an internal counter goes up. Normally, this counter is used as the basis for random numbers – every time the user asks for a random number the counter will have changed value and a new number can be generated. If a modern-day computer runs at 3 GHz, then every single second the clock ticks three billion times!

Entropy pools

More advanced systems use entropy pools to generate random numbers. An entropy pool is a special piece of hardware that uses physical phenomena as a source of random numbers, such as thermal noise. The best entropy pools use quantum phenomena to generate truly random numbers, as the laws of quantum mechanics allow many systems to behave in an unpredictable and random way.

Secure online transactions How are card details used securely online? Large services like PayPal and Google Checkout use special private key encryption to make sure that the session is secure. More recently PayPal has produced a 'key' that generates a temporary code to be used for the duration of the session. In a similar vein, many banks are providing similar units to make online banking safer.

Future Medicine

Many of us today encounter medical techniques unheard of when we were born, such as 'keyhole' surgery, and routine organ transplants. The speed of medical research, especially with the completion of the Human Genome Project (*see page 174*), coincided with other developing technologies, not least pharmaceutical research, but also miniaturization (nanotechnology) and robotics. All of these to some extent depend nowadays on coded computerized technology. In the half century since Watson and Crick identified the DNA code (*see page 170*), our well-being and longevity, for good or ill, rely increasingly on digital technology.

New technologies

The advent of computerization has transformed medical science to the extent that machines such as modified ink-jet printers are now being used to 'build' replacement body tissues while computerized analysis of functions such as sight and hearing among animals are helping to reconstruct damaged systems for the human blind and deaf. On the other hand, DNA-related experiments have shown that rebuilding lost body parts is no longer science fiction, but will soon be science fact.

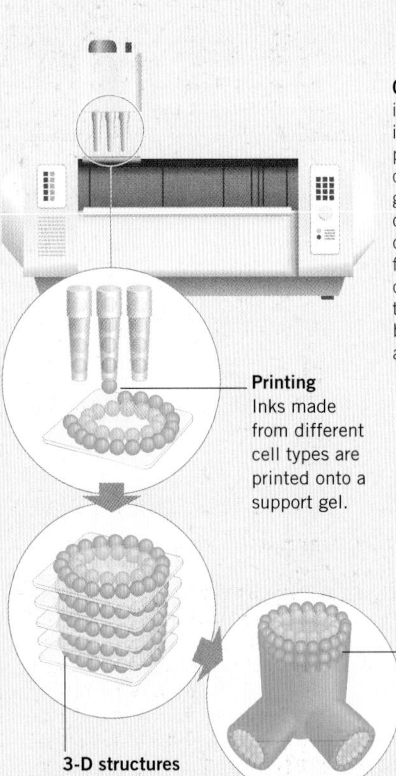

Organ printing Modified ink-jet printers are already in development that may print living tissues and organs, using a support gel for paper and living cell cultures as ink. The cell cultures could be grown from an individual's own cells providing organs that will not be rejected, bypassing the need to find a matching donor.

Ready to use
The cells merge as the support gel relaxes, forming the final tissue. Complete organs could be printed in this manner.

Wearable sensor clothing
Clothing patches can monitor health functions, such as pulse, conductivity, breathing rate, and electrolyte levels in the sweat. They will be able to communicate directly with health centers, giving the medical status and whereabouts of the wearer. This data may be used for general health information and emergency service call-out in the event of severe trauma.

Sensors
Clothing is made by weaving together natural and conductive fibers.

Fast track
Data is received by satellite and forwarded to a monitoring station.

Data collection
An in-built processor collects and sends data to satellite.

The Hippocratic Oath

This venerated code of conduct among doctors worldwide was originated by the clinical physician Hippocrates of Cos (c.460-c.370 BC), whose ideas were recorded by others in the *Corpus*. Across the centuries, in its essentials the code enshrines four main moral precepts:

Tradition The veneration of one's teachers, and a commitment to pass on knowledge to the next generation.
Sanctity of life To offer the best possible medical advice to the patient, and to refuse to give a patient poison if requested (originally extending to refuse to administer abortificants).
Patient confidentiality To never pass on to a third party details of a patient's condition without their consent.
Respect To avoid intimacy with patients.

Physicians today are, however, confronted by an increasing number of challenges as society and scientific research evolves. Abortion on demand remains a heated social, ethical, and often legal issue; arguments for voluntary euthanasia in cases of extreme distress are counterbalanced by moral issues and clinical advances; the problem of innovations in identifying genetic heritage are discomfiting, while genetic engineering, especially in the human sphere, remains an enormously controversial issue for the medical profession.

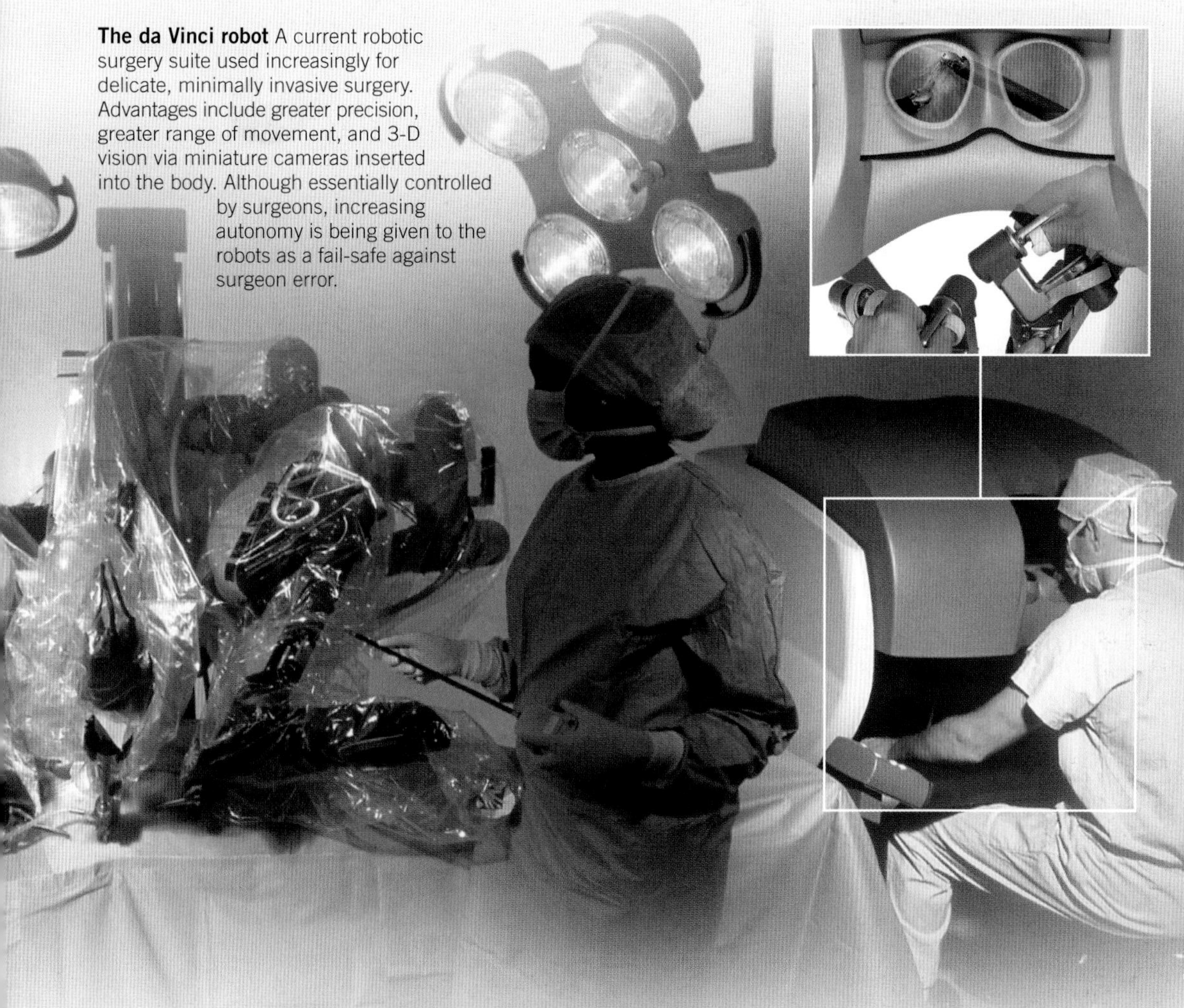

The da Vinci robot A current robotic surgery suite used increasingly for delicate, minimally invasive surgery. Advantages include greater precision, greater range of movement, and 3-D vision via miniature cameras inserted into the body. Although essentially controlled by surgeons, increasing autonomy is being given to the robots as a fail-safe against surgeon error.

Battlefield medicine

Throughout history, the need to treat soldiers injured in battle has provided countless medical breakthroughs. In the future, mobile, robotic surgical units may be deployed to retrieve fallen soldiers and stabilize them prior to evacuation. Named 'trauma pods' after similar machines imagined by Robert Heinlein in his 1957 sci-fi book *Starship Troopers*, these units would provide automatic care during the 'golden hour' (the first hour after injury), crucial to the fate of the injured soldier. Other advances may include regenerative techniques such as spray-on skin to treat burns, blood clot-forming powder, and field dressings chemically impregnated to stop blood loss (the cause of 50% of deaths on the battlefield). These advances are already in limited use.

Small portable anaesthetic devices that shut down pain signals coming from injured regions of the body are also being developed. Ultrasound devices are envisioned that will locate and cauterize internal wounds. Externally, battlefield clothing may remotely inform medics of a soldier's physical status. This early triage will mean the most severely injured will be treated first.

Vital signs These are assessed and in case of emergency a medical station is notified and assistance immediately dispatched.

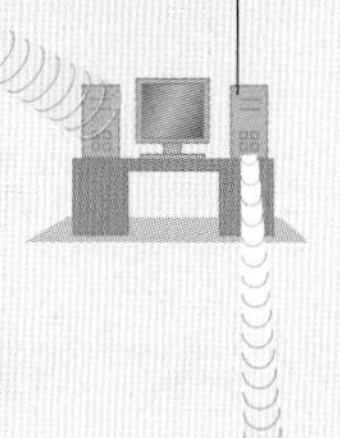

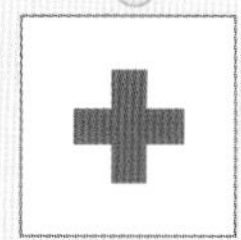

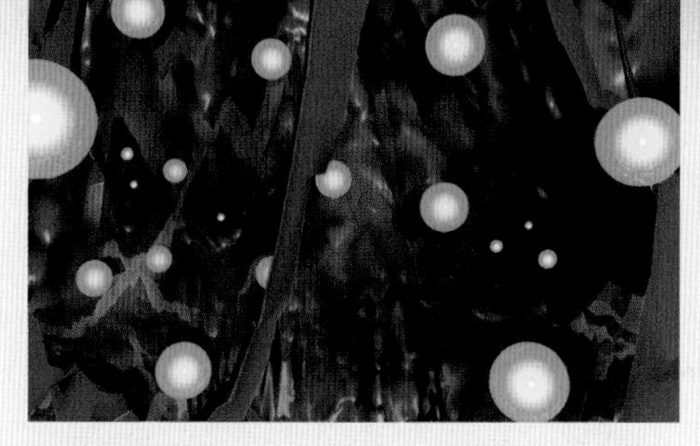

Nanotechnology At the nanoscale (one billionth to 100 billionths of a meter) treatments are being developed that could one day become commonplace. Minuscule molecular balls called 'nanoshells' or 'buckyballs' are beginning to be used to deliver drugs and other therapies to specific sites in the body – particularly useful for the delivery of chemotherapy drugs direct to cancer cells, avoiding normal cells and thereby minimizing side effects. Supports built from nanotubes called 'nanoscaffolds' will help provide a structure for regrowth of damaged tissue such as nerve tissue and as a base for the regrowth of organs.

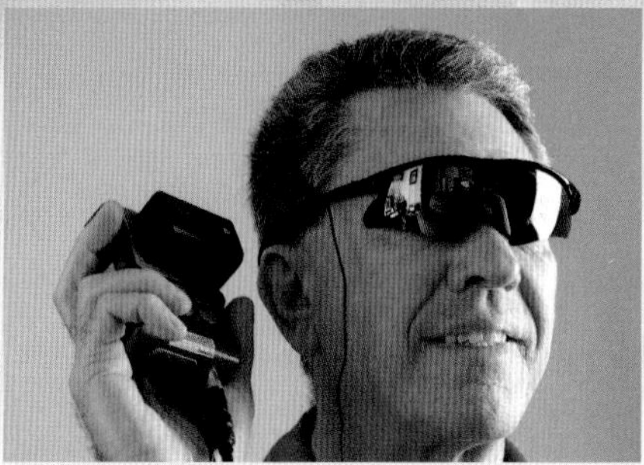

Decoding brainwaves Signals picked up by electrodes inserted into a cat's brain have already been used to recreate hazy images of its visual field. The signals, taken from a point just behind the optic nerve, were interpreted using 'linear decoding technology.' Working in reverse, it may be possible to translate images from a camera into signals that can be fed directly to a person's visual cortex, allowing a blind person to see. Today, algorithms are being researched that can mathematically model the activity of the visual cortex. Ultimately, this may lead to 'seeing' someone else's dreams and imaginings.

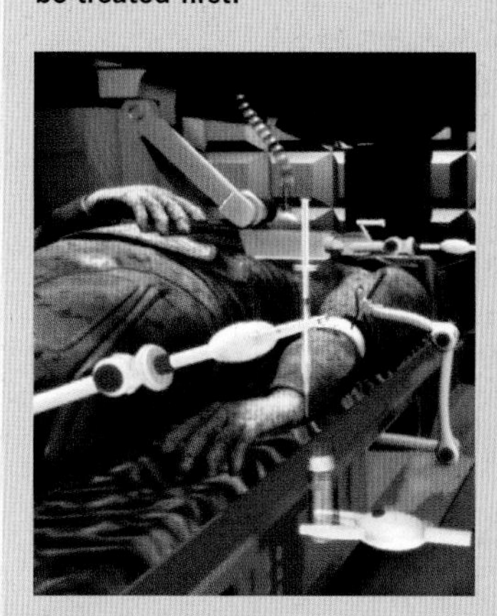

Inside the 'trauma pod' An injured soldier is medically assessed and stabilized robotically prior to evacuation.

Where are codes taking us?

Our finesse in finding coded languages to describe the world and manipulating the results is astounding, yet many believe that the 'digital revolution' has barely begun. Computing power is doubling roughly every two years, according to Moore's Law (*see page 272*), as are improvements in many other aspects of digital technology. Phones, cameras, cars, music systems, televisions, and above all PCs have changed so radically in the last few decades that were we to step back just 20 years their antecedents would seem unfamiliar, if not quaint. Is there a limit to this progress? There are certainly limits to the number of transistors that can be fitted on a silicon wafer. These have formed the basis of microprocessors and thereby computing power for several decades, but overheating and finite size both impose constraints that are increasingly difficult to overcome. Time for something new.

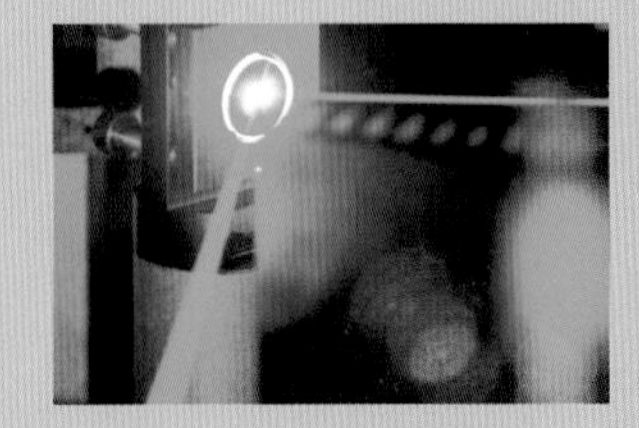

Quantum computers

The most promising research involves harnessing the potential of subatomic particles and the world of quantum physics. When things are very, very small (the size of atoms and subatomic particles), the physical laws that govern them change radically. This is the realm of quantum physics, where particles are also waves, and matter is energy. Researchers are finding ways to exploit this particle/wave duality to build computers with a vast increase in storage capacity and processing speed to solve problems in seconds that would currently take hundreds of years, but there are many problems yet to be overcome in building fully-functioning quantum computers. One quantum effect called 'entanglement,' referred to by Einstein as "spooky action at a distance," has been used to 'teleport' quantum information. This has implications for both quantum computation and data encryption, and will allow for completely secure transfer of data.

The DNA connection

Research into new types of computer points to a future of near infinite computing power and capacity at speeds unheard even with today's supercomputers (*see page 270*). For example, one pound of DNA has more storage capacity than all the silicon-based computers that have ever existed, and DNA is plentiful and relatively cheap. Computers harnessing this potential should be able to perform calculations in parallel rather than linearly (as conventional computers operate). This will vastly increase their speed and reduce their size, enabling computers the size of a raindrop to outperform the fastest of today. Similar ideas are being explored using a 'soup' of different chemicals with operations performed by chemical reactions.

Freeway to flyway

Could flying cars, the promise of so many sci-fi predictions, finally become a reality? Current research suggests that they will be on the market within two decades. Prototypes of private vertical takeoff vehicles already exist. The main drawback involves their control technology, but computer modeling, GPS, and 3-D positional software is solving these problems.

Digital warfare

Technological advances in weapon systems are always at the cutting edge of our knowledge due to the massive expenditure on defense research budgets. In this sometimes surreal world reality can be stranger than science fiction. This image is of a pilot's helmet from the new F-35 Joint Strike Fighter, a state-of-the-art development providing the pilot with unprecedented levels of information and control. In addition to the features detailed below, digital cameras mounted on the exterior of the fighter allow the pilot to access views to the side, above, below, and behind the aircraft.

Twin projectors These beam a range of images onto the interior of the tinted visor.

Vocal commands Most of the digital functions can be activated by voice.

Data cable The digital feed supplies data and relays commands.

Oxygen supply Air is pumped into the pilot's lungs at high pressure.

Earphones These relay radio messages and synthesized voice information from the aircraft's computerized control systems.

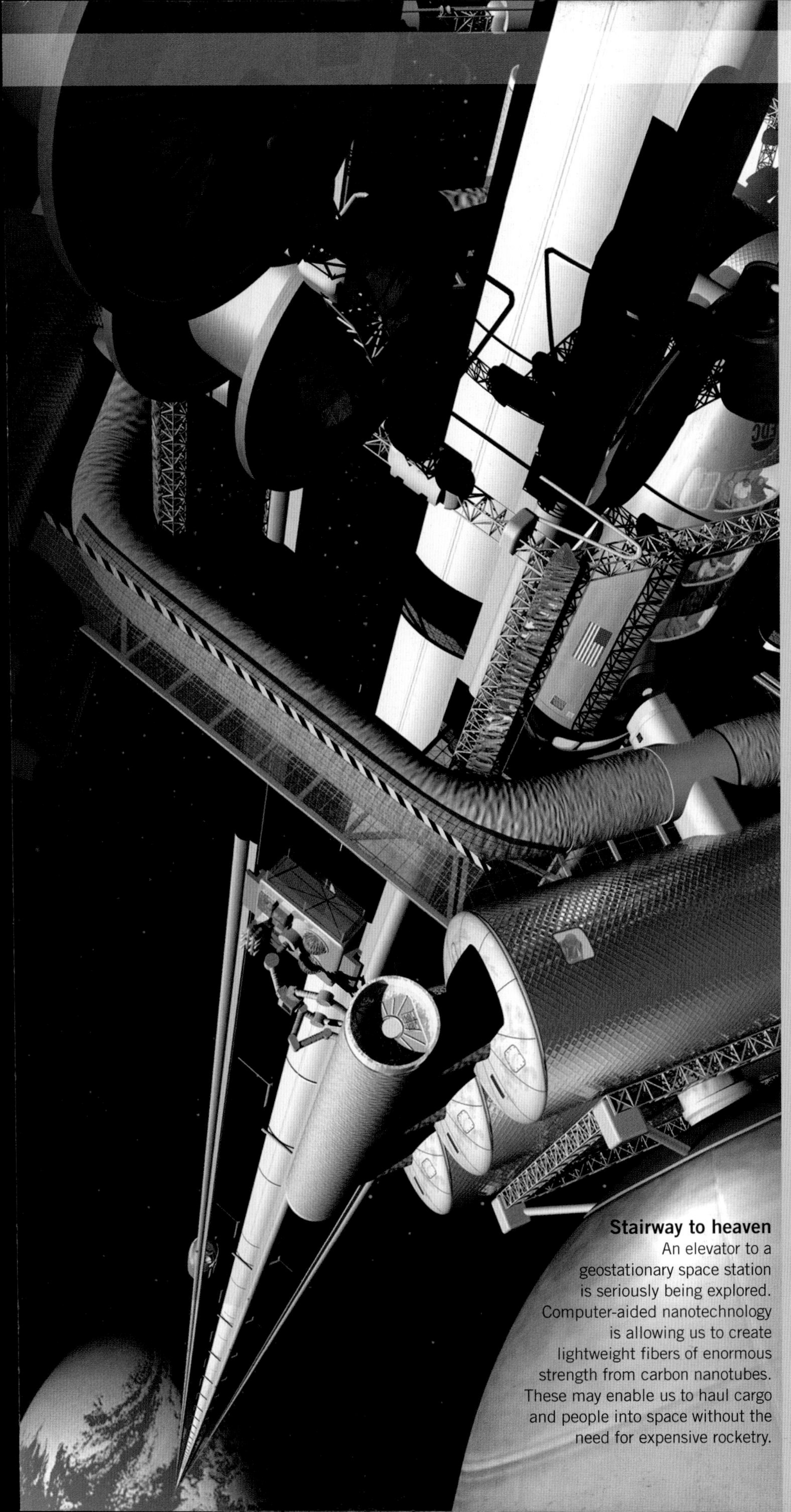

Stairway to heaven
An elevator to a geostationary space station is seriously being explored. Computer-aided nanotechnology is allowing us to create lightweight fibers of enormous strength from carbon nanotubes. These may enable us to haul cargo and people into space without the need for expensive rocketry.

Into the matrix

With processors functioning at the atomic scale, we may soon see a world in which superfast, minute computers are printed onto objects or the skin, communicating through a quantum network. Reductions in scale and cost are already allowing microprocessors to be attached to everyday objects such as RFIDs; with wi-fi technology and sufficient bandwidth they will soon be communicating spontaneously with each other and collecting information via the Internet. Lawn sprinklers will read weather forecasts; children's clothes will reveal their whereabouts via GPS systems; medicine cabinets will automatically identify the drugs they contain and warn of possible adverse interactions; and food packaging will communicate with the oven to tell it how the food it contains is to be cooked.

It is likely that the majority of computing power will be moved out of our homes and offices to vast, remote computer sites. Our terminals will become reduced to a wearable size – wristwatches or perhaps headbands with constant access to the Internet, feeding us information about the world, aware of our location and environment, and able to infer our desires and intentions.

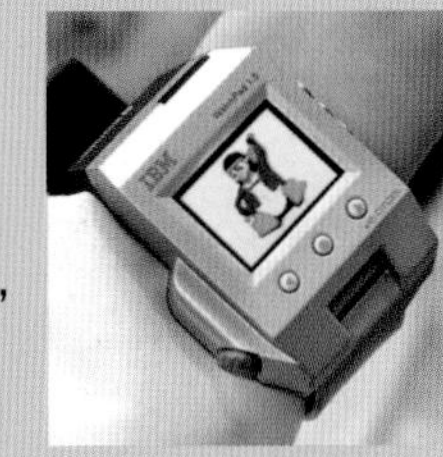

Plugging in

Brain-computer interfaces (BCIs) are becoming increasingly sophisticated, along with our ability to decipher the brain's electrical activity. People can already move cursors and write messages on computer screens by thought alone using sensors mounted on the scalp. Monkeys have been trained to feed themselves using robotic arms linked to brain sensors, and we have seen through cats' eyes by interpreting signals from electrodes placed in their brains (*see page 277*). Given this, we can envisage a world in which humans and machines are married together to enhance our cognitive skills, and keep us healthy, informed, connected, and entertained. Anything and everything could be coupled to the Internet – with dire consequences for our privacy. Will we accept electrodes hardwired into our brains? The only brakes on the process are cultural: do we consider this desirable or an Orwellian nightmare?

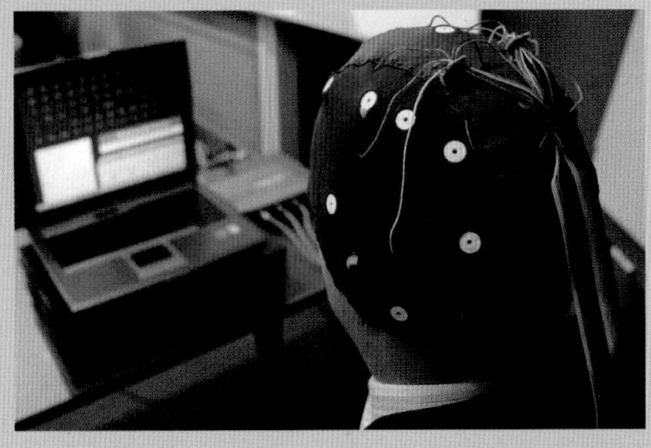

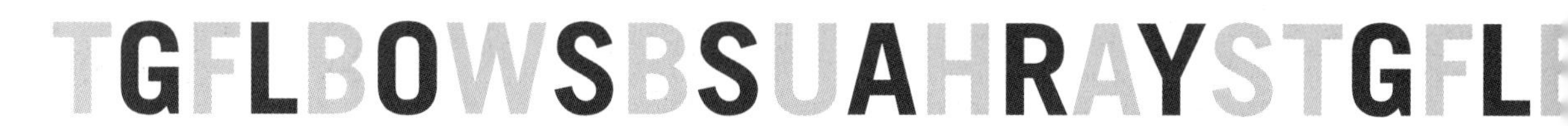

GLOSSARY

Algorithm
A general term for a system used to encrypt a plaintext into a ciphertext. Algorithms are usually made more specific by the use of a key. *See page* 66.

Anagram
A simple transposition cipher involving the rearrangement of the characters of a plaintext word or sentence to produce a different word or sentence. *See page* 66.

ASCII
American Standard Code for Information Interchange, a widely used system which transforms characters into binary numbers. *See page* 273.

Asymmetric key cryptography
A modern form of secret messaging in which two different keys are used, one for encryption and one for decryption. Often referred to as the 'public' key and the 'private' key, this system is typified by the RSA and PGP systems, and is increasingly used for digital, online communications and transactions. *See pages* 274-275.

Book cipher
A cipher in which the key takes the form of an extended text; the Bible and the Declaration of Independence are among the texts which have been used.

Cipher
Any system in which the letters, numbers, or characters of a plaintext have been substituted with another series of letters, numbers, or characters in order to disguise the meaning of the plaintext.

Cipher alphabet
An arrangement of letters, characters, or numbers for which the normal alphabet may be substituted to create a ciphertext.

Ciphertext
The encrypted plaintext.

Cryptanalysis
The science of code breaking, and specifically of detecting the algorithm and key used to create a cipher.

Cryptography
Disguising the meaning of a message. The creation or development of systems for encryption.

Cryptolect
A language within a language, usually spoken rather than written, which has developed as a means of secret communication within select groups. Cryptolects include cants, argots, and languages such as Cockney rhyming slang. *See pages* 99, 128-129, 132-133, 134-135, 146-147.

Cryptology
The study of coded messages and secret writing.

Decipher
To work a ciphertext back through an encipherment system to reveal the plaintext.

Decode
To transform a coded message into the original plaintext.

Decrypt
To decipher or decode a coded message.

Digital signature
A method of validating a digital message usually using asymmetric key cryptography.

Digrams/Digraphs
Commonly occurring combinations of pairs of letters or characters.

Encipher
To transform a plaintext into a ciphertext.

Encode
To transform a plaintext into a coded message.

Encryption
The process of transforming a plaintext into a ciphertext using an algorithm.

Fractionation
The process of transforming the alphabetic letters of a plaintext into numbers or groups of numbers.

Grille
A method of revealing a hidden message in an otherwise innocuous plaintext. Grilles usually take the form of a sheet of paper or card with holes cut out, which, when laid over a plaintext, reveals selected words or letters forming a hidden or secret message.

Homophone
One of a number of letters, numbers, or characters used to replace recurring letters, numbers, or characters in the plaintext. Homophones tend to be used to disguise the more frequently occurring characters or words in monoalphabetic substitution ciphers.

Key
The element in a cipher which, together with the algorithm, enables encryption and decryption by specifying how the algorithm should be used. Key words and key phrases (and sometimes extended texts) may be used in conjunction with a tabula recta to create polyalphabetic ciphers, the most widely known being the de Vigenère cipher, *see page* 104.

Monoalphabetic substitution ciphers
A substitution cipher system in which the replacement of letters, numbers, or characters by other letters, numbers, or characters remains consistent for the length of the message, unless homophones are used. *See pages* 66, 74, 103.

Nomenclator
A combination of a monoalphabetic substitution cipher with a large number of homophones representing recurring letters, characters, numbers, or words. *See pages* 70-71, 74-75, 106-107.

One-time pad
A random key in the form of a codebook, to be used once only. In principle unbreakable, unless the codebook is lost or stolen.

Polyalphabetic substitution ciphers
A substitution cipher system in which the letters, numbers, or characters of the plaintext may be replaced by a variety of other letters, numbers, or characters.

Plaintext
The message before encryption, or after decryption.

Steganography
Ways of hiding or concealing a message which do not involve cryptography.

Symmetric key cryptography
A general term for most premodern cryptographic systems, where the algorithm and key are known by both the sender and receiver, the latter merely reversing the system used by the former.

Tabula recta/tableaux
An encrypting tool usually used for creating polyalphabetic ciphers or ratiocination. The horizontal and vertical axes of a tabular layout, usually in conjunction with a keyword or keyphrase, can be used to create a variety of polyalphabetic ciphers, the most well known being the de Vigenère cipher, *see page* 104-105.

Transposition ciphers
Methods of rearranging the characters of a plaintext to produce a ciphertext. Transposition ciphers include anagrams, and the Rail Fence system. *See page* 66.

Trigrams/Trigraphs
Commonly occurring sequences of three letters or characters, such as 'the' and 'ing' in English.

INDEX

A

B

Publisher's Acknowledgments

This book covers an enormously complex subject, and the publishers would like to express their thanks, in addition to the consultant editors listed on page 4, to the following for their expertise and advice in preparing the title for publication:

Britt Baille, Laura Cowan, Denise Goodey, Amelia Heritage, George Heritage, Julian Mannering, Tim Osborne, Alexander Stone, Caroline Stone, James Stone, and John Sullivan.

With consultants, designers, editors, and friends scattered across the globe, the coded miracle of the Internet deserves a mention.

An outstanding debt of gratitude is due to Christopher Davis, who was determined to see this project reach publication, and whose support helped us all to make it happen.

Any book on this subject owes much to the masterful *The Code Book* by Simon Singh (1999), one of the few truly readable books on the subject.

For further information, the following resources are also recommended:

Books

F.H.Hinsley *British Intelligence in the Second World War* (London, 1975)

David Kahn *The Codebreakers* (New York, 1996)

David E. Newton *The Encyclopedia of Cryptology* (Santa Barbara, 1997)

Lawrence Dwight Smith *Cryptography* (New York, 1943)

Web sites

www.omniglot.com

www.simonsingh.net/The_Black_Chamber.html

www.zodiackiller.com

TPBIWCHTBUMRBEITPCIDCHTBUIRTEB
ATCBKWNHOBWMLBEIDTGCMDEHNBTI

The publisher would like to thank the following for their kind permission to reproduce their photographs.

Key: (a-above; b-below/bottom; bg-background; c-center; f-far; l-left; r-right; t-top)

akg-images: 42tl, 52-53c, 56-57bc, 58tl, 58c, 86-87c, 114cla, 164-165bg, 179tr, 185c, 194br, 232bc; Elie Bernager 59br; Bibliothèque Nationale, Paris 59tl, 104tl, 160tl, 160bl; Collection Archiv für Kunst & Geschichte, Berlin 112-113bc; Erich Lessing 61l, 64b, 186br, 190bl; Musée du Louvre, Paris/Erich Lessing 43clb; Museo Nazionale Archeologico, Naples/Nimatallah 102-103bc; Postmuseum, Berlin 162br; Ullstein Bild 114bc; Victoria & Albert Museum, London/Erich Lessing 201tl.
Alamy: blickwinkel 14fclb; Mike Booth 209cbr; capt.digby 152tl; Classic Image 74br; Phil Degginger 159ca; Javier Etcheverry 19tr; Mary Evans Picture Library 9bc, 47t, 53tr, 58br, 65tr, 72tl, 73bl, 80tl, 97cb, 109tr, 114cr, 124bl, 172-173bc, 197cr, 260tl; Mark Eveleigh 15ftl; Tim Gainey 182b; Duncan Hale-Sutton 95crb; Dennis Hallinan 77tc, 194tl; Nick Hanna 205bl; Peter Horree 264tr; INTERFOTO Pressebildagentur 21tl, 49tc, 73tr, 96tl, 110tl, 189bl; Steven J. Kazlowski 15br; Stan Kujawa 145tc; David Levenson 189tc; Jason Lindsey 188br; The London Art Archive 78tl, 265b; Manor Photography 179tl; Mediacolor's 208tl; Todd Muskopf 244c; Jim Nicholson 15tr; Photo Researchers 156tl; Photos12 204bl; Phototake Inc. 277clb; Pictorial Press Ltd 94bcl, 121br, 125ca; The Print Collector 20ca, 36tl, 55cbl, 70b, 91br, 113tc, 131cl, 189c; PYMCA 146c; Friedrich Saurer 167c; Sherab 187br; Ian Simpson 18-19c; Skyscan Photolibrary 74tl; Stefan Sollfors 188clb; Stockfolio 207ca; Amoret Tanner 224l; Don Tonge 91tc; Genevieve Vallee 15cr; Vario Images GmbH & Co. KG 275br; Mireille Vautier 157bl; Visual Arts Library, London 20br, 33tr, 33cl, 53cb, 172c, 197bc, 242tl; Dave Watts 175tc; Ken Welsh 186cla; Norman Wharton 45br; Tim E. White 120cla; Maciej Wojtkowiak 183tc; World Religions Photo Library 187bl, 187bcl; Konrad Zelazowski 15tc.
***Amhitheatrum Sapientae Aeternae*, by Heinrich Khunrath, 1606:** 52bl.
Apex News & Pictures: 155tr.
Courtesy of **Apple Computer, Inc.:** 98bc, 271br.
The Art Archive: 116clb, 120bl, 211tr; American Museum, Madrid 153c; Archaological Museum Chatillon-sur-Seine/Dagli Orti 102tl; Archaeological Museum, Chora/Dagli Orti 29tl; Archaeologica Museum Sousse, Tunisia/Dagli Orti 43tl; Archives de l'Académie des Sciences, Paris/Marc Charmet 158cb; Ashmolean Museum, Oxford 57tr; Bibliothèque des Arts Décoratifs, Paris/Dagli Orti 28tr; Egyptian Museum, Cairo/Dagli Orti 35cr; Egyptian Museum, Turin/Dagli Orti 20tl; Galerie Christian Gonnet, Louvre des Antiquaires/Dagli Orti 209cb; Heraklion Museum/Dagli Orti 28tl, 30-31; Jan Vinchon Numismatist/Dagli Orti 212tl; Musée Cernuschi, Paris 185br; Musée Condé, Chantilly/Dagli Orti 228tl; Musée du Louvre, Paris/Dagli Orti 46tl; Musée Luxembourgeois, Arlon/Dagli Orti 27cla; Museo della Civilta, Romana, Rome/Dagli Orti 42clb, 152br; Museum of Carthage/Dagli Orti 43cr; National Gallery, London/Eileen Tweedy 195tr, 195b; National Gallery, London/John Webb 196tl; National Maritime Museum/Eileen Tweedy 92-93c; Nationalmuseet, Copenhagen 189br; Dagli Orti 7fbl, 26c, 34bc, 45ca, 85c, 190tl, 196bl; Palenque Site Museum, Chiapas/Dagli Orti 36cra; Private Collection/Marc Charmet 188cra; Ragab Papyrus Institute, Cairo/Dagli Orti 152cl; Science Museum, London/Eileen Tweedy 84cra; Eileen Tweedy 93tc; Mireille Vautier 185cla; Victoria & Albert Museum, London/Sally Chappell 74cla; Laurie Platt Winfrey 85tl.
Philippa Baile: 206tl.
The Bridgeman Art Library: Bonhams, London 198cbl; Centre Historique des Archives Nationales, Paris/Giraudon 49br; Chartres Cathedral 193cra, 193br; Fondazione Giorgio Cini, Venice 82cl; Vadim Gippenreiter 43bc; Look and Learn 257tr; Oriental Museum, Durham University 180bc; Prado, Madrid 242bl; Private Collection 54clb, 210tl; Royal Geographical Society, London 164tl, 164bl; Santa Sabina, Rome 43br; St Paul's Cathedral Library, London 210cr, 210b; Victoria & Albert Museum, London 181t; The Worshipful Company of Clockmakers' Collection, UK 153bc.
Corbis: 6br, 120-121bc, 123tr, 123c, 187t, 214tl; The Art Archive 224tl, 246cl; Asian Art & Archaeology, Inc. 130clb; Alinari Archives 76-77b; Richard Berenholtz 206-207bc; Bettmann 9fbl, 56tl, 77br, 78-79bc, 104b, 124br, 129cr, 134bl, 134-135c, 138clb, 170tl, 179bl, 228-229c, 232bl, 243tl, 264bl; Stefano Bianchetti 148c; Tibor Bognár 226-227c; Andrew Brookes 6fbl, 175bl; Christie's Images 82-83c, 172tl, 200bc; Dean Conger 192bl; Ashley Cooper 220-221c; Gérard Degeorge 186cr; Marc Deville 106-107c; DK Ltd 95tc; EPA 6bl, 246-247bc; Robert Essel NYC 174tl; Kevin Fleming 7br, 38-39c; Michael Freeman 184-185c, 209br; Rick Friedman 248tl; Gallo Images 14tr; Historical Picture Archive 111tc; Angelo Hornak 5br, 192-193c; Hulton-Deutsch Collection 132-133b, 133tc, 254-255tc; Kim Kulish 270bl; Massimo Listri 194cr, 218bl; Araldo de Luca 24cr; Maritime Museum, Barcelona/Ramon Manent 166c; Francis G. Mayer 209l; Gideon Mendel 7fbr, 131bc; Ali Meyer 128-129b; David Muench 18bl; Robert Mulder 45bc; Kazuyoshi Nomachi 22-23b; Charles O'Rear 208bl; Gianni Dagli Orti 20bl, 23c, 234tl; Papilio/Robert Gill 15tl; Steve Raymer 204-205bg; Roger Ressmeyer 119tr; Reuters 173ca; Reuters/Mike Mahoney 148bl; Reuters/Guang Niu 25tr; Reuters/Mike Segar 207tr; H. Armstrong Roberts 261bg; Royal Ontario Museum 209cb; Tony Savino 207tl; Stapleton Collection 57tc, 66-67c; Swim Inc. 2, LLC 122tl, 219tr; Sygma/Denver Post/Kent Meireis 139tr; Sygma/Franck Peret147tr; Sygma/Gaylon Wampler 274cb; Atsuko Tanaka 147tl; Penny Tweedie 19br; TWPhoto 131tr; Underwood & Underwood 16br; Werner Forman Archive 68bl, 186bc, 188tr; Adam Woolfitt 185cl; Michael S. Yamashita 249tc; Zefa/Howard Pyle 147bl; Zefa/Guenter Rossenbach 166tl; Zefa/Christine Schneider 224bc.
***Dogme et Rituel de la Haute Magie*, by Levi Eliphas, 1855:** 44-45c.
Mary Evans Picture Library: 32br.
Eric Gaba: 168tl.
Getty: 175br; AFP 156-157bc, 157br; AFP/Frederic J. Brown 226tl; AFP/Alastair Grant 230-231tc; AFP/Stephane de Sakutin 279br; AFP/Yoshikazu Tsuno 279cr; De Agostini Picture Library 200-201bg; Aurora/David H. Wells 183br; Aurora/Scott Warren 17br; Tim Boyle 265cla; The Bridgeman Art Library/Art Gallery and Museum, Kelvingrove 75bl; The Bridgeman Art Library/Bibliothèque Nationale, Paris 106tl; The Bridgeman Art Library/Instituto da Biblioteca Nacional, Lisbon 27br; The Bridgeman Art Library/National Museum of Karachi, Pakistan 4br, 20cb; The Bridgeman Art Library/Private Collection 5bl, 90cl, 234bl; The Bridgeman Art Library/Royal Geographical Society, London 163br; The Bridgeman Art Library/Stapleton Collection 25bc; Paula Bronstein 9bl, 227tr; Katja Buchholz 146tr; Matt Cardy 148-149c; CBS Photo Archive 263cb; Central Press 135tr; Evening Standard 118b; Christopher Furlong 50clb, 50bc, 50-51tc, 51cr; Cate Gillon 154r; Tim Graham 14clb, 128tl, 230bc; Henry Guttmann 229tc; Dave Hogan 55bl; Hulton Archive 29tr, 81br, 113tr, 205tc, 231bc, 233br; Hulton Archive/Horst Tappe 255cr; IDF 213bl; Image Bank/Ezio Geneletti 262-263c; Image Bank/Martin Puddy 98-99bc Imagno 60br, 191l, 233tc; Kean Collection 260clb; Keystone 235tr, 255bl; London Stereoscopic Company 132tl, Lonely Planet Images/Chris Mellor 32l; Haywood Magee 262c; Ethan Miller 218-219bc; Minden Pictures/Ingo Arndt 171tra, 171trb; Minden Pictures/Michael & Patricia Fogden 249cb, 249b; Minden Pictures/Mark Moffett 249tr; Michael Ochs Archive 94bcr; NEC 270-271c; Panoramic Images 8br, 144-145b; Photographer's Choice/Marvin E. Newman 16-17bg; Photographer's Choice/Bernard Van Berg 125tr; Picture Post/Bert Hardy 222cla; Popperfoto 122-123; Sportschrome/Rob Tringali 222-223c; Stone/Paul Chelsey 146bc; Stone/Frank Gaglione 240-241c; Stone/Jason Hawkes 6fbr, 64-65bg; Stone/Arnulf Husmo 96-97c; Stone/Nicholas Parfitt 13bg; Stone/Stephen Wilkes 144-145tc; Stone/Art Wolfe 13br; Taxi/Christopher Bissell 274-275c; Taxi/FPG 173tl; Taxi/Elizabeth Simpson 98-99tbg; Time & Life Pictures 79tr, 108b, 121tc, 247br, 268tl; Time & Life Pictures/Dorothea Lange 136-137; Roger Viollet 94-95c; Ian Waldie 175tr; Mark Wilson 125bc.
Prof. Jochen Gros/www.icon-language.com: 239br.
Helsinki University Library: 244tl.
Amelia Heritage: 206bl, 225tr, 255tr.
Image courtesy **History of Science Collections, University of Oklahoma Libraries**; copyright the Board of Regents of the University of Oklahoma: 84tl.
***Identification Anthropométrique*, by Alphone Bertillon, 1914:** 138c.
Courtesy of **IKEA:** 239cb.
© 2008 Intuitive Surgical Inc.: 277tc, 277c.
IOC/Olympic Museum Collections: 239cla.
iStockphoto.com: 12tl, 24-25bg, 36-37b, 61br, 91tr, 178c, 224c, 241crb, 245br, 272bc, 273bl; Adrian Beesle 178bl; Nicholas Belton 68-69bg; Daniel Bendjy 212cb; Anthony Brown 9fbr, 247cra; Mikhail Choumiatsky 90-91c; Lev Ezhov 273bc; Markus Gann 152bl; Vladislav Gurfinkel 200ca; Uli Hamacher 272tl; Clint Hild 69tr; Eric Hood 218tl; Hulton Archive 226clb; Gabriele Lechner 272-273bc; Arie J. Jager 211c; Sebastian Kaulitzki 276-277bg; Mark Kostich 248bc; Matej Krajcovic 152-153bg; Arnold Lee 8bl, 240tl; Tryfonov Levgenii 68bc; Marcus Lindström 140bg; Susan Long 98tl; Robyn Mackenzie 213tr; José Marafona 49tr; David Marchal 262-263tbg; Roman Milert 5fbl, 243cla; Vasko Miokovic 210-211bg; S. Greg Panosian 179tc; Joze Pojbic 251tr; Heiko Potthoff 106bl; Achim Prill 12bg; Johan Ramberg 149tr; Stefan Redel 240cra; Amanda Rohde 158br; Emrah Turudu 213c; Smirnov Vasily 256-257c; Krzysztof Zmij 273tr.
Susan Kare LLP: 239cr.
The Kobal Collection: A.I.P. 173tr; Artisan Ent 256clb; Bunuel-Dali 235br; MGM 273tc; Paramount 251crb; Warner Bros 263tr.
Library of Congress, Washington, D.C.: 61tr, 84bl, 84-85b, 112tl, 260cr, 261tr; Edward S. Curtis 258-259t.
***Light for the Blind*, by William Moon, 1877:** 243cra.
Musée Condé, Chantilly: 46bc.
Musée du Louvre, Paris: 34tl.
Museo Nazionale Archaeologico, Naples: 103tl.
Muséum des Sciences Naturelles, Brussels: 26tl.
Museum of Natural History, Manhattan: 169cr.
National Archives, London: 75cra.
National Portrait Gallery, London: 74tr.
NASA: 250bc; ESA and H.E. Bond (STScl) 250tl; ESA and J. Hester (ASU) 250-251bg; JPL 164br, 251bla, 251blb; MSFC 279l.
Courtesy of the **National Security Agency:** 84cb., 85crb, 119tc, 125tc.
Photo12.com: ARJ 199; Pierre-Jean Chalençon 110-111c.
Photolibrary: AGE Fotostock/Esbin-Anderson 227br; Jon Arnold Images 48bc; Jon Arnold Travel/James Montgomery 39br; F1 Online 28-29bc; Garden Picture Library/Dan Rosenholm 168bl; Robert Harding Travel/David Lomax 166-167c; Hemis/Jean-Baptiste Rabouan 64br; Imagestate/Pictor 265tc; Imagestate/The Print Collector 156bl; Pacific Stock/John Hyde 248-249tc; Franklin Viola 38bl.
***Philosophiae Naturalis Principia Mathematica*, by Sir Isaac Newton, 1687:** 156ca.
Private Collection: 22tl, 42bc, 47b, 54r, 55tc, 56bl, 60cr, 64cr, 65tl, 73br, 87tc, 94c, 97bl, 102cl, 107cb, 115cl, 115br, 116cbl, 117bc, 124tl, 129tl, 134tl, 139b, 158tl, 162-163tc, 164cra, 164crb, 165, 168c, 173br, 196br, 197tl, 197c, 201tr, 201bc, 206tr, 212tr, 212clb, 232br, 234-235bc, 238cr, 238bc, 254tl, 254bl, 260bl, 264tl, 264cr, 269crb, 271tc, 271tr, 272cb.
Antonia Reeve Photography: 51bl.
***Relación de las Cosas de Yucatán*, by Diego de Landa, 16thC:** 37tl.
Rex Features: 278bl; Greg Mathieson 65c; Sipa Press 147br; Dan Tuffs 277crb.
Tony Rogers: 96bl.
Royal Swedish Academy of Sciences: 168br.
Science and Society Picture Library: 117r, 119tl, 119cl, 152cr, 268bc, 268-269c, 269tc, 269tr, 269c, 271tc, 271cr.
Science Photo Library: 116tl.
SETI League Photo: Used by Permission 251tc.
SRI International: Image courtesy of DARPA and XVIVO 277br.
Still Pictures: Andia/Zylberyng 179br; Biosphoto/Gunther Michel 184tl; The Medical File/Geoffrey Stewart 171bc; Ullstein/Peters 185tl; Visum/Wolfgang Steche 167br; VISUM/Thomas Pflaum 278tl; WaterFrame.de/Dirscherl 221tr.
Caroline Stone: 7bl, 9br, 198cbr, 201cl.
Tim Streater: 64cl.
Telegraph Media Group: 86bl.
Louise Thomas: 14-15bc, 198tl.
Times of India: 87bc.
University of Pennsylvania: 154tl.
U.S. Air Force: 124-125c.
U.S. Government: 212br.
The U.S. National Archives and Records Administration: 115tc, 115cb.
Courtesy of **VSI:** 278br.
Werner Forman Archive: 32cr; Biblioteca Universitaria, Bologna 27tr; Haiphong Museum, Vietnam 180tr; Museum für Volkerkunde, Vienna 21tc; Museum of Americas, Madrid 153tc; National Gallery, Prague 180tl; National Museum, Kyoto 130cr.
John Wolff, Melbourne: 85tr.
Zodiackiller.com: 5fbr, 140cr, 141, 142-143.